International Political Economy Series

General Editor: Timothy M. Shaw, Professor and Director, Institute of International Relations, The University of the West Indies, Trinidad & Tobago

Titles include:

Leslie Elliott Armijo (*editor*)
FINANCIAL GLOBALIZATION AND DEMOCRACY IN EMERGING MARKETS

Eudine Barriteau
THE POLITICAL ECONOMY OF GENDER IN THE TWENTIETH-CENTURY CARIBBEAN

Gabriel G. Casaburi
DYNAMIC AGROINDUSTRIAL CLUSTERS
The Political Economy of Competitive Sectors in Argentina and Chile

Peter Clegg
THE CARIBBEAN BANANA TRADE
From Colonialism to Globalization

Matt Davies
INTERNATIONAL POLITICAL ECONOMY AND MASS COMMUNICATION IN CHILE
National Intellectuals and Transnational Hegemony

Yvon Grenier
THE EMERGENCE OF INSURGENCY IN EL SALVADOR
Ideology and Political Will

Ivelaw L. Griffith (*editor*)
THE POLITICAL ECONOMY OF DRUGS IN THE CARIBBEAN

Jerry Haar and Anthony T. Bryan (*editors*)
CANADIAN–CARIBBEAN RELATIONS IN TRANSITION
Trade, Sustainable Development and Security

Tricia Juhn
NEGOTIATING PEACE IN EL SALVADOR
Civil–Military Relations and the Conspiracy to End the War

R. Lipsey and P. Meller (*editors*)
WESTERN HEMISPHERE TRADE INTEGRATION
A Canadian–Latin American Dialogue

Gordon Mace, Andrew F. Cooper and Timothy M. Shaw (*editors*)
INTER-AMERICAN COOPERATION AT A CROSSROADS

Don Marshall
CARIBBEAN POLITICAL ECONOMY AT THE CROSSROADS
NAFTA and Regional Developmentalism

Juan Antonio Morales and Gary McMahon (*editors*)
ECONOMIC POLICY AND THE TRANSITION TO DEMOCRACY
The Latin American Experience

Leo Panitch and Martijn Konings (*editors*)
AMERICAN EMPIRE AND THE POLITICAL ECONOMY OF
GLOBAL FINANCE

Eul-Soo Pang
THE INTERNATIONAL POLITICAL ECONOMY OF TRANSFORMATION IN
ARGENTINA, BRAZIL, AND CHILE SINCE 1960

Julia Sagebien and Nicole Marie Lindsay (*editors*)
GOVERNANCE ECOSYSTEMS
CSR in the Latin American Mining Sector

Henry Veltmeyer, James Petras and Steve Vieux
NEOLIBERALISM AND CLASS CONFLICT IN LATIN AMERICA
A Comparative Perspective on the Political Economy of Structural Adjustment

Henry Veltmeyer, James Petras
THE DYNAMICS OF SOCIAL CHANGE IN LATIN AMERICA

International Political Economy Series
Series Standing Order ISBN 978–0–333–71708–0 hardcover
Series Standing Order ISBN 978–0–333–71110–1 paperback
(*outside North America only*)

You can receive future titles in this series as they are published by placing a standing order. Please contact your bookseller or, in case of difficulty, write to us at the address below with your name and address, the title of the series and one of the ISBNs quoted above.

Customer Services Department, Macmillan Distribution Ltd, Houndmills, Basingstoke, Hampshire RG21 6XS, England

Governance Ecosystems
CSR in the Latin American Mining Sector

Edited by

Julia Sagebien
School of Business Administration, Dalhousie University, Canada and Graduate School of Business Administration (EGAE), University of Puerto Rico, Puerto Rico

and

Nicole Marie Lindsay
School of Communication, Simon Fraser University, Canada and Faculty of Management, Royal Roads University, Canada

Introduction, selection and editorial matter © Julia Sagebien and Nicole Marie Lindsay 2011
Individual chapters © contributors 2011

All rights reserved. No reproduction, copy or transmission of this publication may be made without written permission.

No portion of this publication may be reproduced, copied or transmitted save with written permission or in accordance with the provisions of the Copyright, Designs and Patents Act 1988, or under the terms of any licence permitting limited copying issued by the Copyright Licensing Agency, Saffron House, 6-10 Kirby Street, London EC1N 8TS.

Any person who does any unauthorized act in relation to this publication may be liable to criminal prosecution and civil claims for damages.

The authors have asserted their rights to be identified as the authors of this work in accordance with the Copyright, Designs and Patents Act 1988.

First published 2011 by
PALGRAVE MACMILLAN

Palgrave Macmillan in the UK is an imprint of Macmillan Publishers Limited, registered in England, company number 785998, of Houndmills, Basingstoke, Hampshire RG21 6XS.

Palgrave Macmillan in the US is a division of St Martin's Press LLC, 175 Fifth Avenue, New York, NY 10010.

Palgrave Macmillan is the global academic imprint of the above companies and has companies and representatives throughout the world.

Palgrave® and Macmillan® are registered trademarks in the United States, the United Kingdom, Europe and other countries.

ISBN 978-0-230-27764-9

This book is printed on paper suitable for recycling and made from fully managed and sustained forest sources. Logging, pulping and manufacturing processes are expected to conform to the environmental regulations of the country of origin.

A catalogue record for this book is available from the British Library.

Library of Congress Cataloging-in-Publication Data
Governance ecosystems : CSR in the Latin American mining sector / edited by Julia Sagebien, Nicole Marie Lindsay.
   p.   cm.
Includes index.
ISBN 978-0-230-27764-9 (hardback)
  1. Mineral industries—Government policy—Latin America. 2. Mineral industries—Social aspects—Latin America. 3. Mineral industries—Environmental aspects—Latin America. 4. Social responsibility of business—Latin America. 5. Sustainable development—Latin America.
  I. Sagebien, Julia, 1954– II. Lindsay, Nicole Marie, 1975–
HD9506.L252G68 2011
622.068′4—dc23   2011030624

10 9 8 7 6 5 4 3 2 1
20 19 18 17 16 15 14 13 12 11

Printed and bound in the United States of America

Contents

List of Tables, Figures, Maps and Box	vii
Preface	ix
Acknowledgments	x
Notes on Contributors	xii
List of Acronyms	xviii

	Introduction: Companies and the Company They Keep: CSR in a 'Social and Environmental Value Governance Ecosystems' Context *Julia Sagebien and Nicole Marie Lindsay*	1
1	Systemic Causes, Systemic Solutions *Julia Sagebien and Nicole Marie Lindsay*	12
2	Sustainable Development Norms and CSR in the Global Mining Sector *Hevina S. Dashwood*	31
3	CSR and the Law: Learning from the Experience of Canadian Mining Companies in Latin America *Kernaghan Webb*	47
4	The Role of Governments in CSR *Jan Boon*	64
5	Regulatory Frameworks, Issues of Legitimacy, Responsibility, and Accountability: Reflections Drawn from the PERCAN Initiative *Bonnie Campbell, Etienne Roy-Grégoire, and Myriam Laforce*	84
6	Conflict Diamonds: The Kimberley Process and the South American Challenge *Ian Smillie*	102
7	Whose Development? Mining, Local Resistance, and Development Agendas *Catherine Coumans*	114

8 Mining Industry Associations and CSR Discourse: Mapping the Terrain of Sustainable Development Strategies   133
Nicole Marie Lindsay

9 Drivers of Conflict around Large-scale Mining Activity in Latin America: The Case of the Carajás Iron Ore Complex in the Brazilian Amazon   154
Ana Carolina Alvares da Silva, Silvana Costa, and Marcello M. Veiga

10 Community and Government Effects on Mining CSR in Bolivia: The Case of Apex and Empresa Huanuni   170
Robert Cameron

11 Corporate Social Responsibility in the Extractive Industries: The Role of Finance   187
Allen Goss

12 Responsible Investment Case Studies: Newmont and Goldcorp   201
Irene Sosa

13 Anti-corruption: A Realistic Strategy in Latin American Mining?   214
Carol Odell

14 Sustainable Juruti Model: Pluralist Governance, Mining, and Local Development in the Amazon Region   233
Fabio Abdala

15 Energy and CSR in Trinidad and Tobago in the Second Decade of the Twenty-first Century   245
Timothy M. Shaw

16 Mining Companies and Governance in Africa   260
Ralph Hamann, Paul Kapelus, and Ed O'Keefe

17 Conclusion   277
Nicole Marie Lindsay and Julia Sagebien

Appendix: The Latin America–Canada Mining Connection   303

Bibliography   307

Index   341

List of Tables, Figures, Maps and Box

Tables

2.1	Mining companies' CSR reports, by year of first release	40
2.2	Number of major mining companies with sustainable development/sustainability in the title of their reports, by first year of use of term	43
8.1	Preconditions for CSR performance and existing conditions for Canadian mining companies with operations in Latin America	143
8.2	Mining Industry Association CSR initiative comparison chart	145
8.3	Comparison of key themes – industry association CSR Codes/Guidelines and OCMAL declaration	151
10.1	Comparison of communities surrounding the Apex and Empresa Huanuni mines	175
10.2	CSR comparison of Apex Silver and Empresa Huanuni mines	177
12.1	Newmont mining corporation	206
12.2	Goldcorp Inc	208
13.1	Typology of corruption responses for mining companies in Latin America	226
14.1	Sustainable Juruti model	238
14.2	Projects funded by FUNJUS, 2009/2010	241

Figures

1.1	Basic two-tier stakeholder map	21
1.2	The Social and Environmental Value Governance Ecosystem (SEVGE) model	26
4.1	High-level categories for government roles	65
13.1	Corruption perceptions for the Americas	218
13.2	Superposition of corruption perceptions on world values	220
13.3	Corruption and the factors found to increase it in empirical econometric studies	221

viii *List of Tables, Figures, Maps and Box*

14.1	Sustainable Juruti model	237
16.1	A schematic representation of the corporate presence as part of the local governance system in Africa	262
16.2	A schematic illustration of the two-way relationship between organizations and the governance system of which they are a part	264

Maps

North America	xxii
South America	xxiii

Box

6.1	Basic elements of the KPCS	104

Preface

This book is the result of serendipity – a most fortuitous series of coincidences – and of a great deal of work on the part of the authors and all project participants, advisors, and sponsors. It represents the culmination of an International Development Research Centre (IDRC) project entitled 'Both Sides Now – Corporate Social Responsibility in the Latin American Extractives Industries: An Examination of Challenges and Best Practice Opportunities.' The objective of the Both Sides Now project (2008–11) was to foster the formation of a network of individuals with the skills, competencies, and commitment to explore the emerging dynamics of corporate social responsibility in the extractive industries of Latin America from an interdisciplinary research perspective. The research team gathered at a workshop held at Royal Roads University in Victoria, BC in October 2009. This book presents some of the papers presented at the workshop plus a number of additional works we felt were necessary in order to achieve our research objectives.

At least on our part, the IDRC project and this book have been a testament to Zen Master's Shunryu Suzuki Roshi's adage that 'In the beginner's mind there are many possibilities.' Now that the project and the book have both been successfully brought to an end, we close these endeavors with Suzuki Roshi's words: 'Everything is perfect, but there is a lot of room for improvement.'

Julia Sagebien and Nicole Marie Lindsay

Acknowledgments

The editors and authors are indebted to many individuals for their assistance with the completion of this book. Giselle Morin Labatut and Loredana Marchetti of IDRC facilitated the Centre's generous support of this project and guided us through each step of the project development. They were both always available to help us along. Aida DuBois from IDRC kept our budgeting on track. Dr Pedro Marquez, Deborah Zornes, Evelyn Goedhart, and Isabel Cordua-von Specht from Royal Roads University provided us with a helpful, friendly, and efficiently managed 'home' for the workshop and the management of the project. We thank Dr Timothy M. Shaw, Editor, International Political Economy (IPE) Series for Palgrave Macmillan, our champion in the publication quest, and the team at Palgrave Macmillan – Alexandra Webster, Christina Brian, and Renée Takken – for helping us turn a research idea into a publication with international reach. Trish O'Neill's keen eye and editing skills allowed us to submit the manuscript on time.

We would also like to thank the researchers who participated in the workshop but who did not contribute to this volume: Peter Campbell, Policy Analyst, Minerals and Metals Sector, Natural Resources Canada; Oscar Gonzales, Environmental Services Association of Alberta (ARPEL); Mr Andrés Recalde, Latin America Caribbean Development Associates; Dr Robert Boutilier, Associate, Centre for Sustainable Community Development, Simon Fraser University; Dr Sujit Sur, Assistant Professor, Dalhousie University; Yanina Kowszyk, General Manager of Forum Empresa; Melissa Whellams, Senior CSR Advisor, Canadian Business for Social Responsibility; Patricia Ferreira, SjD candidate Faculty of Law, University of Toronto; Steven Schnoor, PhD Candidate, Communication & Culture, York/Ryerson Universities; Jessica Dillabough, PhD Student, University of Calgary, International Resource Industries and Sustainability Centre (IRIS); and Circe Niezen, PhD Candidate in International Business and Assistant Professor, Industrial Engineering Department, Polytechnic University of Puerto Rico.

A special thanks to Yolanda Banks of EDC for funding a separate research project on Peru, which served as a very powerful mining industry CSR learning lab for Nicole Marie Lindsay and Julia Sagebien, and to Dr Marketa Evans, the Government of Canada's first Corporate Social

Responsibility Counselor for the Extractive Sector, who was the original co-principal investigator in the IDRC project proposal while Director of the Munk Centre, University of Toronto.

Julia Sagebien wishes to thank Drs Paul Latortue and Emilio Pantojas from the University of Puerto Rico (UPR), as well as Dean Peggy Cunningham from the Faculty of Management at Dalhousie University, for providing the research environment that facilitated the production of this volume. She would also like to thank Circe Niezen and Alba Brugueras, doctoral candidates at UPR, for their research assistance. On a personal basis, she wishes to thank her mother Olivette (Beba) Sagebien as well as Ms. Sharon Meadows for their love, care, and moral support. Thanks also to Veronica Fernandez for providing her with the original inspiration to pursue this research interest, and Monica Perez Nevarez for helping her edit some of the work and keeping her sane during the process. She thanks her teachers, the Vidyadhara Chogyam Trungpa Rinpoche and Venerable Kenchung Thrangu Rinpoche, for showing her that a better world is possible. Cachita the Puerto Rican rescue 'sata' Schnauzer mutt brought many hours of playful joy and an excuse to embark upon much-needed exercise.

Nicole Marie Lindsay wishes to thank Dr Pedro Marquez for his support and insights through a directed studies course on the history of neoliberalism in Latin America, and Drs Shane Gunster, Robert Anderson, and Yuezhi Zhao, her supervisors at Simon Fraser University, for their expert and knowledgeable guidance. Adam Molnar, PhD candidate in Political Science at the University of Victoria, provided moral support and an intellectual sounding board. Nicole is most grateful to her family: her parents Don and Susan Lindsay provided love and support from the ground up, and her sons Noah and Solomon hope for a better future.

Notes on Contributors

Fabio Abdala is a political scientist by training, a PhD in International Relations (University of Brasilia). Currently, Fabio is Sustainability Manager to Mining Operations for Alcoa Latin America and the Caribbean region. He was Executive Secretary of Amazon Working Group, the largest community-based networks of Brazilian rainforest, for four years. Before that Fabio worked for the Brazilian Ministry of Environment, where he was responsible for coordinating studies and dissemination on Brazilian rainforest conservation; and was project manager for the Sustainable Development Program of Amapá State.

Jan Boon's career includes university-level teaching of physical chemistry, oil sands research, and management of provincial and federal geological survey organizations. After his retirement, he took an MA in Globalization and International Development and he is currently enrolled in a PhD program in the Department of Sociology and Anthropology, Carleton University, Ottawa, Canada. His MA and PhD research focus is on corporate social responsibility in the mineral exploration and mining industry. He has published articles on 'globalization, mining and Peru' and on whether corporate social responsibility can replace development aid.

Robert Cameron recently graduated with a MA in International Development from Saint Mary's University in Canada, where he studied the relationship between mining CSR and development in Latin America. As part of his thesis, Robert conducted fieldwork in Bolivia and presented at a number of international development conferences. Robert has also worked as a socio-economic consultant for an oil project in Africa and was involved in a university–community partnership, researching and promoting community-based natural resource management. Currently, Robert is a Traditional Knowledge and Land Use Study Lead in the Community and Aboriginal Affairs Division of Stantec Consulting, Inc.

Bonnie Campbell is Professor of Political Economy at the Department of Political Science at the University of Quebec in Montreal, where she heads the C.-A. Poissant Research Chair on Governance and Aid for Development. She is also the Director of the Groupe de recherche sur les activités minières en Afrique (GRAMA). Bonnie Campbell has written extensively on issues related to international development, governance, and mining and is the author of many journal articles, and author, editor, or co-editor of more than ten volumes including *Mining in Africa. Regulation and Development* (2009) [French version *Ressources minières en Afrique. Quelle réglementation pour le développement?* (2010)], *Qu'allons-nous faire des pauvres?* (2005), *Regulating Mining in Africa: For whose Benefit?* (2004) [*Enjeux des nouvelles réglementations minières en Afrique* (2004)], *Restructuring in Global Aluminium* (with Magnus Ericsson) (1996), *Structural Adjustment in Africa* (with John Loxley) (1989), and *Political Dimensions of the International Debt Crisis* (1989).

Catherine Coumans is Research Coordinator and Coordinator of the Asia-Pacific Program at MiningWatch Canada. She holds an MSc (London School of Economics) and a PhD (McMaster University) in Cultural Anthropology and carried out postdoctoral research at Cornell University. Her recent publications include: 'Indigenous peoples and NGOs in the contested terrains of mining and corporate accountability' in *Earth Matters: Indigenous Peoples, the Extractive Industries and Corporate Social Responsibility*, 2008; 'Alternative accountability mechanisms and mining: the problems of effective impunity, human rights and agency' in *Canadian Journal of Development Studies* 2010; 'Occupying spaces created by conflict: anthropologists, development NGOs, responsible investment and mining' in *Current Anthropology* (April 2011).

Silvana Costa is a community planner by training. She completed her PhD at the University of British Columbia (UBC), Norman B. Keevil Institute of Mining Engineering in early 2008. Dr Costa was a UBC Bridge Program Doctoral Fellow – a program designed to strengthen links between engineering, policy, and public health research. Her career interests include a variety of aspects of the concepts of Sustainability and Corporate Social Responsibility (CSR) within the natural resources sector in Canada and abroad.

Hevina S. Dashwood is Associate Professor of Political Science at Brock University, Canada. She is completing a book manuscript on corporate social responsibility, mining, and the dissemination of global norms, to be published by Cambridge University Press. She is conducting case study research on Canadian mining companies in Africa, and is the author of numerous journal articles and book chapters on CSR in the mining sector.

Ana Carolina Alvares da Silva's academic and professional experience includes several projects for junior exploration and large mining companies, industry associations, and local communities in Brazil, Canada, Chile, Panama, and Peru on issues of indigenous rights, artisanal mining, community development, and social impact assessment. Her PhD work at the University of British Columbia, Canada, focused on the development of stakeholder consultation approaches and sustainability indicators for better life cycle management of metals. Dr da Silva has authored many academic publications and presented extensively on community-based indicators, corporate social responsibility, and stakeholder consultation processes.

Allen Goss is Assistant Professor of Finance at the Ted Rogers School of Management at Ryerson University in Toronto, Canada. His current research looks at the intersection of corporate social responsibility and financial performance. His articles on CSR and socially responsible investing have appeared in the *Journal of Banking and Finance* and the *Journal of Portfolio Management*.

Ralph Hamann is Associate Professor and Research Director at the University of Cape Town Graduate School of Business. His research is on strategic change in organizations and governance systems in response to complex socio-ecological problems, with a focus on food security, climate change, and human rights issues in extractive industries. Ralph is also a founding director of FutureMeasure, a company that provides an Internet-based sustainability performance measurement system

Paul Kapelus was a founding Director of the African Institute of Corporate Citizenship. He has worked in the field of Corporate Social Responsibility in Africa for the past 13 years as a researcher, consultant, facilitator, and as an employee at a platinum mine. He has a Masters in Social Anthropology from the University of Sussex, UK, where he specialized in multinational mining companies and corporate social responsibility. Paul is a member of the Global Reporting Initiative Stakeholder Council and is on the AccountAbility (AA1000) Council. He has written extensively on the subject of corporate–community relations and participated in conferences and projects around the world with the aim of ensuring that an African perspective on corporate responsibility is considered in both policy and practice.

Myriam Laforce is a researcher at the Groupe de recherche sur les activités minières en Afrique (GRAMA), part of the C.-A. Poissant Research Chair on Governance and Aid for Development at the University of Quebec in Montreal. In 2006 she completed a Masters in Political Science at the same university. Her recent research has focused on the political processes surrounding the management of foreign mining activity impacts in Peru and on the main implications of present forms of liberalization in the mining sector. Since 2006 she has been involved in a GRAMA research project funded by the Social Sciences and Humanities Research Council (SSHRC), based on the analysis of the regulation processes that led to the signing of specific agreements between mining companies and aboriginal communities in two regions of Canada in the 1990s.

Nicole Marie Lindsay's research is focused on the discourses of corporate social responsibility and sustainability in the mining industry in Latin America. She is an associate faculty member at Royal Roads University in the Faculty of Management and the School of Communication and Culture, where she has taught courses on communication, rhetoric, and sustainability. She is a PhD candidate in the School of Communication at Simon Fraser University, and is currently researching CSR and sustainability in Mexico's mining industry at the Centro de Investigaciones Interdisciplinarias en Ciencias y Humanidades (CEIICH) at the Universidad Nacional Autónoma de México (UNAM).

Carol Odell is a PhD candidate at the Norman B. Keevil Institute of Mining Engineering at the University of British Columbia, where she was a Fellow of the Bridge Program. Her doctoral research focused on the interface between mining company practices, social context, policy, and community health and well-being in the Ancash Region of Peru and was partially financed by the IDRC through a Doctoral Award. Carol has also consulted for mining companies and industry associations, governments, and local communities in Canada and in Latin America, specializing in the socio-environmental management of mining operations and the effectiveness of Corporate Social Responsibility in action. Her teaching experience includes course module development at UBC, and she teaches a course on Effective Mine Social Management at the University of Chile. Her specific areas of academic interest include Social

Impact Assessment, Policy Analysis, Risk Communication, and the Integration of Sustainability issues within Mine Design and Corporate Social Responsibility, all focused on the mining industry in the Americas.

Ed O'Keefe is Director at Synergy Global Consulting. He has worked on extractive industry and community development issues in over 20 countries over the past decade, including working with organizations such as AngloGold Ashanti, Rio Tinto, Shell, BG, IFC, DFID and Fauna & Flora International.

Etienne Roy-Grégoire holds a Masters in political science from the Université du Québec à Montréal and a degree in anthropology from the Université de Montréal. His work focuses on Canadian foreign policy relative to extractive activities in conflict and post-conflict countries in the Americas. He is a member of the Groupe de recherche sur les activités minières en Afrique (GRAMA) of the Université du Québec à Montréal.

Julia Sagebien is Associate Professor at the School of Business Administration and Adjunct Professor in the International Development Studies program and the College of Sustainability at Dalhousie University in Halifax, Nova Scotia, Canada. She is also a Catedratica (Full Professor) at the Escuela Graduada de Administracion de Empresa of the University of Puerto Rico. She served as Senior Fellow for the Canadian Foundation for the Americas (FOCAL) and for Forum Empresa.

Timothy M. Shaw is Professor and Director of the Institute of International Relations at The University of the West Indies, St Augustine, Trinidad & Tobago, Professor Emeritus, University of London, Visiting Professor, Mbarara University of Science & Technology & Makerere University Business School, Uganda, Senior Fellow, Centre for International Governance Innovation, Waterloo, Canada, and Editor, International Political Economy (IPE) Series.

Ian Smillie was a founder of the Canadian NGO, Inter Pares, was Executive Director of CUSO and is a long-time foreign aid watcher and critic. He is the author of several books, including *Blood on the Stone: Greed, Corruption and War in the Global Diamond Trade* (2010). Ian helped develop the 55-government 'Kimberley Process,' a global certification system to halt the traffic in conflict diamonds. He was the first witness at Charles Taylor's war crimes trial in The Hague and he chairs the Diamond Development Initiative.

Notes on Contributors xvii

Irene Sosa is Senior Analyst at Jantzi-Sustainalytics, an independent sustainability research and analysis firm. As the lead analyst in the mining sector, she is responsible for identifying key risks for investors related to poor environmental management systems or strained community relations. She engages regularly with mining companies, NGOs, and local communities and benchmarks companies against best practices in the sector. Recently, Irene contributed to Jantzi-Sustainalytics' 'The Link between Sustainability Performance and Competitiveness in the Canadian Natural Resource Sector.' Irene has a Masters in Environmental Studies degree from York University, Toronto.

Marcello M. Veiga has for the past 32 years worked as a metallurgical engineer and environmental geochemist for mining and consulting companies in Brazil, Canada, Chile, China, Colombia, Costa Rica, Ecuador, Guinea, Guyana, Indonesia, Laos, Mozambique, Peru, South Africa, Sudan, Suriname, Tanzania, the US, Venezuela, and Zimbabwe. He has worked extensively on environmental and social issues related to mining. As an Associate Professor at the Norman B. Keevil Institute of Mining Engineering at the University of British Columbia (UBC) since 1997, his research and teaching topics include sustainable development in mining, conflicts with communities, mine closure and reclamation, remedial procedures for metal pollution (in particular, mercury pollution), bioaccumulation of metals, environmental and health impacts of metals and cyanide, acid rock drainage, and mineral processing. Dr Veiga has published more than 260 technical papers and has directly supervised 28 Masters and PhD theses.

Kernaghan Webb is Associate Professor in the Department of Law and Business at Ryerson University's Ted Rogers School of Management, where he is also the Founding Director of the Institute for the Study of Corporate Social Responsibility (CSR). Dr Webb has published widely on innovative regulatory approaches and his work has been cited and followed by the Supreme Court of Canada. He is Special Advisor to the United Nations Global Compact on the ISO 26000 Social Responsibility Standard, Project Leader of the SSHRC-supported Multi-Perspective Collaborative CSR Case Study Project, and is the Law Editor for the *Journal of Business Ethics*.

List of Acronyms

ALBA	Alianza Bolivariana para los Pueblos de Nuestra America
ANC	African National Congress
AGA	AngloGold Ashanti
the BRICs	Brazil, Russia, India, and China
FUNBIO	Brazilian Biodiversity Fund
IBAMA	Brazilian Institute for Renewable Natural Resources
IBRAM	Brazilian Institute of Mining
MST	Brazilian Landless Workers' Movement (*Movimento Sem Terra*)
CalPERS	California public employees retirement system
CBSR	Canadian Business for Social Responsibility
CIM	Canadian Institute of Mining
CIDA	Canadian International Development Agency
CIFA	Canadian Investment Fund for Africa
CMALA	Canadian mining activity in Latin America
CPP	Canadian Pension Plan
CTPL	Center for Trade Policy and Law
CAFTA	Central American Free Trade Agreement
CERLAC	Centre for Research on Latin America and the Caribbean
CMA	Chemical Manufacturers Association
CVRD	Companhia Vale do Rio Doce
CCC	corporate codes of conduct
CSI	corporate social investment
CR	Corporate responsibility
CFPOA	Corruption of Foreign Public Officials Act
CSR	Corporate social responsibility
CPI	Corruption Perceptions Index
DRC	Democratic Republic of Congo
DDS	development diamond standards
DI	Devonshire Initiative
DFAIT	Department of Foreign Affairs and International Trade (Canada)
DDII	Diamond Development Initiative International
FRE	Economic Reconstruction Fund
ESTTA	Energy Security Through Transparency Act

EIA	Environmental impact assessment
EMA	Environmental Management Authority
ESE	environmental, socia,l and economic
ESG	environmental, social, and governance
EDC	Export Development Canada
EITI	Extractive Industries Transparency Initiative
EIR	Extractive Industry Review
FDI	foreign direct investment
FPIC	free, prior, and informed consent
FTA	Free Trade Agreements
FNI	Front des Nationalistes et Intégrationnistes
FUNAI	Fundação Nacional do Indio
GMI	Global Mining Initiative
GRI	Global Reporting Initiative
GES	Government of El Salvador
GCP	Great Carajás Project
FGV	GV Foundation
SERNA	Honduran Ministry of Natural Resources and Environment
HRIA	human rights impact assessment
IBA	impact-benefit agreement
IACHR	Inter-American Commission on Human Rights
ICME	International Council on Mining and the Environment
(ICMM)	International Council on Mining and Metals
IDRC	International Development Research Center
IFC	International Finance Corporation
IFIs	International financial organizations
ILO	International Labour Organization
IMF	International Monetary Fund
ISO	International Organization for Standardization
ISA	International Studies Association Conference
KP	Kimberley Process
KPCS	Kimberley Process Certification Scheme
OCMAL	Latin American Observatory of Mining Conflicts
LNG	liquefied natural gas
M&A	merger and acquisition
MDG	Millennium Development Goals
MAC	Mining Association of Canada
MMSD	Mining, Metals and Sustainable Development
MEM	Ministry of Energy and Mines (Peru)
MONUC	Mission des Nations Unies en Republique Democratique du Congo

MIGA	Multilateral Investment Guarantee Agency
MNC	Multinational corporation
NIC	Newly industrializing countries
NGO	Non-governmental organization
NBK	Norman B. Keevil Institute of Mining Engineering
OECD	Organisation for Economic Co-operation and Development
OTs	Overseas territories
PAC	Partnership Africa Canada
PERCAN	Peru–Canada Mineral Resources Reform Project
PDAC	Prospectors and Developers Association of Canada
PPPs	Public–private partnerships
RI	responsible investing
RJC	Responsible Jewellery Council
RCMP	Royal Canadian Mounted Police
SWUP	Save the Wild UP
SIDS	small island developing states
SEV	social and environmental value
SEVGE	social and environmental value governance ecosystem
SLO	Social license to operate
SRI	Socially responsible investment
SAHRC	South African Human Rights Commission
STCIC	South Trinidad Chamber of Commerce
SWF	Sovereign Wealth Fund
SCC	Supreme Court of Canada
SG	Sustainable governance
CONJUS	Sustainable Juruti Council
FUNJUS	Sustainable Juruti Fund
TSX	Toronto Stock Exchange
TSM	Toward Sustainable Mining
T&T	Trinidad and Tobago
UNCTAD	United National Conference on Trade and Development
UN	United Nations
UNEP	United Nations Environment Program
UNEP FI	United Nations Environment Programme's Finance Initiative
(UNPRI) (PRI)	United Nations Principles for Responsible Investment
SRSG	United Nations Special Representative of the Secretary General
UWIO	University of the West Indies
EPA	U.S. Environmental Protection Agency

WMC	Western Mining Company
WMI	Whitehorse Mining Initiative
WB	World Bank
WBCSD	World Business Council on Sustainable Development
WCED	World Commission on Environment and Development
WDC	World Diamond Council
WVS	World Values Study
WWF	World Wildlife Fund

North America

South America

Introduction: Companies and the Company They Keep: CSR in a 'Social and Environmental Value Governance Ecosystems' Context

Julia Sagebien and Nicole Marie Lindsay

The primary purpose of this book is to share the results of a collective exploration by the authors and by several other research project partners into the way in which a wide variety of public, private, and civil society actors impact the process of designing, implementing, and evaluating corporate responsibility in the mining sector of Latin America. By taking a systemic approach that reveals the political economy surrounding the mining industry and its Corporate Social Responsibility (CSR) efforts, we hope to begin to provide a broader view of the myriad contextual relationships and dynamics that can potentially enable or disable the balance of economic, social, and environmental value that CSR strategies pursue. Through this systemic approach, we hope to add a new variant to debates about whether mining is good or bad for communities and countries and whether CSR is good or bad for mining firms and their stakeholders. Rather, by asking how a variety of different social, economic, and political actors negotiate their conflicting interests surrounding large-scale extractive projects, we hope to gain insight into the complex relations bound up in the practices of mining and CSR.

Being an initial exploration, the chapters in this book draw from an eclectic set of multidisciplinary approaches, rather than a single unified approach to examination of the systemic dynamics that surround CSR and mining. In terms of root disciplines, the chapters include insights from business management, finance, law, mining engineering, political science, discourse analysis, public policy, Non-governmental Organization (NGO) advocacy, and international development studies. The chapters focus largely on Latin American countries as 'host' states, with an emphasis on Peru and Brazil. All but one chapter focuses on Canada and its government and overseas development assistance institutions as the 'home' state. The vast majority of companies discussed in the chapters are either Canadian or have a strong association with Canada. In order to provide a geographic contrast to Latin America, Chapter 16 presents reflections from Africa.

Why a systemic approach to CSR?

The chapters in this collection in some way examine the four basic research premises articulated by the editors in Chapter 1. The premises are that (1) poverty, social exclusion, and environmental degradation, while often exacerbated by irresponsible business, exist within a broader local and global political economic context in which irresponsibility and lack of accountability are built into the system; (2) CSR provides firms with a strategic response to the risks that systemic dynamics present, by addressing governance gaps that can, in turn, increase the potential for obtaining a 'social license to operate'; however, (3) firm-centred CSR as currently conceptualized cannot be expected to bring about the long-term, transformative change needed to address multi-actor, system-wide issues; and (4) given this, new analytical models that can capture system-wide dynamics and put CSR into context within a broader governance system should be used as a complement to traditional stakeholder-management CSR planning tools.

In sum, CSR may be helpful in gaining a company a 'social license to operate,' but it cannot always truly foster 'sustainable' development because the isolated, piecemeal efforts that characterize most CSR policies do not address, either directly or indirectly, the dynamics of the dysfunctional, risk-ridden, and unequal systems that are at the root of the problems and issues targeted by CSR programs. Moreover, the potential and limitation of CSR as a strategic policy tool for sustainable development depend on the capacity to analyze and understand the political economic context in which transnational firms operate. In other words, while the act of developing and implementing CSR initiatives lies firmly within the firm's domain and its primary intent is to strategically advance corporate interests, the sustainable, long-term impact of these initiatives depends to a great extent upon the enabling or disabling dynamics resulting from the actions and interactions of a host of actors outside the boundaries of the firm.

In order to better understand the mixed results of CSR policy and programs, this collection of chapters explores the multiple and cross-cutting relationships between major actors and institutions involved in negotiations over mining development in search of ways in which they can make more meaningful contributions to sustainable development. Companies must begin to learn how to better factor into their CSR planning, implementation, and evaluation process 'the company they keep.'

The chapters presented in this volume explore a range of actors and interactions affecting the political economy of CSR in Latin America's mining industry, engaging head-on with the complexity of competing interests in a dynamic system of interrelationships. In doing so, they explore the contours of debates over mining development and shed some light both on the limitations and potential offered by CSR as a provisional and practical strategy for addressing long-term, systemic problems. This collection of chapters seeks to question *how, under what conditions, and enabled by which actors* CSR in the mining industry might meaningfully contribute social and environmental value (SEV) to communities and countries dependent on mining.

In Chapter 1, Sagebien and Lindsay take the arguments presented in the four premises further by proposing that CSR theory and practice be contextualized within a broader 'ecosystem' of governance that prioritizes the creation (and protection) of social and environmental value. The social and environmental value governance ecosystem (SEVGE) model encourages a systemic approach to conceptualizing the multiple actors and institutions involved in negotiations over development, highlighting contextual and relational dynamics that either enable or inhibit creation of social and environmental value. We argue that SEVGE facilitates a more complex analysis of the contextual and relational dynamics shaping the system as a whole, considering in particular the specific actor/institutional objectives, goals, abilities, initiatives, mechanisms, and relationships that influence and are influenced by firm CSR.

Though there is general agreement on the descriptive and diagnostic value of the four premises, the authors in this collection do not all necessarily agree with, nor do they specifically adopt, the SEVGE model in their work. There is, however, general agreement between the editors and the authors that there is still much theoretical research and fieldwork to be done in order for the SEVGE model proposal to serve as a useful modeling and planning tool. The concluding chapter weaves together the findings presented in this book with a more in-depth examination of the specific components of the SEVGE model.

Organization of the book

In order to better understand the mixed results of CSR policy and programs, the authors in this collection focus their attention on the complex dynamics that shape relationships between key actors and institutions involved in negotiations mining and sustainable development. The chapters in this book are not meant to be read in any particular order. However, we have sequenced them in such a way as to connect concepts in the interests of building a number of progressive arguments, as well as in highlighting points of digression and controversy.

In Chapter 1, 'Systemic Causes, Systemic Solutions,' Sagebien and Lindsay outline the four premises that frame the collection of chapters presented in this book. We argue that systemic problems require systemic solutions and that company CSR programs alone cannot make the substantial contributions to sustainable development they claim to make. In part, this is because the CSR planning models do not accurately represent the context in which CSR operates. We argue that stakeholders are legitimate independent political actors interacting in a governance ecosystem, with goals related to social and environmental value that are much larger than any triple bottom line can capture. Incorporating this perspective into CSR planning, implementation, and evaluation in an iterative fashion with existing CSR stakeholder models should allow companies to better leverage the enabling dynamics of the system. We propose a planning model that we believe can more accurately capture the systemic dynamics in which companies and their CSR programs and policies are imbedded. Using the SEVGE model, companies must begin to learn how to more accurately factor in the contextual and relational dynamics of the broader governance system in which they operate.

Hevina Dashwood's 'Sustainable Development Norms and CSR in the Global Mining Sector' takes a broad, system-wide perspective, highlighting how a number of actors both internal and external to the firm influence the definition of individual corporate CSR goals. Specifically, Dashwood examines the influences that have led to the near-universal adoption by large mining multinationals of 'sustainable development' as a conceptual framework for their CSR policies. She concludes that although external pressures at the global level, such as advocacy efforts and the dissemination of global norms, have been important influences, the domestic institutional context in which mining companies operate and the role of some mining companies as norms entrepreneurs are also important in explaining why mining companies have accepted the normative validity of sustainable development.

There is considerable controversy worldwide concerning the extent and adequacy of the national (home/host) and international legal framework in which mining companies operate, and about the relation between those laws and corporate social responsibility as a voluntary and discretionary activity. Kernaghan Webb's 'CSR and the Law: Learning from the Experience of Canadian Mining Companies in Latin America' addresses some of the most profound questions that mining activity raises about how societies make decisions and how power is wielded. Webb's analysis reveals that the legal framework, as well as the concept and content of corporate social responsibility, are in a state of considerable flux and that we are currently witnessing a 'seismic shift' away from largely pro-economic development approaches, toward more balanced approaches in which the interests of stakeholders other than the mining companies are beginning to be more fully integrated into decision-making processes. Webb concludes that those firms that proactively address their environmental, social, and economic impacts through CSR commitments and activities that extend beyond current legal requirements may be better able to remain compliant with both present and future laws, and may be more likely to be seen as acceptable to affected stakeholders.

In the entire debate about corporate responsibility, the 'governance' buck stops at the state's desk. States develop legal and regulatory frameworks, are entrusted by citizens and communities with the protection of public goods (civic order, environment protection, and so on), and the mediation of conflict between constituencies. They also design the enabling environments for economic activity and redistribute gains through fiscal policies and state investments. Jan Boon's chapter, 'The Role of Governments in CSR,' demonstrates the relative paucity of CSR and government studies, despite the fact that such a role is generally 'implied' in the CSR literature. He examines how the nature of the relation between the state and its communities and corporations, the state's vulnerability to international pressures, transparency and the availability of information, and the enforcement and accessibility of a legal framework are key factors affecting governments' abilities to live up to the expectations outlined above and influence CSR development initiatives undertaken by the extractive industry. Boon's chapter provides a categorization of possible government roles in CSR and illustrates the issues using examples from a series of interviews with key stakeholders in Canada and Peru. The case study demonstrates that both host government and home government have either failed to see or have not yet been able to take full advantage of opportunities presented by CSR in the mining sector.

Bonnie Campbell, Etienne Roy-Grégoire, and Myriam Laforce's chapter, 'Regulatory Frameworks, Issues of Legitimacy, Responsibility and Accountability: Reflections Drawn from the PERCAN Initiative,' weaves many of the issues raised in the previous chapters by Dashwood, Webb, and Boon into the examination of a specific home-host state cooperation program, The Peru–Canada Mineral Resources Reform Project (PERCAN). Campbell, Roy-Gregoire, and Laforce's chapter adopts a historical perspective in order to explore the problems PERCAN seeks to address, examining the roles played by regulatory frameworks and both public or private actors in shaping the environment in which mining activities take place. In doing so, they explore the issues of legitimacy, responsibility, and accountability that are often at the origin of conflicts. The chapter analyzes the strategies implemented by PERCAN in order to determine the extent to which they address the problems that they set out to resolve. The example permits deepening understanding of the complex and sometimes contradictory roles of various actors in current initiatives to address host country governance issues.

Though, in principle the 'governance buck' stops at the state's desk, in reality this is not often the case. Ian Smillie's chapter, 'Conflict Diamonds: The Kimberley Process and the South American Challenge,' elucidates the consequences of systemic governance gaps and the challenges faced by multi-actor collaborations intended to address them. Africa's diamond wars took countless lives between the end of the Cold War and the early part of this century's first decade. To end the phenomenon, governments, industry, and campaigning NGOs created a legally binding global certification system called the Kimberley Process (KP). The diamond industry, completely unregulated and firmly rooted in nineteenth-century notions of CSR, was obliged to change dramatically as the twenty-first century dawned. The first and largest challenges for the KP were not in Africa, however: they were in South America – in Brazil, Guyana, and Venezuela – and where KP effectiveness is concerned, the jury is still out. The industry, having relied on the KP to improve its tarnished image, is now developing other mechanisms aimed at providing consumers with the confidence that governments either cannot or will not.

Catherine Coumans' chapter 'Whose Development? Mining, Local Resistance, and Development Agendas' focuses the reader's attention on what is perhaps the primary driver of the extraordinarily high levels of CSR activity in the mining industry – the increase in local opposition to mining and the international support and exposure that this opposition has received. She describes the concerted corporate response to this opposition through campaigns directing attention to mining's positive contributions to human development through employment, taxes, and royalties, and local CSR projects, while at the same time seeking to involve governments and civil society actors in development projects at mine sites. The chapter examines the interests of various stakeholders in the debate about 'mining and development' in the context of ongoing struggles by community members to protect the basis of their livelihoods and to determine their own futures. This chapter calls for greater transparency and public dialog on the issue of mining and local level development, particularly on the part of Canadian government agencies and development NGOs who seek to partner with mining companies.

Nicole Marie Lindsay's 'Mining Industry Associations and CSR Discourse: Mapping the Terrain of Sustainable Development Strategies' addresses a fundamental question in the analysis of the perspectives explored in this book – 'are we speaking the same language?' The chapter examines the discourse of CSR in the mining industry as a mode of speaking and thinking about the relationships between mining companies, the environment, and the communities in which they operate. Highlighting the ways in which discourse both shapes practices on social and institutional levels, and is shaped by the broader political economic context in which it emerges, she argues that the relationship between CSR and the political–economic project of neoliberalism warrants greater attention. Further, in analyzing the discourse of CSR as articulated by three prominent mining industry association codes and guidelines, and comparing industry CSR discourse with one example of a discourse of resistance to mining, Lindsay's chapter seeks to show the profound differences in issue framing and worldview represented by diverse discourse communities engaged in negotiations over mining development.

Ana Carolina Alvares da Silva, Silvana Costa, and Marcello M. Veiga's chapter,' 'Drivers of Conflict around Large-scale Mining Activity in Latin America: The Case of the Carajás Iron Ore Complex in the Brazilian Amazon,' delves further into the drivers of conflict surrounding mining, as well as into the corporate practices and social and political factors that influence governance dynamics in the region. Historically, large-scale mining activities in the Brazilian Amazon have been associated with conflicts with artisanal gold miners, who compete for the same resources; aboriginal people, who do not have autonomy to exercise their mineral rights; and the widespread landless people movement, MST, which advocates and takes action for more equitable distribution of land and resources. da Silva, Costa, and Veiga discuss these actors and their relationships to key drivers of conflict situations. They demonstrate how challenging the management of a mining operation can be in an ecologically sensitive area with restricted governance, political instability, and a social legacy of conflict. The results of their study provide lessons learned for implementation of firm-level CSR policies on the ground.

Robert Cameron's 'Community and Government Effects on Mining CSR in Bolivia: The Case of Apex and Empresa Huanuni' case study examines the potential for mining operations and their CSR to make meaningful contributions to sustainable development, particularly in a context where legacies of poverty have left communities with low capacity for governance and desperation for sources of income creates internal divisions. This context of poverty is further complicated where government social services and regulatory capacity are largely absent or of low quality. Cameron compares the CSR practices of two companies, a state-owned tin mining company and a transnational silver mining company, with a focus on how specific factors in the Bolivian government and in the local communities motivated, assisted, or impeded the contributions that CSR could make to sustainable development.

As the 2008 global financial meltdown clearly demonstrated, finance is not only about money – it is about country-specific laws, industry-wide regulations and norms, watchdog enforcement agencies, public policy alternatives, economic theories, corporate governance practices, and so on. Allen Goss' chapter, 'Corporate Social Responsibility in the Extractive Industries: The Role of Finance,' provides a primer on the kinds of finance available at different stages of the mining development process and sheds light on the potential for finance to more positively shape the conduct of firms. Drawing on theoretical and empirical research, Goss explores the potential for both public and private debt and equity to influence corporate behavior, asking how sensitive investors are to the activities of the firms they invest in, whether investor boycotts influence corporate behavior, and whether banks and bondholders care about the environmental, social, and governance records of the firms to which they lend.

Irene Sosa's chapter, 'Responsible Investment Case Studies: Newmont and Goldcorp,' provides a bridge between the social and environmental concerns of affected communities and advocacy NGOs as highlighted by authors such as Campbell et al., Cameron, and Coumans and the environmental, social, and governance (ESG) concerns of investors and corporate managers as presented by Goss. The chapter explores the impact of active ownership and responsible investment (RI) on CSR. Sosa argues, for example, that since the interests of responsible investors are aligned with the company's financial performance, shareholders can establish a dialog with management that other stakeholders such as NGOs, often labeled 'anti-mining,' might find difficult to achieve. Moreover, shareholder engagement can give marginalized groups a new route of access to companies. Though currently a small and modestly powerful trend, Sosa expects that RI groups will continue growing, especially though coalition building. The chapter examines two case studies of responsible investors engaging with two large mining companies active in Latin America through shareholder resolutions: Goldcorp Inc. and Newmont Mining Corporation. She concludes that, as a case of active ownership, Newmont's can be considered more successful than Goldcorp's. Sosa examines the contextual issues that have fostered the growth of RI, as well as the factors that might enable the success of shareholder activism. Her chapter concludes with an assessment of the potential and limitations of RI to influence CSR in the mining industry.

Corruption – private, public, and personal – may present the greatest 'disabler' to the achievement of social and environmental value. Carol Odell's chapter, 'Anti-corruption: A Realistic Strategy in Latin American mining?,' explores the relationships between corruption and mining in Latin America. Drawing from emerging economic research, Odell examines the extent and effectiveness of corporate anti-corruption strategies in enabling mining companies operating in Latin America to pursue their business objectives and mitigate escalating risks while contributing to social value. Odell explores the costs, benefits, and risks of a range of corporate responses to corruption for different types of companies in a variety of Latin American mining contexts, as well as identifying opportunities for and threats to continued anti-corruption efforts in the region.

The authors in the book have argued to some extent that CSR may have some potential, but alone is not enough. It might provide a license to operate, but it won't deliver sustainable development. Thus, multi-actor models of long-term and broad-based development and collective governance over resources are being actively sought. Fabio Abdala's chapter, 'Sustainable Juruti Model: Pluralist Governance, Mining, and Local Development in the Amazon Region,' presents one such model.

Alcoa is spearheading the Sustainable Juruti project at the site of its bauxite mine near Juruti City in the Brazilian Amazon region. The Sustainable Juruti model is based on a multi-institutional governance partnership aimed at providing mutual benefits for companies, communities, and local government. Its components include the Sustainable Juruti Council, a permanent forum for dialog and collective action among the parties, considering a long-term agenda; sustainability indicators used to monitor the development of Juruti and to provide the Council with accurate and valid information; and the Sustainable Juruti Fund, a vehicle through which to finance the activities prioritized by the Council to mobilize the resources needed to generate an endowment fund for present and future generations.

Timothy M. Shaw's 'Energy and CSR in Trinidad and Tobago in the Second Decade of Twenty-first Century' brings the arguments back to an underlying question – can resource exploitation provide a route to development? Shaw recounts how at the height of the energy boom toward the end of the last decade, Trinidad and Tobago (T&T) claimed to be next in line to become a developed state, but questions whether it can transform its offshore gas reserves into such a political economy over the next decade. On the positive side, after a hundred years of oil, T&T has established two major industrial estates at Points Lisas and Fortin, the latter for liquefied natural gas (LNG). On the negative side, its infrastructure is deficient, its ecological status problematic, and its level of violence is high. Its developmental prospects are augmented by its diaspora and their demand for cultural and other service sector exports. And whilst T&T is a formal democracy, it is a 'flawed democracy' with limits on its advocacy of CSR initiatives such as the Extractives Industries Transparency Initiative (EITI). In short, the plausibility, let alone sustainability, of T&T as an aspiring industrial economy is questionable, with implications for comparative analysis and practice in meso- and macro-regions and beyond. T&T may remain the quintessential example of a rentier economy, state, and society.

Since the African continent represents the second largest recipient of Canadian mining FDI (Foreign Direct Investment; CAN$15 billion), the editors felt that that the last chapter of the book should provide a window into the African context. The editors invited Ralph Hamann, Paul Kapelus, and Ed O'Keefe to provide this 'reflection from Africa,' in part because of their intimate knowledge of the sector and of the continent, and in part, because they were pioneers in the conception of CSR as an element of a complex multi-actor governance system. 'Mining Companies and Governance in Africa' contributes to the overarching theme of this book with a focus on the African context, arguing that corporate responsibilities can be better understood and acted upon when seeing corporations as part of a broader governance system. The African case studies suggest that companies do not just respond to their external governance context, but because of a range of operational and reputational drivers, they also feel the need to contribute to more effective and legitimate governance arrangements beyond the firm boundary. In other words, they seek – and are trying to learn how – to contribute to relationships between stakeholders, rather than just relationships between themselves and stakeholders. The authors situate this analysis theoretically by explaining the need for companies to do so as a result of 'limited statehood' in their operating environment in many parts of Africa – that is, because political institutions are too weak to hierarchically adopt and enforce collectively binding rules. In such circumstances, companies recognize the state's inability to enforce collectively binding rules and to provide collective goods as a prominent risk to effective operations or corporate reputation. One of the many challenges for companies becoming involved in governance relates to the important and difficult task of finding or even establishing legitimate stakeholder representation structures.

In the concluding chapter, Nicole Marie Lindsay and Julia Sagebien tie together the main arguments presented in previous chapters, providing a fuller analysis of components of the proposed SEVGE model and suggesting lines of inquiry for future research.

Lastly, the Appendix provides a brief overview of the level of mining activity in the region, Canadian exploration efforts, and direct investment in the region's mining sector.

The present volume is an interdisciplinary and an exploratory work, and as such there are limitations. For example, from a thematic point of view, there are many under-explored issues in this volume on Canada–Latin American mining CSR. Among them, four issues stand out as priorities for future research: first, the role of gender in development; second, the importance of environmental imperatives; third, the impact of Chinese mining FDI in Latin America; and fourth, the role of media and especially electronic media in information sharing and in the formation of climates (and sub-climates) of opinion. Further comparisons with Africa, Asia, and other BRIC countries besides Brazil are also promising research areas.

1
Systemic Causes, Systemic Solutions

Julia Sagebien and Nicole Marie Lindsay

Introduction

This chapter presents the basic arguments informing the approach taken by the editors in organizing this book. First, we argue that poverty, social exclusion, and environmental degradation, while often clearly exacerbated by irresponsible business, exist within a broader local and global political economic context in which irresponsibility and lack of accountability are built into the system. Second, Corporate Social Responsibility (CSR) strategies conceptualized as a set of discretionary or voluntary actions originating within a company can provide firms with a strategic response to some of the risks that systemic dynamics present, especially in the developing world. By addressing governance gaps, systemic risk is lowered, and firms increase their potential of obtaining a 'social license to operate.' However, third, we argue that firm-centred CSR as currently conceptualized cannot be expected to bring about the long-term, transformative change needed to address multi-actor, system-wide issues. Fourth and finally, given these limitations, new analytical models that can capture system-wide dynamics and put CSR into context within a broader governance system should be used as a complement to traditional stakeholder-management CSR planning tools.

The individual authors collected in this book adhere to most of the tenets presented in the four premises outlined above, although to varying degrees and with different points of emphasis. However, in this chapter the editors take one additional step on which there is no particular consensus or agreement (although we hope this collection will stimulate discussion leading toward further insights and agreement). We introduce an analytical model that, we suggest, might better capture and represent the systemic dynamics influencing development outcomes than traditional modes of understanding CSR as 'stakeholder management.' By bringing into focus a broader range of actors, goals, capabilities, and strategies involved in debate, negotiation and/or struggle over sustainability, development, and governance in the mining industry, the 'Social and Environmental Value Governance Ecosystem' (SEVGE) model can provide finer-grained understandings of the complex dynamics surrounding CSR and development issues. It can be used alone or as a complement and iteratively with more traditional stakeholder planning models, and although it has been developed as a tool for better understanding the political economy of mining in the developing world, we suggest that it has application in other sectors and geographic locations.

Unlike firm-centric CSR stakeholder planning models, the SEVGE model has social and environmental value (SEV) as its hub. The hub is surrounded by an array of political actors (which may or may not be characterized as firm 'stakeholders'), of which the firm is but one. These actors can either enable or disable the overall system's capacity to protect or enhance social and environmental value in any given geographic region through actor-specific tools and capabilities such as, in the case of the mining firm, CSR programs financed by internal profits. Since the various actors are generally 'role bound,' and thus more capable in some types of strategies and with some types of tools than others, for those engaged in development planning it is crucial to conceptualize and design strategies that optimize individual actor contributions as well as leverage the contributions of other actors.

The SEVGE model maps a system of *governance* because the actors in the system are political actors involved in making decisions over the management of resources – decisions that impact the collective social and environmental value of the entire system. We argue that the relationships forming these interlocking acts of governance can be conceptualized as an *ecosystem* – a community of interacting organisms and their physical environment – where each organism and specific interaction between components shapes not only the organisms primarily involved in the interaction, but also the entire system itself.

For those involved in CSR planning from a corporate perspective, we propose a 3-step planning process for CSR strategy development that uses both stakeholder and SEVGE models iteratively. First, stakeholder models would be used to conduct a situation analysis and to establish a baseline for the firm's relationship with its stakeholders as traditionally defined. Second, a broader analysis of the issues affecting a particular locale, or a particular firm at a particular time would be conducted using the SEVGE model. This would help firms better understand the point of view of each actor and how SEV might be best obtained in a given context by focusing attention on how various actors define or understand SEV, what mechanisms various actors can use efficaciously to obtain it, and how CSR programs impact these objectives or their efficacy, and vice versa. Firms would need to consider whether other actors using different tools might better achieve SEV goals, and if so, how firm activities might aid or hamper such efforts. Lastly, firm CSR strategists could use the stakeholder model again to focus their CSR strategy on the most important issues, stakeholders, and/or projects from a *systemic* perspective.

Using both models in an iterative fashion takes into account the fact that while the act of developing and implementing CSR initiatives lies firmly within the firm's domain and its intent is to strategically advance corporate interests, the impact of these initiatives on SEV depends to a great extent upon the enabling or disabling dynamics resulting from the actions and interactions of a host of actors outside the boundaries of the firm. Further, the efficacy of firm CSR depends as well as on the capacity of all systemic actors to conceive of a set of parallel or aligned objectives, even if conceived as opposing viewpoints (for example, Non-governmental Organization (NGO) critiques of specific industry externalities and the inadequacy of CSR responses usually contain well-founded insights that could be used to improve the positive impact of CSR).

We acknowledge that the proposed model has limitations. For example, some actors may not easily fit into categories shown in the model, nor are these categories definitive. Furthermore, the model should be seen as dynamic and multi-dimensional, with actors forming and reforming a variety of strong or weak relationships with other actors, affected by a greater or lesser degree by contextual forces, and offering varying levels of enabling or constraining influences on SEV outcomes. A refined assessment of the actors/institutions involved in any given development context, along with a finer-grained explanation of the full range of their enabling or disabling dynamics and interactions requires further study as well as adaptation to specific circumstances. It is, nevertheless, our hope that this introduction and overview might provide the impetus for researchers working in a broad range of industries, fields of knowledge, and thematic areas to take up this challenge.

The four premises in detail

Premise 1: Poverty, social exclusion, and environmental degradation, while often clearly exacerbated by irresponsible business, exist within a broader local and global political economic context in which irresponsibility and lack of accountability are built into the system.

This is by no means a new observation, and it is one that has led many to suggest that poverty and ecological degradation are likely or inevitable outcomes of a political economic system that is built around (and perpetuates) fundamentally asymmetrical power dynamics between social actors, as well as a subordination of social and environmental well-being to economic gain (Escobar, 1995; Banerjee, 2003, 2008). In other words, the limitations of the 'the market for virtue' (Vogel, 2005) are rooted in the way in which business–society relations are structured. Unless CSR planners and managers can alter these structures and dynamics, CSR programs and policies will remain open to charges of 'greenwashing' in which surface-level and short-term benefits are seen to be exchanged in a veiled attempt to secure a social license to operate and enhance corporate gain.

In Latin America and other parts of the developing world, the tensions and dynamics associated with neoliberal globalization are particularly visible. A broad sweep of the political economic history of Latin America highlights the deeply conflicted and asymmetrical nature of power relations between the primary actors engaged in negotiations over the shape and trajectory of economic development in the region (see, for example, Cardoso and Faletto, 1979; Harris and Nef, 2008). Historically speaking, poor rural communities have had little input into the decisions and policies that have drastically affected their lives and livelihoods, most often to the greater benefit of elites making the decisions than to the communities who must live with them (Escobar, 1995). The more recent dynamics observed in Latin America's history of asymmetrical development could perhaps be ascribed to the same types of processes that have shaped globalization as a whole.

According to Stiglitz (2006), while globalization has been driven by economics, it has been shaped by a politics that furthers the interests of powerful groups of economic actors. These actors have been influential in defining the rules of the game, and they have not sought to create fair rules, let alone a set of rules that would promote the well-being of those in the poorest countries of the world. Rather, globalization has been used to advance an extreme version of market economics that is more reflective of corporate interests than those of society as a whole.

Different economic sectors and different nations have shaped and been shaped by these processes in various ways. In the 1990s, for example, mineral-rich nations such as Peru, Chile, and Mexico undertook dramatic structural reforms in the interests of creating an investment-friendly climate aimed to attract foreign capital into a number of sectors, most notably the mining sector (Bridge, 2004). Endowed with some of the most promising mineral reserves in the western hemisphere and backed by international financial organizations (IFIs) such as the World Bank (WB) and the International Finance Corporation (IFC), these countries succeeded in turning their formerly languishing state-run mining industries over to the highly efficient and exceedingly profitable enterprises that characterize mineral extraction operations throughout Latin America today.

The growth in mining investment and mineral exports, however, has largely failed to translate into real gains for the vast majority of ordinary citizens in mining-dependent economies, especially in developing nations (Pegg, 2006). This is particularly true for the rural communities most directly affected by mining who also bear the brunt of heightened environmental risks associated with the industry (Bebbington et al., 2008a). Host government corruption and lack of capacity is frequently cited as the primary obstacle to adequate 'trickling down' of mining revenues to the local level (see Bastida et al., 2005, for an example of 'trickle down' problems in Peru). However, an analysis of the broader political economic context of mining investment in the developing world demonstrates that the problems are much more complex than this would suggest. For example, Campbell (2003) points out that regulatory reforms in Tanzania, Mali, and Madagascar undertaken during the 1990s at the behest of the World Bank reduced state governance capacity to the point that the development goals of these nations were seriously compromised. Bridge (2004) shows that the economic policies adopted by many Latin American countries during the 1980s and 1990s increased not only the extensity (number of new areas) of mining development, but also the intensity of these developments, in terms of the scale and value of the individual operations. Further, Coumans argues in this volume (Chapter 7) that an over-focus on 'governance gaps' in the capacity for host governments to effectively regulate the industry given the increased numbers and scales of new mining projects risks veiling efforts to shift too much responsibility for the negative environmental and social impacts (and costs) resulting from these operations onto governments and development NGOs.

Taking into account Coumans' observations in Chapter 7 regarding the risk of focusing too much on governance gaps defined narrowly in terms of a host government's capacity to deal with the negative effects of mining, there remains a great deal of evidence showing that a retrenchment of (or relative weakness in) state governance capacity often exists alongside rapid expansion of mining development. In such contexts of weak governance, failed development, broken promises, social division, environmental and health impacts, and legacies of conflict and violence are as closely identified with the mining industry as are economic and infrastructure development and job-creation (Muradian et al., 2003; Bebbington et al., 2008a, 2008b; Bury, 2008; Calvano, 2008; Slack, 2009; UNEP, 2009; see also MiningWatch Canada (www.miningwatch.ca) and Rights Action (www.rightsaction.org) for NGO accounts of mining impacts.

Although corruption and lack of government capacity is certainly an issue that needs continued attention, as Coumans (Chapter 7 in this volume), Campbell (2003), and others have pointed out (Escobar, 1995; Banerjee, 2003, 2008), a more radical re-conceptualization of the logic informing development economics may be necessary to address the root causes of problems such as deepening rural poverty, environmental degradation, and so on.

Premise 2: CSR strategies conceptualized as a set of discretionary or voluntary actions originating within a company can provide firms with strategic response to some of the risks that systemic dynamics present, especially in the developing world. By addressing governance gaps, systemic risk is lowered, and firms increase their potential of obtaining a 'social license to operate.'

The governance gaps exacerbated by neoliberal globalization present significant operational risks for not only for business, but also for nations and communities. As John Ruggie, the Special Representative of the United Nations Secretary-General on the issue of human rights and transnational corporations, pointed out in his 2008 report:

> The root cause of the business and human rights predicament today lies in the governance gaps created by globalization – between the scope and impact of economic forces and actors, and the capacity of societies to manage their adverse consequences. These governance gaps provide the permissive environment for wrongful acts by companies of all kinds without adequate sanctioning or reparation. How to narrow and ultimately bridge the gaps in relation to human rights is our fundamental challenge. (Ruggie, 2008, p. 189)

Many other risks plague the mining industry. Mining involves difficult terrain, dangerous procedures, harmful substances, and an abundance of hazards to employees. Adjacent communities suffer great threats to their natural environment, livelihoods, social structure, and health while often garnering only a miniscule part of the return from mining operations. Historically poor environmental, social, and labor practices in the mining sector, which have been the norm and not the exception, have aggravated the risks. For example, Peru is littered with *pasivos ambientales*, abandoned mines that continue to pollute groundwater with leeching heavy metals, as well as with *pasivos sociales*, communities whose livelihoods and health have been gravely affected by irresponsible mining operation, leaving a testament to 'dirty' mining.

Boom/bust cycles make mining finance and operations very risky. Mining is also capital intensive, geographically fixed, and long-term, requiring that companies either successfully mitigate significant risks or else face potential delay, increased expenditures, or even closure resulting from conflict with local communities and/or host governments. The ores are also finite, so there is a limited time window in which mineral-rich nations can convert mining revenues into development that is sustainable once the mineral reserves can no longer be exploited. For states, the risk of skewed economic development and the 'resource curse' is always close at hand. Further, the metals and ores are often found in jurisdictions with high political or human rights risk, thin rule of law, and high levels of corruption. A glance at any Maplecroft's Risk Maps or AON's Political Risk Maps offers a sobering view of the risks inherent in Latin American operations.[1]

In Latin America, another risk for the extractives industry is in 'resource nationalism' driven both by national governments' desire to benefit from the high prices of commodities, as well as by perceptions that the mining industry has not contributed enough to society. In the first decade of the twenty-first century, populist governments in several Latin American countries (Ecuador, Venezuela, Bolivia) either threatened nationalization of extractive sectors, or undertook legislative reforms moving in this direction. Recently the Peruvian government established a not very voluntary 'Voluntary Fund' (*Aporte Voluntario*) in order to garner additional royalties from mining companies without having to revise the mining law (called *Canon Minero*) formulas. While the 2010 Argentinean law protecting glaciers from mining exploitation, and the Ecuadorian decision to temporarily halt oil production in the Yasuni National Park jungle area in exchange for United Nations financial support, are certainly extraordinary wins for environmentalists (and we would argue for humanity), for firms planning to exploit these resources (for example, Barrick in Argentina), this type of legislation presents a risk to corporate bottom lines.

CSR has become the primary tool used by many multinational mining companies to address this wide-ranging set of risks and associated governance gaps in developing nations. A review of the CSR activities of Canadian mining companies in Latin America suggests that a significant proportion of them profess adherence to CSR policies, highlight their CSR performance in regular reporting, and engage in various voluntary regulation and reporting initiatives (Sagebien et al., 2008). The activities undertaken as a part of these CSR initiatives include community consultations, infrastructure development (for example, building housing, schools, roads, electricity), employment and economic development programs, training and jobs, health care, direct funding of community projects, and environmental initiatives designed to remediate the impacts of mining operations (ibid.).

Premise 3: CSR as currently conceptualized cannot be expected to bring about the long-term, transformative change needed to address multi-actor, system-wide issues.

Debates about the efficacy and usefulness of CSR, both from critical and from mainstream management perspectives, continue to rage. In evaluating the 'dark side' of success in the corporate citizenship movement, Waddock (2007) highlights several 'disconnects' between intent and practice in her review of the potential and limitations of CSR policies. She argues that the dark side of CSR policies includes the following four tendencies:

> (1) The short-term orientation on which both companies and financial markets operate and the long-term societal issues that short-term thinking creates; (2) an overly narrow focus on corporate citizenship as explicitly doing good, while ignoring other effects of company behaviour; (3) the gap between rhetoric and reality of many companies' corporate citizenship; and (4) the reality that most corporate citizenship agendas, even when quite broadly stated, fail to deal with the significant risks, impacts, and practices of companies that result from their business models. (Waddock, 2007, p. 255)

Development practitioners and scholars, like business executives and NGO staff, are also concerned about the lack of robust empirical evidence supporting the claims of CSR's contributions to development (see Sagebien and Whellams (2011) for a summary of the debates on 'whether CSR is good or bad development'). Despite the alleged and real benefits experienced by many communities through firm CSR policies and practices, there is also still much debate about the long-term impacts of mining itself, as well as concern around the paternalistic, firm-dependent relationships that can emerge when the company takes a leading role in community development through its CSR policies.

In the case of mining interests in developing countries, the gap between the intended outcomes of CSR and its actual outcomes is, in great part, due to the reality that obtaining a license to operate by addressing the governance gaps on the one hand and achieving sustainable development on the other are in often practice two very different things.

In sum, given the limits of CSR, as well as the fact that in developing countries responsibility and accountability have rarely been built in a meaningful way into political systems, the isolated, piecemeal efforts that have characterized CSR practices fall short in terms of addressing the dysfunctional, risk-ridden and unequal systems that are at the root of the issues that CSR programs seek to address.

Premise 4: New analytical models that can capture system-wide dynamics and put CSR into context within a broader governance system should be used as a complement to traditional stakeholder-management CSR planning tools.

Since the earliest iterations of stakeholder theory emerged in the management literature with Freeman's (1984) assertion that business should be understood in terms of its relationships among various groups of actors with interests at stake regarding business impacts, CSR discourse has modeled the dynamics of the business/society relationship as one of firm/stakeholder relations. In CSR strategy development, Freeman's stakeholder model has been used as a way to define CSR as a set of discretionary or voluntary actions originating within a company as a response to stakeholder pressure or market opportunities and risk. Stakeholder analysis and planning models of CSR are ubiquitous in the academic and practitioner literature.[2] They may vary in their finer details, but they tend to have a number of central elements in common: (1) the firm is at the center of the business–society relationship; (2) entities which interact with the firm and have some impact upon it are conceptualized as 'stakeholders' of either primary or secondary importance; and (3) CSR is used as a 'stakeholder management tool,' the ultimate purpose of which is to add value to the firm.

In their recent work, Freeman et al. (2007) have built on the concept of 'stakeholder management' to develop instead an approach that they call 'managing for stakeholders.' They argue that:

> In the business world of the twenty-first century the very purpose of a business in society is connected with creating value for stakeholders. We can better understand business by seeing it as an institution for stakeholder interaction. Corporations are just the vehicles by which

stakeholders are engaged in a joint cooperative enterprise for creating value for each other. Capitalism, in this view, is primarily a cooperative system of innovation, value creation and exchange. (p. 6)

Clear graphic representations of explanatory models are useful management tools because they help conceptualize complex issues in simple ways, and can thus help solve problems expediently and plan, implement and control competitive strategies. Freeman's recently updated model shown in Figure 1.1 has already become a CSR classic for its heuristic power.

Critiques of this model have begun to emerge in recent years. For example, Greenwood (2007) argues that the conflation of stakeholder engagement with 'CSR in action' and responsible practice is problematic. She points out that high levels of stakeholder engagement do not mean high levels of responsibility because 'engagement' is a morally neutral practice. She proposes a model that allows for the coincidence of stakeholder engagement with corporate responsibility and the moral treatment of stakeholders, as well as for the coincidence of stakeholder

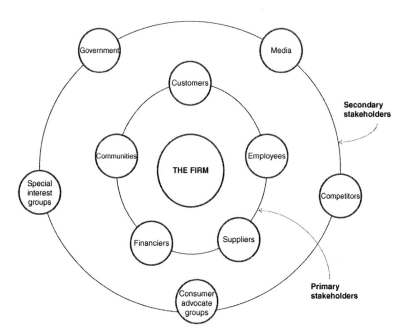

Figure 1.1 Basic two-tier stakeholder map
Source: Adapted from Freeman et al. (2007).

management and corporate irresponsibility and immoral behavior. In other words, high stakeholder engagement does not automatically result in responsible or moral outcomes.

If stakeholder engagement is to go beyond moral neutrality, the CSR practices that derive from this engagement must be truly responsible and result in an increase in collective social and environmental value. Banerjee (2008) suggests that 'alternate visions' of CSR must reject the standard emphasis on win–win situations and must recognize that, if it is to be genuine, CSR might entail 'giving back' power to subaltern groups who have historically been marginalized in economic decision-making processes. Kurucz et al. (2008) echo the recognition that traditional firm-centric stakeholder and business-case views of CSR are inadequate to meet the challenges presented by a changing business–society relationship. They advocate for a reworking of the CSR business case along lines that acknowledge complexity, move away from an organization-centric view of CSR, and build integrative capacity for a more holistic understanding of the business–society relationship. They frame the organization as firmly embedded within a broader social context – that is, a part of an 'integral commons' that foregrounds broader social needs rather than just the needs of either shareholders or stakeholders as conventionally defined. This, they point out, requires a 're-imagination of business strategies' (Kurucz et al., 2008, p. 102) that takes into account the interdependent relationship between business and society – and, we would add, the planet's finite resources.

The limits of the firm and its CSR programs as a macro-level sustainable development agent is perhaps one of the reasons why there has been a pronounced increase toward collaborations and partnerships between companies, governments, and civil society. Waddock (2008) refers to the emergence of a 'corporate responsibility (CR) infrastructure,' which encompasses a range of actors and institutions working either in partnership or unilaterally to enhance CSR performance through a variety of mechanisms and processes. This CSR/CR infrastructure, according to Waddock (2008), includes market/business sector initiatives such as codes, standards, reporting and monitoring/verification, along with CR consulting, accreditation, and responsible investing institutions. Civil society and multi-sectoral initiatives include groups that form over specific issue-areas to exert influence over the CR of business firms. Under this category of CR infrastructure, Waddock points to watchdog and activist groups that monitor corporate behavior and engage in information sharing, reporting, and naming and shaming campaigns. Other civil society and multi-sectoral initiatives include media initiatives (investigative journalism), and ranking and rating organizations. State/government initiatives include at the national level legislation such as the U.S. Sarbanes–Oxley Act or the Alien Claims Tort Act, and at the international level, agreements such as the Kyoto Accord.

While the empirical outcomes of the emerging 'CR infrastructure' will require further analysis to determine the degree of change actually taking place, Waddock (2008) points out that the fact that many multinational corporations are responding to increased demands for CSR 'signals potential fundamental shifts in the rules of the game that companies abide by' (p. 107).

Scherer and Palazzo (2007) argue that more than the rules of the game are changing – in fact, the game itself has fundamentally changed in ways that alter the roles of business firms and the business–society relationship. They point out that both the global expansion of the corporation, as well as the kinds of partnership addressed above have led to the politicization of the corporate role in society and the politicization of CSR, resulting in serious questions regarding the appropriate role of business firms in democratic societies.

Scherer, Palazzo and Matten (2009) point out that the process of globalization has undermined two fundamental assumptions regarding the role of business in society. The first assumption is based on a notion of the 'division of labor' between politics and governance (the realm of state governments) on the one hand, and economics on the other. This view holds that business firms should only be concerned with economic matters – that is, creating economic wealth. The second outmoded view is that governments have the capacity (and jurisdiction) required to control and regulate corporate behavior for the common good. However, as the authors point out:

> Today, businesses do not necessarily operate within the borders of a clearly defined legal system and a more or less homogenous set of social expectations. Instead many operations are shifted offshore and beyond the reach of the rule of law or the enforcement of taxes or regulations... In addition, nation state institutions face social and environmental challenges that have transnational origins and cannot be regulated or compensated unilaterally by national governance. (Scherer et al., 2009, p. 332)

Zadek (2006) and Zadek and Radovich (2006) point out that partnerships are a form of collaborative governance, which they argue has become the most important emergent institutional form in global governance. Though some of these multi-stakeholder partnerships are likely to remain a simple 'institutional patch designed to overcome technical glitches,' some have moved from being experiments to becoming the mainstream institutional foundations for the delivery of services, resource transfer, and rule setting. Midtun (2008) also suggests that the confluence of four factors – CSR-based self-regulation, public policy oriented toward engagement with business, civil society organizations capable of establishing 'moral rights' as credible voices in 'just causes,' and a 'media driven communicative society' (p. 406) – is creating new forms of partnered global governance that can address governance gaps. In a similar vein, Boxembaum (2006) has suggested that CSR is an eclectic and malleable 'institutional hybrid' composed of a combination of heterogeneous institutions that can be adapted to suit specific local and institutional circumstances.

Though the notion of private sector agents voluntarily addressing governance gaps, alone or in partnership, is laudable, it also somewhat questionable not just from a veiled self-interest point of view (how else to obtain a social license to operate?) but also from a political legitimacy point of view. Utting (2005) suggests that CSR risks involving a 'transfer of regulatory authority to largely unaccountable agents and renders more stable and palatable a model of capitalism that generates or reinforces widespread social exclusion, inequality and environmental degradation' (p. 23). He argues that this is likely to take place in instances when the CSR agenda ignores or marginalizes issues related to empowerment, redistribution, and strong public sector and civil society organizations. Utting concludes that CSR should be both activist and self-reflexive, with an orientation toward social justice, ecological stewardship, and participatory development alternatives.

As the above review shows, we are not alone in suggesting that new analytical tools are needed in order to account for and better understand new business–society relationships and the CSR/CR infrastructure developing around it. Thus, the current challenge is how to model the relationship between business and society in a manner that can foster not just better long-term corporate value though better CSR, but also foster broader social and environmental value beyond the scope of that which directly benefits the business firm. More importantly, we need to be able to model the system of 'collective governance' in which the firm/CSR is embedded in order to better manage it, and to find the specific interventions that can more effectively foster system-wide social and environmental value-enhancing dynamics. In other words, we need to 'politicize' the discourse of CSR, while at the same time being careful to respect issues of legitimacy. We would caution against private sector forays into the public space beyond discretely defined spaces legitimately agreed to through consensual processes that involve the polity and the state. We are not suggesting a return to the 'company town,' but rather a complex and well-governed set of overlapping communities of interest where actors negotiate optimum systemic value.

Hamann et al. (2005) conducted a case study-based analysis of mining in South Africa, Mali, and Zambia that led them to acknowledge the importance of such shifts in traditional CSR approaches:

> Implementing corporate citizenships at a local level may require support for more sustainable patters of local governance, based on proactive and creative approaches to enhancing collaboration and responding to complexity.... [In these complex environments] traditional corporate citizenship activities based on unilateral company action and stakeholder engagement are unlikely to meet their objectives... [and] companies have learned that a more proactive involvement in moving local governance towards accountability and inclusiveness is necessary. (p. 18)

The authors map out these complex interactions and identify places of possible collaboration. In the following section of this chapter, we propose a planning model that follows along similar lines of thought as Hamann et al. (2005). We argue in the next section of this chapter that this systemic approach is useful in all circumstances, local or global, given the complexity of developing and implementing effective mining CSR.

Proposal: Social and environmental value governance ecosystem

We have argued above that, particularly in developing nations in the global South, CSR policies and practices directed toward immediate stakeholder groups and in response to local governance gaps, while potentially beneficial to the firm and to its selected stakeholders, possess limited potential to transform the systemic political and economic structures that create the conditions in which inequities and injustices persist, despite the best intentions and efforts of any corporate actor. In this section we introduce and describe a dynamic systems-level model that can help managers better understand and plan for collective governance.

26 *Systemic Causes, Systemic Solutions*

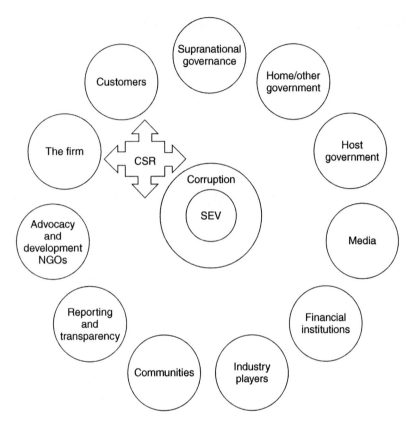

Figure 1.2 The Social and Environmental Value Governance Ecosystem (SEVGE) model

In order to address some of the limitations and shortcomings of current thinking about the role of CSR in the business and society relationship and of the models that are used as aids to conceptualization and planning, we introduce the Social and Environmental Value Governance Ecosystem (SEVGE) model (represented in Figure 1.2). This model conceptualizes the following:

(1) A collectively defined central *goal* as the hub of the system: that is, the creation, enhancement and protection of *social and environmental value*;
(2) The firm as just one of many role-bound *actors* embedded in a complex political system of conflicting and/or synergistic interests;

(3) CSR programs and strategies as just one of the *mechanisms* available to 'govern' this system, with other mechanisms available to other actors in the *collective governance* of the system;
(4) System-wide *relationships* between actors and *interactions* of system dynamics (including actions and inactions) that can either *disable* or *enable* multi-actor, multi-mechanism governance efforts; and finally
(5) *Corruption* as a corrosive, system-wide disabling factor.

In order to illustrate the differences between traditional stakeholder models and the SEVGE model, we draw an analogy from the natural sciences. Stakeholder management CSR strategy development models/tools are conceptualized in a *Ptolemaic* manner, with the firm as the center of the system, around which stakeholders revolve. Thus, stakeholder management CSR models assume a natural centrality of private sector agents in driving social/environmental goal setting. While this Ptolemaic corporate mental map/planning model has proven extremely useful in facilitating CSR strategies that address the primacy of the firm, even in the so-called 'win–win' situations, we argue that because the model does not capture the contextual and relational system-wide dynamics in which the firm and its CSR are embedded, it is insufficient and must be complemented by an analytical tool and CSR planning model that is not firm-centered.

We argue that any model that attempts to realistically capture the complex dynamics of interrelations present in a multi-actor system cannot have as the center or hub of the system the firm and its objectives. Instead, the shared and/or aligned aims of all actors are what, realistically, either contribute toward or detract from the attainment of a collectively defined social and environmental value, a notion that lies much beyond what any triple bottom-line, firm-centered stakeholder model can possibly capture. Thus the SEVGE model provides a more *Copernican* view of CSR stakeholder relations; it places at the hub of the system social and environmental value, surrounded by multiple actors – of which the firm represents only one. Each actor interacts with each other and with the hub mainly though specific role-bound tools and mechanisms.

Though the fact that firms are not the center of the universe is obvious, the conceptual frameworks of many corporate planning models generally assume the centrality of the firm. However, there are also frameworks that do not put the company at the centre of the universe.

For example, the *e3PLUS: A Framework for Responsible Exploration, Community Engagement Guide* (PDAC, 2009a) is careful to warn:

> **Mine/project centred thinking:** You must avoid thinking of your project as the centre of all activity and thought, with relationships arranged radially around the project. This perspective is usually not the reality experienced by the community, which had many functioning relationships before the project arrived and will continue to have them, and others, outside of the project. You must understand these relationships, and see your project or mine as one node in a more complex network of relationships. By this means, it becomes possible for you to place the project or mine within the community, rather than imposing it upon the community. (PDAC, 2009a, p. 8)

We use the word *governance* because the actors in the system are political actors involved in making decisions that affect the collective social and environmental value. Issues of power and legitimacy form part of this notion. We also use the concept of an *ecosystem* – a community of interacting organisms and their physical environment – because in such systems each organism and specific interaction between components shapes not only the organisms primarily involved in the interaction, but also the entire system itself.

The chapters that follow address a significant number of the actors and dynamics that the SEVGE approach models. In the conclusion, we examine in more detail a select number of actor groups, drawing insights from the authors collected in this volume to illustrate the complexities of the relationships, interactions, strategies, and tools that make up the governance ecosystem, serving sometimes to enable and other times to disable the protection and preservation of social and environmental value.

Conclusion

In the late 1990s and in the first decade of the new millennium, the poorest and most marginalized populations of the developing world have suffered the dire consequences of extreme forms of capitalism. The governance gaps that resulted from the simultaneous expansion of industrial/commercial activity and the retrenchment of the state that accompanies free-market globalization have stalled the process of effectively addressing these 'externalities.' In order to address the risks of 'cowboy capitalism,' companies have embarked on a myriad of CSR projects. However, these piecemeal, isolated actions cannot truly address the systemic causes of the problems they are trying to redress. The SEVGE model is an aid in mapping the systems thought necessary for creating more effective solutions.

Without systemic support, the reorientation of corporate behavior required to address deep systemic problems seems almost unthinkable. However, with a systemic view it is possible to strategically leverage the capacities of a variety of actors in order to move synergistically in a new direction. For example, supranational organizations and institutions have a significant role to play in setting norms and standards that place human rights, environment, and social inclusion at the top of the agenda; trade agreements can be built around equity instead of advantage; and public sector governance capacity can be developed along lines that increase accountability, responsiveness, and meaningful participation of civil society, at the same time as decreasing inefficiencies and corruption. Civil society needs to rise to the challenge of organizing, educating, and mobilizing to bring about viable development alternatives that are truly environmentally and socially sustainable. All of this, we argue, can be supported and enabled as a part of the CSR strategies of firms, but to do so will require a fundamental shift in values and understandings of the 'way things really work' in a non-firm-centric world.

Besides a possible element in a CSR management 'tool kit,' we see the SEVGE model as an attempt to capture the 'prisoner's dilemma' dynamics that mining and other resource industry multi-actor governance systems are facing. It highlights the rapid process of institutional learning that is taking place by a multitude of actors, concurrently and at a phenomenal rate, in order to even conceive of an optimum collective strategy. As is the case in prisoner's dilemma situations, though actors may be able to obtain optimum benefit from collective 'cooperate' moves, they iteratively also make 'defect' moves before there is sufficient institutional learning (if ever) as to create an equilibrium where all parties cooperate and can thus achieve optimum systemic collective benefit. Moreover, as in any game, credible sanctions for illegal behavior are needed in order for players to obey the rules. Sanctions are a part of learning, and the debate about going beyond voluntary and discretionary codes of ethics and codes of behavior into instituting state-mandated regulations and laws with enforceable sanctions is likely to intensify until they are achieved.

This chapter and this book are both primarily focused on the conflicts over mining in Latin America. However, we believe that this high level of conflict and the unprecedented effort to address it through CSR activity is akin to a 'canary in the mine' – an early warning system that all is not well in the global governance of resources. There are similar struggles over other natural resource industries in many home/host jurisdictions, and we are likely to see struggles over many more resources (water, energy, and food, to name three primary examples). It is thus imperative to find tools that can help provide parts of the solution. Modeling these interactions and dependencies is crucial, and we submit that the SEVGE model can provide a simple, intuitive start.

Notes

1. See http://www.maplecroft.com/portfolio/ and http://www.aon.com/risk-services/political-risk-map2/index.html, date accessed 16 February 2011.
2. For an extensive literature review, see Jamali (2008).

2
Sustainable Development Norms and CSR in the Global Mining Sector

Hevina S. Dashwood

Introduction

This chapter seeks to account for the process by which sustainable development emerged as a norm that informs the policies and practices of mining companies. As recently as 15 years ago, only a small number of major mining companies were reporting on their corporate social responsibility (CSR) policies, and fewer still had come to grapple with the notion of sustainable development. Yet, at the end of the first decade of the twenty-first century, the vast majority of major mining companies frame their CSR policies in terms of sustainable development, reporting on the economic, social, and environmental aspects of their operations. Since the mid-2000s, an increasing number of mid-tier companies are doing the same.[1]

The pattern whereby an initially small and then growing number of mining companies came to frame their CSR policies in terms of sustainable development is consistent with findings in the International Relations literature on the 'life-cycle' of global norms (Finnemore and Sikkink, 1998; Keck and Sikkink, 1998; Risse et al., 1999). Although that literature focused on the influence of emerging global norms on states, it can be readily applied to firms, which along with non-governmental organizations (NGOs), have come to play an increasingly important role in global governance processes alongside states (Ruggie, 2004). To explain the timing and impetus behind the adoption by mining companies of sustainable development as a normative framework for their CSR policies, insights from historical institutionalism are drawn upon (Campbell, 2006).

By the mid-1990s, major mining companies around the world were experiencing pressures from a variety of sources to improve their environmental performance and minimize the negative social impact of their operations. Environmental NGOs raised awareness of the serious environmental and social consequences of mining, and a series of major environmental disasters cemented the bad reputation of mining companies in the public's mind. Faced with public pressure, governments in advanced industrialized economies tightened environmental regulations pertaining to mining and began to restrict access to certain ecologically sensitive areas to mining. By the mid- to late 1990s, the mining industry had reached a 'critical juncture' (Pierson, 2004), which provided the impetus for the adoption of sustainable development as a normative underpinning of CSR policies.

This chapter traces the evolution of sustainable development in the mining sector over time, showing how reporting on the contribution of mining to economic performance, environmental protection, and social progress has become an accepted norm of business practice. The next section introduces the theoretical literature that helps to explain the emergence of sustainable development as a global norm. It explains how and why a normative concept can be applied to instrumental actors such as firms. It then explains the methodology for demonstrating how the sustainable development norm has evolved and strengthened over time. Section three traces the external pressures exerted on major mining companies in the advanced industrialized democracies and accounts for how a small number of mining companies (early movers) came to push globally for better practices within the mining industry. Section four outlines the results of these efforts, providing evidence for the widely accepted normative validity of sustainable development within the mining sector. The concluding section will reflect on the implications of these developments in the context of the themes of this book and provide suggestions for future research.

Theoretical and conceptual approach

The global normative context

To explain how it is that by the mid-2000s, the majority of major mining companies were framing their CSR policies in terms of sustainable development, it is necessary to consider the global normative context in which they were operating. Norms can be defined as collectively held understandings of acceptable behavior within a given realm (Finnemore, 1996). Global norms around human rights and protection of the environment have evolved quite significantly over the past few decades, as reflected in international human rights and environmental treaties. These treaties are binding on states, not firms, so the challenge has always been how to induce companies to accept wider obligations to society in the areas of human rights and the environment (one that the Special Representative of the UN Secretary General has sought to address; see Ruggie, 2008).

Much of the consolidation of global norms in the areas of human rights and the environment reflects prior and/or parallel developments in the advanced industrialized economies, from whence the initial major impetus in the evolution of global human rights and environmental norms originated. Within mining, the early pressure on companies to meet their social and environmental obligations came from their home governments, NGOs, and the public. The domestic context in which major mining companies from countries such as Canada, the US, Japan, the UK, and Australia were based produced pressures on these companies to improve their own CSR practices (Jenkins and Yakovleva, 2006), and to work globally to develop CSR norms. Such companies were acting strategically in response to external pressures, but also came to accept the normative validity of sustainable development, and were therefore responsive to larger shifts in the global normative environment (Dashwood, 2007a).

The norms life-cycle

According to the life-cycle theory of norms development, domestic support/pressure is crucial in the early stages of norms development (Finnemore and Sikkink, 1998). The literature on the dissemination of human rights norms shows that the initial pressure came from transnational advocacy networks of NGOs with a global reach but based in the 'West.' The fact that early efforts to promote CSR on the part of mining companies came from those that were headquartered in the advanced industrialized economies bears this observation out. By this logic, considerable explanatory weight must be assigned to the domestic (or state-based) sources of norms that eventually become accepted at the global (or international level) (Katzenstein, 1996).

Early efforts to promote CSR on the part of mining companies came from those that were headquartered in the advanced industrialized economies. At a critical point in time, a 'tipping point,' or norms cascade, occurred, such that a critical mass of companies came to accept sustainable development as a means to frame their CSR policies. This norms cascade occurred around the mid-2000s, when most major mining companies converged around sustainable development. This process is consistent with how sustainable development came to be an accepted norm among states, with the key difference that sustainable development achieved global normative status among states by the early 1990s, a full decade before major mining companies came onside.

Evidence of the acceptance of sustainable development norms comes both from discourse and patterns of behavior that are consistent with their prescriptions (Finnemore, 1996, p. 23). It is important to note the distinction between the acceptance of norms as discourse and the action required to follow through on them (Dingwerth, 2008). It is suggested here that the fact that mining companies have adopted sustainable development to frame their CSR policies is evidence of acceptance of the normative validity of sustainable development at the discursive level. While reference to sustainable development in their CSR reports may be dismissed as a mere public relations exercise, the acceptance of sustainable development at the discursive level is a significant development. As recently as the mid-1990s, the majority of mining companies were reactive and defensive about their environmental and social responsibilities. Mining companies lagged behind states in adopting sustainable development, and the norm was strongly contested in the context of mining (where minerals are understood to be a non-renewable resource). Although the challenge of developing suitable sustainable development indicators for mining is ongoing, mining companies' public pronouncements today demonstrate that they understand this norm to be an appropriate basis for action.

The literature on norms socialization helps to clarify the distinction between the acceptance of norms at the discursive level and actual behavior to promote them. Sustainable development can be said to have achieved 'prescriptive status' by the mid-2000s, when the validity of the norm was no longer challenged (Risse et al., 1999). At this stage, actual environmental and social performance may still be inconsistent with sustainable development norms, but firms would need to demonstrate that they are making a sustained effort to improve their CSR policies and practices. Major mining companies have undergone significant organizational changes (through for example, creating executive positions such as VP, Sustainable Development) in an effort to institutionalize sustainable development in their operations.

The final stage of norms socialization is 'rule-consistent behavior,' but few, if any, mining companies have reached this stage because of the inherently damaging nature of their operations to the environment, and the challenging social context in which mining operations are located. All the same, major mining companies have played a leading role in promoting global CSR standards. Mining companies, together with NGOs, have borne much of the responsibility for driving global governance in this area (with states lagging behind for the most part). Such global actors have filled some of the governance gaps with respect to CSR – what Ruggie refers to as a 'reconstituted global public domain' (Ruggie, 2004).

Methodology

Conceptual approach

How is it that some major mining companies played a leadership role in promoting sustainable development both nationally and globally? It is clear that the domestic institutional context in which companies operate is important to explaining how a small number of mining companies from the advanced industrialized economies came to promote CSR, and later, sustainable development. Institutionalist approaches are useful for understanding how and why firms are responsive to evolving global norms, and would be willing to promote global CSR norms through collaborative efforts at industry self-regulation (Campbell, 2006). Firms are understood to be embedded in a larger social environment beyond the marketplace, which induces them to conform to societal norms. The 'new' institutionalism in organization theory, which emphasizes cognitive processes and the normative environment, highlights the ways in which firms interpret their external environment (Powell and DiMaggio, 1991; Hoffman, 1997). Norms entrepreneurs play a critical role in promoting new ideas within firms around their environmental and social responsibilities (Dashwood, 2007a).

Historical institutionalism is useful in explaining the timing whereby a growing number of mining companies came to frame their CSR policies in terms of sustainable development. The mining industry had reached a critical juncture by the mid-1990s and, consistent with historical institutionalism (Pierson, 2004), a number of distinct, yet interrelated developments came together at this time to produce a crisis for the mining industry. Major environmental disasters associated with mining had badly damaged the industry's reputation, and the legitimacy of the industry as a whole was called into question. This critical juncture provided an opportunity for senior management in a number of mining companies from the advanced industrialized economies to take on a leadership role in the late 1990s (Kingdon, 1995; Pierson, 2004).

Combining historical institutionalism and institutionalism in organization theory serves as a useful corrective to ahistorical approaches to CSR that tend to ignore its historical development over time, and the critical role of institutional settings in influencing CSR's development (see Utting and Marques, 2010).

Method

To trace how this process occurred, the major mining companies with headquarters in Canada, the US, the UK, Australia, and Japan were identified, drawing on data from Corporate Register. The date that these companies first started issuing stand-alone reports on their CSR policies was noted, and those that were the first to start reporting beginning in the early to mid-1990s are referred to as the 'early movers.' Although there are other media through which companies can report on their CSR policies and practices, the production of stand-alone reports is inferred to reflect a stronger commitment to CSR.

The early movers tended to be the same companies that, by the mid- to late 1990s, began pushing for global CSR standards and adopting sustainable development as a means to frame their own policies. As such, early movers played two different roles simultaneously: that of norm 'leaders,' by seeking to improve their own practices as an example to others; and that of norm 'entrepreneurs,' by seeking to influence other mining companies and collaborating to establish new normative standards for the mining industry (Flohr et al., 2010). Companies were tracked according to when they first began adopting 'sustainable development' reports, in order to ascertain when a sustainable development norms cascade occurred.

Although the release of CSR reports might be dismissed as being merely strategic, they are a reasonable indicator of the acceptance of sustainable development at the discursive level. By the same token, although the adoption by many mining companies of the norm of sustainable development might be the result of 'strategic framing,' it does reflect the acceptance of sustainable development as a valid norm according to which CSR responsibilities should be understood.

Impetus behind the adoption of sustainable development

External pressures

All major mining companies with global operations encountered in the mid-1990s common constraints in their external environment. Around that time, a number of widely publicized environmental disasters cemented the bad reputation of mining companies (for a detailed list, see Hamann, 2003). In 1996, Placer Dome's (now Barrick) Marcopper mine in the Philippines suffered a major accident when the plug to the tailings dam gave way, causing serious damage to rivers downstream and disrupting the livelihoods and health of people living nearby. In another example, in 1998 the tailings dam at the Aznacollar mine (owned by Swedish–Canadian Boliden–Apirsa) in Spain collapsed, killing almost all life in the river, and threatening the nearby Donana National Park, a UN World Heritage site. In one legendary case, at the Summitville mine in Colorado, USA, Galactic Resources (a Canadian company) declared bankruptcy and walked away from a tailings pond failure that was deemed to be imminent by the U.S. Environmental Protection Agency (EPA). Such well-publicized disasters served to damage the already tarnished reputation of all mining companies, not just those directly implicated in such events.

Role of NGOs

Environmental NGOs such as Greenpeace and the World Wildlife Fund (WWF), operating nationally and globally, helped to mobilize public opinion against mining. Hostility to new mines in industrialized countries, together with improved investment climates in developing countries and the former Soviet states, prompted mining companies to significantly expand their operations globally. Whatever truth to the assumption that companies operating in developing countries or countries in transition were seeking to evade their environmental responsibilities, mining companies found that NGOs were able to adeptly use information technologies to raise awareness about mining company operations in far-off places, and to mobilize local community opposition to mining.

Role of international organizations

A number of other developments at the global level affected mining companies directly or indirectly. Early pressure came from the United Nations Environment Program (UNEP, created in 1974), which, through various programs initiated in the 1990s, sought to improve the environmental performance of the mining sector through engagement with national governments and the mining industry (Yakovleva, 2005, p. 42). National governments that had signed on to international environmental treaties began to restrict mining companies' access to mining sites in an attempt to meet their treaty commitments. The World Heritage Convention (1972), for example, can restrict access to areas that could negatively impact natural or cultural values, which is of particular significance to companies wishing to extract ore from lands claimed by indigenous peoples. The Biodiversity Convention (1992) restricts access to land with fragile ecosystems (Balkau and Parsons, 1999, cited in Yakovleva, 2005, p. 43). The Basel Convention (1989) restricted the use of certain metals and minerals in consumer products, affecting mining companies' (especially those engaged in smelting) access to markets, particularly in Europe.

Conditions attached to financing

In the mid-1990s, under pressure from Northern-based NGOs, the World Bank began to strengthen its environmental criteria in its lending. Commercial banks, concerned about liability risks, began to attach environmental conditions to their loans through the Equator Principles. The Extractive Industry Review (EIR), launched in the late 1990s, was a multi-stakeholder series of negotiations that resulted in the International Finance Corporation's (IFC) Policy and Performance Standards on Social and Environmental Sustainability (IFC, 2006). In light of the dependence of mining companies on financing for their large, capital-intensive projects, conditionalities attached to lending were a major source of concern.

Domestic institutional context

These global initiatives were reinforced at the national level in the advanced industrialized economies. In the 1990s, governments tightened up existing environmental regulations affecting various aspects of mining, pertaining to safeguarding the quality of land, air, and water affected by mining and mineral processing (Dias and Begg, 1994). New regulations were introduced, such as environmental impact assessments for new projects, which set limits on discharges of waste and contaminants to water, required monitoring, and incorporated legal penalties in the event of violation of permit conditions. In countries such as Canada, Australia, the US, the UK, and Japan, national parks and wildlife management areas were established that were off limits to mining. For example, in 1996 Noranda (now Xstrata) had to walk away from its New World Mine project due to public outcries over the development of a mine only three miles from Yellowstone National Park.

In short, a variety of developments at the global and national levels conspired to restrict mining companies' access to mining sites, to markets, and to finance. Due to the fact that they must operate where the ore is, mining companies are vulnerable to community opposition to their operations. The combination of widely publicized environmental disasters, restricted access to new mine sites, tighter regulations in developed countries, the closing-off of markets, sustained NGO targeting, and the conditionalities attached to loans by private and development banks created a critical juncture for mining companies in the late 1990s. The coming together of these disparate developments provided the opportunity for change agents/norms entrepreneurs to overcome internal resistance and push for better practices both within their own companies and, through global initiatives, among all companies. Growing awareness of the need for a 'social license to operate' (Gunningham et al., 2003), coupled with ongoing environmental challenges, made sustainable development an increasingly attractive means for mining companies to address their reputational problems.

The normative validity of sustainable development in mining

By the early 1990s, global initiatives that brought the environment and sustainable development onto the international agenda and efforts at the national level among the advanced industrialized economies led to a consensus on the norm of sustainable development among the majority of states. The establishment in 1983 of the World Commission on Environment and Development (WCED) set the stage for a lengthy process of consultation that culminated in the report, *Our Common Future* (better known as the Brundtland Report). The report defined sustainable development as 'development that meets the needs of the present without compromising the ability of future generations to meet their own needs' (WCED, 1987, p. 43). While much ink has been spilled in attempts to clarify what this means in practice, sustainable development refers to the economic, social, and environmental dimensions of development.

The 1992 United Nations Conference on Environment and Development (better known as the Rio Conference) set out principles for achieving sustainable development, and thereby cemented the normative validity of sustainable development as a means to guide action on the part of governments, international organizations such as the World Bank, and global civil society as represented by NGOs. The mining industry, however, had not accepted sustainable development, and with the exception of a small number of leaders, companies were resistant and defensive about their environmental responsibilities (Dashwood, 2007a). Aside from the definitional problems associated with applying sustainable development to a finite resource, the majority of mining companies had not taken serious steps to address the environmental and social impacts of their operations.

Response of early movers

In the early 1990s, a small number of mining and metals companies from the advanced industrialized countries began reporting on their environmental policies and practices (Scott, 2000, cited in Yakovleva, 2005, p. 52). Most reporting in the early to mid-1990s focused on environmental aspects, and efforts to report on the social impacts of their operations did not begin until the late 1990s. The fact that mining companies began to publicly report on their environmental, and later, social performance, is significant because it marked a shift away from denial and toward a willingness to publically acknowledge their environmental responsibilities. The publication of CSR reports means that internal organizational and policy changes have occurred, reflecting the existence of internal mechanisms capable of reporting regularly on environmental issues. While information in the reports themselves cannot be taken at face value in the absence of external verification (Guthrie and Parker, 1990; Neu et al., 1998, cited in Yakovleva, 2005, p. 53), they are important indicators of the normative affirmation by mining companies of their environmental and social obligations.

Early movers are those companies that began releasing CSR reports during the 1990s. Normally, such reports were labeled as environment, health, and safety (EHS) reports, although there was and remains considerable variation in the maturity of content of reporting and styles (Jenkins and Yakovleva, 2006). Table 2.1 shows the number of major companies by the date that they *first* began CSR reporting, in five-year cohorts. In 2009, there were 56 major (revenues over US$500m) mining companies with HQ in Canada, the US, the UK, Australia, and Japan, of which 40 published CSR reports. The data show a significant increase in the number of companies reporting on CSR for the first time from the late 1990s to the present.

Table 2.1 Mining companies' CSR reports, by year of first release

	1990–1994	1995–1999	2000–2004	2005–2009
No. of companies	5	10	15	10

Note: Merger and acquisition (M&A) activity means that some of the companies included above no longer exist. The combination of M&A and bankruptcies also means that the total number of major mining companies has changed, making it difficult to express accurately the above numbers as a percentage of the total number of major mining companies.
Source: Corporate Register: www.corporateregister.com.

It was the early movers that began in the mid-to late 1990s to push for greater responsibility on the part of mining companies, to advocate for consistent voluntary global standards against which environmental and social performance could be measured, and to promote sustainable development as a means to frame CSR policies. The timing of social reporting corresponded with efforts on the part of a small number of mining companies to incorporate sustainable development into their CSR reporting. These early movers recognized before most mining companies that a normative consensus had emerged around sustainable development among states and within international organizations that would be foolhardy to ignore.

Norms entrepreneurs within these early mover companies played a key role in promoting the norm of sustainable development, both within their companies and through international collaboration (Dashwood, 2007b). Internally, norms entrepreneurs played a critical role in gaining internal acceptance of sustainable development as an appropriate means to frame their CSR policies.

The key internal driver for mining companies in accepting the normative validity of sustainable development flowed from the recognition that mining companies needed to take dramatic steps in order to improve their reputations and acquire a 'social license to operate.' In addition to these strategic-driven considerations, early movers sought to differentiate themselves from poor performers, so as to avoid being painted with the same negative reputational brush. In some cases, early movers were driven not simply by strategic interests, but out of normative conviction that sustainable development should serve as a guiding principle or value underpinning business conduct (Dashwood, 2007b). As such, individual norms entrepreneurs relied on *rationalist* appeals drawing on market considerations in order to win internal consensus, as well as on appeals to the *intrinsic* value of sustainable development in its own right (see Flohr et al., 2010).

For mining companies committed to improving their environmental and social performance, sustainable development served as a useful means to frame the growing range and complexity of reporting, in particular with respect to the social dimensions. Especially important was the need to address the growing risks associated with developing mines in populated areas in developing countries, as well as with indigenous peoples around the world. Internally, sustainable development served as a means for companies to integrate the traditional EHS approach with policies respecting social performance, allowing for a more holistic approach on the part of business (Hamann, 2003). For early movers committed to engaging with external stakeholders such as governments and NGOs, sustainable development served as a common language to address the full range of economic, social, and environmental issues associated with mining (Dashwood, 2007b).

The push on the part of early movers to promote sustainable development came initially within the International Council on Mining and the Environment (ICME) (established with HQ in Canada in 1991). The ICME sought to address the challenge of Principle 10 of the Agenda 21 declaration emanating from the Rio Conference (1992), calling for the disclosure and dissemination of information by industry on environmental performance (Yakovleva, 2005, p. 52). After lengthy, contentious, and controversial internal debates, the ICME ultimately succeeded in developing the Sustainable Development Charter, released in 1998 (ICME, 1998).

It was clear to the early movers that more needed to be done than the release of sustainable development principles, as significant a step as the Charter was in reflecting an emerging normative consensus within the mining industry around sustainable development. Many of the early movers were active in the wider industry association, the World Business Council on Sustainable Development (WBCSD). Under the auspices of the WBCSD, leading mining companies launched the Global Mining Initiative (GMI) in 1998. Wishing to address the reputational issues affecting their operations, companies such as Rio Tinto, Placer Dome, Noranda, Alcoa, Western Mining Company (WMC), and Mitsubishi sought to promote global voluntary standard setting in the mining sector. The most significant development to emerge from the GMI was the launch of the Mining, Metals, and Sustainable Development (MMSD) research and consultation project, which culminated with a major conference held in Toronto in 2002 (Dashwood, 2005).

A key outcome of the GMI was the creation in 2001 of the International Council on Mining and Metals (ICMM), which superseded the ICME. The ICMM serves as the global institutional nucleus for the promotion of sustainable development in mining, and of global standards for addressing problems common to mining. Through the ICMM, the Mining and Metals supplement to the GRI was developed, an important step in identifying what indicators are relevant to sustainable development in mining on which companies can report. The promotion by the ICMM of standards applicable to sustainable development in mining, together with the MMSD process, has been helpful in strengthening the norm of sustainable development amongst mining companies.

Sustainable development norms cascade

The early to mid-2000s can be said to be the time when a norms cascade was reached among mining companies around sustainable development. An important indicator of the normative strength of sustainable development is the extent to which companies' CSR reports are framed as such. As the following table reveals, a growing number of major mining companies came to refer to their CSR reports as 'sustainable development' reports, or 'sustainability' reports. Although some of these companies, such as Placer Dome, released their first report as a 'sustainable development' report, a number of the companies counted here later changed the names of already-existing reports to sustainable development. For example, Noranda, which first began EHS reporting in 1990, changed the name of its report to 'sustainable development' in 1999. In addition to the 30 companies tallied in Table 2.2 below, virtually all mining companies refer to sustainable development within their reports.

Table 2.2 reveals that there was a significant increase in the number of companies framing their CSR policies as sustainable development in the early to mid-2000s. Of the 56 major mining companies with headquarters in Canada, the US, Australia, the UK, and Japan in 2009, 40 published CSR reports. Although a few of the 30 companies recorded above have merged or been taken over, it can safely be observed that close to 75 per cent of major mining companies reporting on their CSR policies frame them in terms of sustainable development. This is confirmed by KPMG's (2006) *Global Mining Reporting Survey*, which found that 40 out of the world's 44 major global mining companies counted were reporting according to sustainable development indicators.

By the late 1990s, a growing number of mining companies became sensitized to the need to promote and sustain relations with the local communities in which they were operating. The notion of stakeholder engagement began to creep into the lexicon of mining companies as

Table 2.2 Number of major mining companies with sustainable development/ sustainability in the title of their reports, by first year of use of term

	1998–2000	2001–2003	2004–2006	2007–2009
No. of companies	3	9	14	4

Source: Corporate Register: www.corporateregister.com.

they came to realize the potential for local communities (aided by NGOs) to deny them a 'social license to operate' (Gunningham et al., 2003; Wheeler et al., 2003). By the mid-2000s, there was general acceptance on the part of mining companies of the need to not only consult with local communities, but maintain ongoing dialogue (Dashwood, 2007b). It was in this context that mining companies came to see the value of framing their CSR policies in terms of sustainable development. Doing so served the external purpose of showing that companies had integrated both environmental and social aspects into their business strategies, and had the internal purpose of helping to integrate social aspects into already-existing environmental systems and to raise awareness of employees across the organization (Dashwood, 2007b).

Conclusion

Writing in 1999, Richard Falk noted the emergence of the norm of sustainable development among states, but feared they would only pay lip service to the consensus about the need to protect the environment while pursuing economic development. Notwithstanding the disappointing outcome of the December, 2009, UN climate change summit in Copenhagen, states are taking important steps toward halting climate change. In constructivist language, most states have been sufficiently socialized to the norm of sustainable development that they are endeavoring to approach 'rule-consistent behavior.'

Over a decade since they first starting acknowledging the need to grapple with sustainable development in their operations, the key question with respect to mining companies is whether they are merely paying lip service to the norm. The distinction between the acceptance of the validity of a norm at the discursive level and actions consistent with the norm is important in answering this question. The data presented above point to the acceptance of the validity of the sustainable development norm at the discursive level. This is an important development, as it signifies a significant shift in attitudes on the part of mining company executives, and it suggests that firms are receptive to normative changes in the global environment. Acceptance of the validity of sustainable development as a norm guiding their operations is a critical first step. Many companies have achieved 'prescriptive status' in that they are seeking to improve their CSR policies and practices, even though their actual behavior may not yet be fully consistent with sustainable development norms. Falk's fear that states (and companies) would pay only lip service to sustainable development does not appear to have been entirely borne out, given that some (but certainly not all) leading companies have demonstrated a strong commitment to sustainable development.

The pattern whereby initially major mining companies headquartered in the advanced industrialized economies were the early movers in promoting norms globally is consistent with the life-cycle theory of norms, where initiatives at the domestic level are important in the early phase of norms socialization. Once a global norms cascade is reached, it is expected that pressure for change will increasingly come from the global level, such that individual state- or firm-level initiatives become less important. This has been partially borne out in that global pressures from transnational NGOs, international organizations such as the World Bank, and private international financial institutions played a critical role in producing the 'critical juncture' that induced mining companies to address their bad reputation. The growing number and influence of global CSR standards pertaining to mining have influenced major and mid-tier mining companies, as well as developing-country governments, to strengthen their environmental regulatory and monitoring capacity. By looking at macro-level trends, this study has been able to establish that a growing number of major mining companies have accepted the normative validity of sustainable development. To fully understand the degree of a firm's commitment to sustainable development, however, it is necessary to look at the micro level of the firm, which the case studies in this volume show is still uneven and inconsistent in terms of operation-level adherence to sustainable development practices.

The life-cycle theory's prediction that domestic-level initiatives become less important once a norm reaches a global degree of acceptance (norms cascade) is not entirely borne out, however. Constructivist theory is weak in terms of explaining how, once norms reach a degree of global acceptance, they then filter back down to the level of states, are interpreted, and then acted upon. This point can be seen most clearly in the case of the emerging economies of Brazil, China, and India, whose governments are only now attempting to alter their policies to adhere to sustainable development norms. How those countries go about achieving this will, in turn, influence the behavior of major mining companies headquartered there, which to date have weak CSR records. An important question for future research, then, is to explore how and to what extent global CSR norms are being interpreted and applied in these countries, and the impact this is having on mining companies headquartered there.

Acknowledgments

I would like to thank the editors, Julia Sagebien and Nicole Marie Lindsay, for their very helpful feedback on this chapter, as well as Bill Buenar Puplampu, Tim Shaw, and Kernaghan Webb for their insightful comments. The research assistance of Chris Hann (Honours B.A., Brock University) is gratefully acknowledged. This research was funded by a Standard Research Grant from the Social Sciences and Humanities Research Council of Canada.

Note

1. Major mining companies are defined as those whose annual revenues exceed US$500m, and mid-tier companies are those whose annual revenues are between US$50m and 500m.

3
CSR and the Law: Learning from the Experience of Canadian Mining Companies in Latin America

Kernaghan Webb

There is considerable controversy about the nature, extent, and adequacy of Canadian, Latin American, and international laws used to address the environmental, social, and economic (ESE) impacts of Canadian mining activity in Latin America (CMALA), and concerning how corporate social responsibility (CSR) and the law are and could be used to address ESE issues associated with CMALA (for example, SCFAIT, 2005, 2010; Imai et al., 2007; McGee, 2009; Lemieux, 2010a). This chapter draws on publicly available accounts concerning ESE-related disputes pertaining to CMALA in order to better understand the actual and potential roles of CSR and the law in addressing ESE dimensions of CMALA. While the veracity of assertions made in publicly available accounts is open to challenge, the disputes are taken as points of departure for discussion of the sorts of ESE challenges that are associated with CMALA.

Mining activity is difficult enough to undertake in a sustainable, socially responsible way in stable democracies where citizens are comparatively affluent and educated, the social and physical infrastructure is well developed, there is political stability and respect for the rule of law, and political, judicial, and regulatory regimes is widespread, and corruption is minimal. Unfortunately, in many countries of Latin America, these conditions are often absent (see for example, discussion of the governance conditions in Guatemala described in Imai et al., 2007). For better or worse, this is often the operating environment for CMALA. It is in this context that CSR, as part of a broader range of actions, may be particularly valuable, as discussed below. Thus, the suggestion made here is that the law and the CSR should not be viewed as polar opposites, but rather, as complementary approaches intended to address similar ESE challenges.

The chapter is organized as follows: first, the possible and actual application of Canadian law to CMALA ESE challenges is examined, as are the CSR implications. Then, an example of application of international law to CMALA ESE issues and CSR dimensions is discussed. Next, the use of Latin American law in the context of CMALA ESE issues and CSR aspects is explored. Finally, some conclusions and a suggested path forward are provided.

Use of Canadian law to address allegedly harmful ESE aspects of CMALA

One option to address allegedly harmful ESE-related misconduct of CMALA is through reliance on elements of Canadian law. Examples of recourse to Canadian legal mechanisms to challenge Latin American activity of Canadian firms provide us with a test of the viability of such 'long arm' (extra-territorial) application of Canadian law. From a legal standpoint, the fact that a particular mining firm has its headquarters in Canada and/or is incorporated in Canada provides at least a theoretical foundation for Canadian law-based actions to address overseas behavior, as discussed below.

Private litigation launched in Canada

Research conducted for this chapter suggests that plaintiffs have rarely brought private lawsuits through the Canadian courts to address ESE aspects of CMALA, and that doing so presents numerous challenges. Major hurdles to be overcome by plaintiffs bringing private law actions in Canada to address CMALA include persuasion of the courts that: there are justiciable legal obligations recognized in Canadian law that are owed by Canadian-based firms concerning activities of the firms or their subsidiaries overseas; there is a real and substantial connection to Canada for events that are largely taking place elsewhere; and Canadian courts are appropriate fora for addressing the issues in question, when compared with the possibility of legal actions brought through the courts in the Latin American jurisdictions where the alleged harm has taken place (see Scott and Wai, 2004).

In 1998, a Quebec class action brought on behalf of Guyanese plaintiffs who had allegedly been harmed by a 1995 tailings dam collapse at the operations of Omai Gold Mines Limited, a Guyanese-based subsidiary of Quebec-based Cambior Inc., was dismissed (the following discussion is derived from Business Wire, 1998; Preville, 1997; Mordant,

2003; Kosich, 2006; Stikeman Elliot, 1999; and Scott and Wai, 2004). The Quebec Superior Court ruled that although Quebec courts had joint jurisdiction with the courts of Guyana, neither the victims nor their action had any real connection with Quebec, and therefore Guyana was the more appropriate forum for redress. Factors that seemed to count against the plaintiffs included the following: the fact that the plaintiffs brought the action against Cambior rather than Omai when it was Omai that was responsible for the design, construction, and operation of the mine; the allegedly harmed parties and the witnesses were based in Guyana, not in Canada; the location of the elements of proof (the mining operation, the river, the plans and documents, the medical records, and so on) were all located in Guyana; the alleged harm took place in Guyana; and testimony was given to show that the Guyanese justice system was not so deficient that the plaintiffs would be denied justice if the case proceeded in Guyana. A lawsuit to address the mining spill was later filed in Guyana, but was eventually dismissed.

The second private legal action pertains to a lawsuit concerning an Ecuadorian mining project brought in the Ontario courts by three Ecuadorian community members against TSX-listed and BC-headquartered Copper Mesa Mining Corporation, two of its individual directors, and the Toronto Stock Exchange, concerning a Bahamian subsidiary of Copper Mesa (Klippensteins, 2010a). In March 2010, the Ontario Superior Court struck out this action on the grounds that it failed to disclose a reasonable cause of action. The decision has been appealed. The plaintiffs in this action were villagers who opposed the proposed Copper Mesa mine in Ecuador, including through blockades of roads. Essentially, the plaintiffs asserted that the defendants were made aware of local opposition to and concerns about activities associated with Copper Mesa (and the existence of and potential for continued conflict and violence); were made aware of the concerns through communications that took place on Canadian soil with Ecuadorian community members (there was a direct Canadian connection on this issue); and failed to take steps to address the issue.

Ontario Superior Court Justice Colin Campbell held that '[o]ther than having its shares registered on the...TSX...Copper Mesa does not appear to have any connection with Ontario. The [named directors who were defendants in the case]...are residents of Ontario who in 2006 and 2007 became non-management directors of Copper Mesa' (*Piedra v. Copper Mesa Corporation*, 2010).

The plaintiffs asserted that the defendants knew or ought to have known that violence would ensue once Copper Mesa was listed and financed through the TSX, unless Copper Mesa took steps to ensure that its agents, employees, and contractors were instructed to prevent violence. Justice Campbell concluded that there was nothing 'that satisfies me that the conduct alleged is of the type of personal conduct by a director that could ground personal liability' and that the plaintiffs had not 'made out the necessary connection for foreseeability and duty in respect of the two individuals who are non-management directors.' With respect to TSX, Justice Campbell held that there was 'no connection between the Plaintiffs and the TSX Defendants.' The limitations of using private law actions undertaken through the Canadian legal system to address overseas corporate behavior are apparent in this judgment, although this is not meant to suggest that there is no place for such actions, or that such actions will not succeed in the future. As this chapter was going to press, another Canadian-based private law action was launched against Canadian-based HudBay Minerals Inc. and related companies, concerning the death of a community activist associated with a proposed mining operation in Guatemala (Klippensteins, 2010b). The approach to the lawsuit is similar to that in the Copper Mesa case.

From a CSR standpoint, the three Canadian private lawsuits can be characterized as cautionary tales, pointing to the value of firms putting in place proactive and preventative ESE management approaches for both employees and contractors, that would decrease the likelihood of the alleged incidents arising in the first place, coupled with strong efforts to partner with governments and local communities to develop benefits sharing and ongoing monitoring of mining projects. This is often not an easy task. To provide an illustration of the extent firms might need to go to in order to avoid being associated with illegality and violence, the author recalls a speaker at a recent conference describing a mining company that had purposely put all of its vehicles 'on blocks' (thus disabling them) in anticipation of and as a way of avoiding becoming complicit in illicit activity by state-based security forces who it feared would commandeer the vehicles in order to engage in wrongdoing and violence against a community. The mining firm had worked out a form of intelligence network with local community organizations whereby the firm and the community organizations would alert each other if rogue security force activity was sighted in the area. The point here is that much problematic activity can be anticipated and addressed by firms – and they may find allies amongst communities in doing so. And if and when conflicts or problems do arise, firms would be well advised to explore the potential use of redress mechanisms (for example, mutually trusted, impartial third party services) to attempt to find solutions that reduce tensions and re-engage the parties on a constructive path.

The use of shareholder proposals under Canadian law

Another way in which Canadian law can be and has been used to address the ESE aspects of CMALA is through shareholder proposals brought pursuant to Canadian corporate law statutes. An attractive feature of shareholder proposals in this context is the prospect of directors and shareholders (and others) directly engaging on ESE issues. Before specifically discussing the shareholder proposal mechanism, it is useful to review more generally the legal obligations of directors vis-à-vis stakeholders, as understood in Canadian corporate law.

In *BCE Inc. v. 1976 Debentureholders et al.* (2008), the Supreme Court of Canada (SCC) held that pursuant to the Canada Business Corporations Act (CBCA, 1985), directors are obliged, in the course of meeting their fiduciary duties, to protect the long-term best interests of the corporation and to consider a broad set of stakeholder interests, including those of shareholders, employees, creditors, consumers, government, and the environment. The Court held that directors 'need to treat affected stakeholders in a fair manner, commensurate with the corporation's duties as a responsible corporate citizen.' While the exact meaning and extent of corporate duties associated with acting as a 'responsible corporate citizen' is unclear at this point, it is significant that the SCC in this decision is explicitly linking the 'fiduciary duties of directors' with consideration of stakeholder interests, the need to consider the long-term best interests of corporations, and corporate duties to be responsible corporate citizens. The Court also stressed the need to be deferential to board decisions: if board directors are properly informed and act in good faith, courts will generally defer to the director decisions, so long as the decisions lie within a range of reasonableness.

One mechanism that shareholders can use to bring issues to the attention of corporate directors that is not available to other stakeholders is the shareholder proposal mechanism. The shareholder proposal mechanism has been made more accessible to a broader range of shareholder proposals as a result of amendments to the CBCA that were passed in 2001 (van Duzer, 2003). The situation of Goldcorp, Inc. and its Marlin Mine operation in Guatemala is provided as an illustration of how the shareholder proposal mechanism has been used to address ESE aspects of CMALA (the following discussion is derived from Coumans, 2008b; Dhir, 2010; Goldcorp Steering Committee, 2010; Hoffman, 2008; Mining Watch, 2010). Montana Exploradora de

Guatemala S.A., a wholly owned subsidiary of Goldcorp, operates the Marlin Mine in the western highlands of Guatemala. Following claims of human rights abuses at the Marlin Mine, a group of socially responsible-oriented Goldcorp shareholders submitted a shareholder proposal that suggested that the company should undertake an independent human rights impact assessment (HRIA) of the firm's Marlin Mine operation. The proposal was subsequently withdrawn after Goldcorp agreed to commission such an assessment (framed in the form of a Memorandum of Understanding between Goldcorp and the shareholders). The Steering Committee that directed the implementation of the human rights assessment included a company representative, a shareholder representative, and a Guatemalan representative. However, subsequent to Goldcorp agreeing to submit itself to the HRIA, a range of Canadian non-governmental organizations (NGOs) and Guatemalan community spokespersons expressed their significant concerns with the assessment (Coumans, 2008b). Most notably, the claim was made that the assessment did not meaningfully involve and did not obtain the free prior and informed consent of the locally affected communities.

While not by any means a panacea, the Canadian law concerning shareholder proposals has the potential to address certain ESE impacts of CMALA, but the controversy concerning the Goldcorp HRIA suggests the need for activist shareholders and mining companies to work closely with governments and affected communities and NGOs to ensure that shareholder proposals and HRIAs are seen from the outset to be of value and credible to all parties concerned. By doing so, Canadian ESE-oriented shareholder proposals can be optimally useful.

Application of Canadian anti-bribery/Corruption law

When Canada and other countries agree to, ratify, and implement international treaties concerning a particular activity, and those treaties include provisions for enforcement by one country concerning actions taking place in other country's jurisdiction, there is a more widely acceptable basis for domestic laws having a 'long arm' effect in other countries. Such is the case with respect to the federal Corruption of Foreign Public Officials Act (CFPOA) (1998), which is Canadian legislation intended to implement Canada's obligations under the OECD *Convention on Combating Bribery of Foreign Public Officials in International Business Transactions* (1997). Mexico has also ratified and implemented this convention – a point that becomes relevant in light of the discussion of alleged bribery activities of Blackfire Exploration, Ltd (Blackfire), a Calgary mining company, concerning operations of a subsidiary in

Mexico. The Blackfire situation provides an illustration of how Canadian public law that implements international treaty obligations could apply to activities of Canadian corporations that take place outside of Canada (the following information is derived from Business and Human Rights, 2010; Popplewell, 2010; and Mines and Communities, 2010).

In March, 2010, a coalition of Canadian NGOs brought a complaint to the attention of the Royal Canadian Mounted Police (RCMP) concerning activities taking place at Blackfire's Mexican operations located near the town of Chicomuselo in the state of Chiapas, claiming that actions contrary to the Corruption of Foreign Public Officials Act (CFPOA) had taken place (Mines and Communities, 2010). Blackfire has recently made a statement refuting the allegations of bribery:

> Blackfire Exploration has never provided inappropriate funds to a Mexican official. In this case, we contributed funds to the local town fair and to assist the municipal government in limited ongoing expenses. Cheques were addressed to the mayor as an elected official because the town has no banks. This is considered an appropriate practice in the State of Chiapas. Unfortunately the contributions were not used for their intended purpose and we reported this fact to the authorities immediately. Blackfire never sought nor obtained any benefit or service from the mayor or any other official different than the services provided and available to all citizens. We will seek other avenues for charitable giving in Chicomuselo in the future as we grow and support the community. (Business and Human Rights, 2010)

It is perhaps not surprising that mining firms often engage in a wide variety of informal (not legally required) CSR activities, such as investing in local infrastructure (roads improvement, hospital/school construction, water systems improvement, and so on), provision of worker training, or support for local projects. There is clearly a need for mining firms to exercise diligence in ensuring that these activities are not characterized as running contrary to the CFPOA. Support for projects will often involve interactions with local government officials, and activity taking the form of a 'loan, reward, advantage or benefit of any kind' (CFPOA, 1998) to those local officials can potentially be problematic.

The Blackfire situation points to the possible application of Canadian public law to overseas activity of Canadian firms where it can be characterized as non-compliant with the CFPOA, and it points to the value of firms having anti-bribery/corruption systems in place to ensure that no activities of the company or anyone associated with the company can be construed as running contrary to the terms of the CFPOA. An antibribery policy, a senior official accountable on bribery issues, training for employees and contractors, and creation of an anonymous tip-line so that possible problematic activity can be brought to the attention of management as early as possible, are examples of CSR approaches to preventing problematic bribery activity from taking place.

The investment protection provisions of free trade agreements

The main objective of free trade agreements is to encourage an unimpeded flow of economic activity between the countries that are party to the agreement (Gantz, 2004). Although it is typical for international trade law agreements to include some language and mechanisms pertaining to environmental and social protection, it is the investment protection provisions – and in particular, the 'private enforcement' ability of companies using these provisions to bring claims against participating FTA governments – that have attracted considerable attention (as discussed below). Typically, private companies using these investment protections have alleged that governments have not accorded the same treatment to them as they have to domestic companies.

In light of the fact that large-scale mining projects such as those developed by Canadian mining firms involve enormous investments, and that many countries in Latin America have historically been politically unstable, the existence of investor protections in free trade agreements provides some assurance to the companies in question that their investments will not be directly or indirectly expropriated. One example of a trade investment case involving CMALA is the situation of Canadian-based Pacific Rim Mining Corporation (hereafter 'PacRim'), with respect to its proposed El Dorado mine in El Salvador, and the use by PacRim of the Central American Free Trade Agreement's (CAFTA, 2005) investor provisions, to launch an action for compensation against the Government of El Salvador (the following information is derived from Anderson et al., 2010; Pacific Rim Mining Corp, 2008; *Pac Rim Cayman LLC v. Republic of El Salvador* (2009); and Public Citizen/Global Trade Watch, 2010). While PacRim is a Canadian-based company, Canada is not a signatory to CAFTA; however, both the US and El Salvador are CAFTA signatories. In December, 2007, PacRim incorporated a Cayman Islands subsidiary as a Nevada (US-based) corporation. The newly reincorporated PacRim US subsidiary initiated a CAFTA foreign investor action against the Government of El Salvador (GES), and in May, 2010, the first hearing on the case commenced at the International Centre for Settlement of International Disputes in Washington, DC.

PacRim has claimed that Salvadoran mining law guarantees a right for it to operate a gold mine in the country once the 'existence of mineable deposits has been demonstrated' and the firm fulfills all administrative requirements for obtaining an exploration permit. PacRim also claims that the Salvadoran government capriciously halted progress on approving the proposed mine for political reasons. One account of the situation states that '[t]his proposed project as well as applications filed by various companies for 28 other gold and silver mines, generated a major national debate about the health and environmental implications of mining in El Salvador, a densely populated country the size of Massachusetts with limited water resources' (Public Citizen/Global Trade Watch, 2010). In the face of calls for a reformed mining law and calls for a ban on intensive gold and silver mining, successive Salvadoran governments have not issued any mining permits, and have launched legislative studies of proposed projects and mining law reforms. PacRim argues that in its actions the GES violates the National Treatment, Most Favoured Nation, Minimum Standard of Treatment and Expropriation, and Compensation Provisions of CAFTA.

GES has countered that it has a right to set the terms for mining in its country, that the company failed to meet these legal obligations in order to be eligible for an exploitation permit, that PacRim was misinterpreting Salvadoran mining law, and that PacRim failed to fulfill critical legal requirements, including submission of the final feasibility study. If PacRim is successful in its allegations of unfair treatment, this could lead to compensatory damages being provided by the governments in question to PacRim. One can speculate as to whether lengthy, expensive, and uncertain FTA litigation of this type could be avoided altogether through proactive CSR activity, by engaging early and often with governments and affected stakeholders in order to demonstrate how the proposed mining activity can be undertaken with benefits to the community and with safeguards to address their concerns.

CMALA and the laws pertaining to mining in Latin America

In this part of the chapter we examine how Latin American mining laws can and have been used in relation to CMALA, and the potential and actual application of CSR practices to those operations. First, the environmental laws of Latin America are discussed, followed by an examination of Latin American indigenous protection laws, and laws and activities pertaining to community referendums.

CMALA and Latin American environmental laws

All Latin American countries have at least rudimentary environmental protection regulatory regimes – although it is certainly true that these laws may not be as comprehensive or as up to date as those found in Canada and the US, nor may they be effectively enforced. As an example of the problematic nature of Latin American laws to address the environmental dimensions of major mining projects, Professor David Szablowski of York University has provided useful analysis of the Peruvian environmental protection regime (Szablowski, 2007). Among other things, he notes that 'the capability of the Peruvian government to play an effective regulatory role with regard to environmental management in the mining sector remains exceedingly weak,' pointing to 'an overloaded administrative apparatus,' the 'shortage of funds dedicated to environmental matters,' the fact that 'the unit in charge of environmental issues is severely understaffed,' and the inherent conflict of interest of the Peruvian agency charged with approvals and enforcement also being responsible for promoting investment (Szablowski, 2007, p. 54).

The point to be made here is that simply complying with the EIA laws of Latin American countries may lead to problems down the line for companies and governments, as affected communities that distrust or perhaps were not fully involved in the EIA process become concerned with this or that aspect of a proposed project, and mount significant opposition to it. This is the sort of scenario that appears to have manifested itself with respect to the Canadian mining company Manhattan Minerals, with respect to a proposed mining project near the town of Tambogrande in northern Peru in the period 1999–2003 (Business Wire, 2003; McGee, 2009). Available accounts suggest that the company had acquired concessions and prepared an EIA as required by Peruvian law, and submitted the EIA to the Peruvian authorities, but that overwhelming local opposition to the project (as evidenced in a community referendum) eventually led the company to abandon the project (community referendums, and this particular case, are discussed in a later section of the chapter).

While discussion so far might suggest that Latin American environmental regimes are simply inadequate, it is worth pointing out that there have been enforcement actions against CMALA on environmental issues. In April 2007, the Honduran Ministry of Natural Resources and Environment (SERNA) fined a wholly owned subsidiary of Canadian-based Goldcorp for polluting water supplies with cyanide and arsenic above permissible levels, although the Goldcorp subsidiary has appealed the fine, claiming that the tests were not conducted properly (Sustainalytics, 2006). In November 2010, a Costa Rican court annulled the mining concession previously granted to a Costa Rican subsidiary of Canadian-based Infinito Gold, ruling that environmental studies required to grant the concession were incomplete, and the mining contract was therefore illegal, although the company has indicated its intention to appeal (McDonald, 2010).

From a CSR perspective, a good argument can be made that Canadian mining companies might want to exceed whatever requirements are set out in Latin American environmental regimes in order to better ensure that governments and communities are fully involved from the outset in decisions concerning mining projects in their area, and thereby reduce the likelihood of problems arising later. And once a mining project is underway, mining firms might want to have their operations certified as being in compliance with voluntary international standards, such as the Cyanide Code (2005) and ISO 14001 (International Organization for Standardization, n.d.). While certification standards are not without their share of limitations (see also Szablowski, 2007), and are supplements, not replacements for laws, use of the standards provides a modicum of assurance that operations meet international benchmarks of acceptable behavior, not as judged by the firms themselves, but as judged by qualified third party auditors. Use of such approaches can assist in identifying environmental issues in the ongoing operations of mines before they become problematic – particularly useful in contexts where the likelihood of government inspections is low. Moreover, if serious environmental problems do subsequently occur, the fact that a firm has gone to the trouble and expense of complying with third party standards can go some way toward demonstrating (in courts, and elsewhere) that the company in question at least tried to 'go the extra distance' to prevent that problem from occurring (Webb, 1999).

To sum up, Latin American environmental legal regimes can be seen as points of departure for Canadian mining firms to proactively address environmental issues in ways that go beyond what is required by law, including through the use of third party audited certification standards.

CMALA and Latin American indigenous protection laws

Frequently, CMALA will involve significant interactions with affected indigenous communities. In four South American countries, more than 40 per cent of the population is indigenous, while in another nine countries the indigenous population ranges from 5 to 20 per cent of the total (Pando, 1990). Whenever large-scale mining projects take place near indigenous communities, there is considerable potential for problems to arise. The law pertaining to mining-indigenous interactions in Latin America, and the practices and activities of mining companies and indigenous groups, is very much in flux at this time. The focus of discussion here is on ILO Convention 169, the Indigenous and Tribal Peoples Convention, which has been ratified by 14 Latin American countries, and on CSR approaches to addressing the mining-indigenous group interaction (International Labour Organization, 1989).

The ILO Convention mandates numerous protections for indigenous communities, specifically in the context of natural resource use, including the right to participate in the use, management, and conservation of these resources. In cases in which the state retains the resource ownership, governments are to implement consultation procedures before any exploration or exploitation. Relocation from lands occupied by indigenous peoples should only be undertaken as an exceptional measure and shall take place only with their 'free and informed consent.' Where such consent cannot be obtained, relocation can only take place following appropriate procedures established by national laws and regulations. Non-indigenous persons are to be prevented from taking advantage of indigenous customs or of lack of understanding of the laws to secure ownership, possession, or use of indigenous land.

Implementation of the ILO Convention is uneven across Latin America. In September, 2010, Peru's Constitutional Tribunal was reported to have ordered the Peruvian Ministry of Energy and Mines to fully comply as soon as possible with Peru's ILO Convention 169 obligations: the Tribunal is reported to have said that current measures 'don't fulfil the elements of being a prior consultation culturally conditioned to indigenous communities' (Dow Jones Newswire, 2010). In Chile, in 2009 a Chilean court referred with approval to ILO 169 in support of a decision granting water rights to two Chilean indigenous groups (Valeriote, 2009), and a regional government of Chile is reported to be experiencing 'delays' in Convention implementation (Zarnikow, 2010). In 2007, in a decision concerning the Goldcorp's Marlin Mine in Guatemala, the Guatemalan Constitutional Court (at the behest of members of a local indigenous community who brought the lawsuit) urged the government to make effective the requirement for consultation with indigenous peoples found under the ILO Convention 169 by legislating a consultation process for the future (Imai et al., 2007).

From a CSR perspective, attempts by firms to proactively and meaningfully engage now with governments and indigenous populations as partners in ways that are consistent with the above-described principles found in ILO Convention 169 represent a prudent course of action, thereby developing allies in the indigenous community and, at the same time, being better prepared for the indigenous-protective legal regimes that are likely to emerge. The use of impact–benefit agreements between mining companies and indigenous groups in Canada represents an example of mining companies going beyond legal requirements to develop ongoing operational arrangements with affected indigenous communities (Galbraith et al., 2007). The Kinross Gold Maricunga–Colla Protocol in Chile can perhaps be described as a 'proto' example of a Canadian company engaging in a Latin American version of an impact–benefit agreement (IBA) (Webb, 2010).

CMALA and Latin American community referendums

Since 2002, individuals in a number of Latin American communities have resorted to use of the community referendum to address issues associated with mining projects (the following information is derived from McGee, 2009). In a sense, the community referendum mechanism has the potential to 'jump over' the conventional focus on regulatory approvals, and instead tends to put the issue into a less formally structured public space. In 2002, the town of Tambogrande in northern Peru held a referendum in which the residents voted in reportedly overwhelming numbers against a gold and copper mine proposal of Canadian-based Manhattan Minerals. While before, during, and after the holding of the referendum, there was considerable dispute and controversy concerning the legality and status of the referendum, the upshot was that ultimately, the company was unable to find a partner to meet the US$100 million assets requirement for such projects, and in 2005 the project was abandoned, with Manhattan having spent US$60 million and having 'nothing to show for it' (McGee, 2009, p. 680). Manhattan's president is reported to have said '[t]he trouble for us was that we weren't able to find a partner...because the social conditions were so anti-mining that nobody wanted to touch it' (McGee, 2009, p. 680). Community referenda have been invoked by Latin American communities in resistance to Canadian resource projects on at least four occasions since 2002: in Argentina, Guatemala, Mexico, and for a second mining operation in Peru. In all of these documented situations, the referendum has led to reports of high community disapproval concerning the mining project in question, and on more than one occasion, projects have eventually been suspended or terminated.

Once the idea of a referendum has been raised as a possibility in a community, it has not been unusual for companies and national governments to challenge it, on a number of grounds, including the following:

- challenging the legal authority of the local community to hold the referendum. For example, does a local municipality have the legislative power to hold plebiscites on issues of this sort?
- challenging the status of the referendum. For example, even if a community does have the authority to hold a referendum, in some situations it has been ultimately been concluded that the result is not binding, or does not alter a nationally issued mining approval.
- pointing to possible problems with the wording of the referendum. For example, issues will be raised as to whether the referendum question was fair and objective or was framed as a leading question, and hence not an accurate representation of community views.
- raising questions concerning how the referendum was held. For example, doubts might be expressed about how many of the possible total electorate responded to the referendum, or whether there was adequate public notice prior to the holding of the referendum, or whether the way in which votes was counted was valid, or whether the population taking the referendum encompassed all of the actual affected communities in the right geographic areas, or only parts of them.

In particular situations, the validity of one or more these points of challenge has been confirmed (for example, the Guatemalan Constitutional Court, in a lawsuit brought by members of an indigenous community, confirmed that the community could hold a referendum, but that the results were not binding on and therefore did not in themselves alter a national regulatory approval) (Imai et al., 2007). However, as a way of swaying public opinion, the referendum mechanism has been highly effective. It is worth pointing out that a community can hold a referendum entirely outside of a legal framework. Granted, the results of such a referendum will not hold any immediate legal status, but it can still be a highly effective community voicing technique. Elsewhere, the author has suggested that private sector- and NGO-driven codes and standards have represented a significant governance innovation, providing a voicing mechanism for the private sector and NGOs (Webb, 2004). Arguably, the extra-legal community referendum represents another significant governance innovation and voicing mechanism, but this time, the innovation is from the community side.

In the long run, the solution might be to fully integrate meaningful and substantive local community participation into reformed national mining approval processes, in keeping with ILO Convention 169 obligations. In the immediate term, for CMALA, the most promising course of action would appear to be to adopt early, proactive, meaningful, transparent CSR stakeholder engagement approaches undertaken in close cooperation with affected governments and communities, beyond any legal requirements to do so, so that the engagement is aligned with and is seen to be officially supported by all parties.

Conclusions

In this chapter, we have reviewed a number of Canadian, Latin American, and international laws that have been used in relation to ESE dimensions of CMALA. The analysis undertaken suggests that these laws hold considerable potential to address the interests of a wide number of parties, but also reveals some deficiencies in both the laws themselves and their implementation. In addition, the discussion has revealed that in a number of situations, the legal regimes are evolving in significant ways, and on the whole the movement is away from a straightforward 'pro-development' orientation, toward a more balanced approach in which other interests (environmental, worker, indigenous, communities) are being more meaningfully considered and integrated. The examples of courts in several Latin American jurisdictions 'reminding' or constraining the legislative and executive branches concerning their legal obligations, often at the behest of communities, seem to suggest increasing promise for domestic (as opposed to Canadian or US) law-based solutions to CMALA ESE issues.

At the same time, because many of the current laws and/or their implementation are often deficient in important respects at the present time, and because use of formal legal approaches tends to polarize positions and lead to expensive, protracted, and uncertain litigation, and because there is evidence of considerable movement toward reform to Latin American legal regimes, the chapter has suggested numerous ways in which proactive environmental and social CSR practices by Canadian mining companies can potentially address some of the concerns of all parties. The CSR approaches that have been discussed in the chapter are directed at ensuring compliance with current laws, and oftentimes to taking on environmental and social obligations that exceed legal requirements, through direct partnerships and meaningful involvement with affected stakeholders, as well as through use of third party certification standards. To be sure, in some situations, CSR activities may not provide an answer to a significant challenge. Moreover, it is frequently 'easier said than done' to, for example, find common ground between governments, mining firms, and communities. In other situations, however, use of proactive CSR approaches can be the difference between a project proceeding in a manner seen acceptable to all major stakeholders and the project not proceeding at all.

In this sense, CSR can be seen as a bridge that spans some of the gaps between the formal legal system, with all of its strengths and deficiencies, and the 'real world,' with its complex and frequent conflicting sets of demands and interests. In an approach referred to as 'sustainable governance' (SG), the author has elsewhere suggested that effective solutions to twenty-first century ESE challenges will require a combination of rule instruments (laws and voluntary standards), institutions, and processes, systematically developed and/or implemented by governments, the private sector, and civil society (Webb, 2005). In some cases this will involve collaboration among parties, and in others, a certain amount of built-in 'friction' or 'check and balance' or competition among parties will be involved.

The SG approach seems well suited to CMALA, with CSR being part of the business contribution. Arguably, the emerging domestic mining governance approach in Canada – with a wide range of evolving federal and provincial laws and other mechanisms and processes administered by government agencies, combined with self-regulatory CSR mechanisms developed and implemented by mining associations, a memorandum of understanding between peak indigenous and industry groups, impact and benefit agreements between firms and affected communities, and pressure from NGOs and the social responsibility investment community – could be considered an emerging example of SG. In saying this, it is fully recognized that all aspects of the Canadian model are 'works in progress' and are far from perfect. The suggestion made here is that it would be useful if some variation on the emerging Canadian mining SG approach could also be developed and implemented in Latin American countries.

In addition, Canadian governments, mining firms and associations, investment entities (including stock exchanges), indigenous groups, and civil society actors are also well positioned to exercise leadership in development of a coordinated multilateral, multi-stakeholder SG framework for the mining sector, working in close cooperation with their counterparts in other mining-oriented jurisdictions, and relevant international actors. By developing a multilateral, multi-stakeholder SG framework of this sort through a grouping of like-minded entities, the disincentives associated with any one actor unilaterally taking on obligations that then puts them at a competitive disadvantage vis-à-vis their counterparts in other jurisdictions could potentially be minimized.

Acknowledgment

In preparation of this chapter, the author drew on research that was financially supported by the Social Sciences and Humanities Research Council of Canada, the Canadian Department of Foreign Affairs and International Trade, and the Canadian Business Ethics Research Network. The author gratefully acknowledges this support.

4
The Role of Governments in CSR
Jan Boon

Introduction

The term 'government' encompasses the state governance apparatus of a country: its political system, bureaucracies and institutions, as well as its sublevels. 'Home government' is the government of the country where a transnational company is registered and 'host government' is that of any other country where it is conducting operations. Citizens expect their government to promote peace, order, and good governance, thereby creating conditions for prosperity. They have a duty to protect their citizens against human rights abuses by third parties, including business (Ruggie, 2008). The nature of the relation between the state and its communities and corporations, the state's vulnerability to international pressures, transparency and the availability of information, and the enforcement and accessibility of a legal framework are key factors affecting governments' abilities to live up to these expectations and influence Corporate Social Responsibility (CSR) development initiatives undertaken by the extractive industry. This chapter provides a categorization of possible government roles in CSR and illustrates the issues, using examples from a series of interviews with key stakeholders in Canada and Peru (Boon, 2009). The case study demonstrates that both the host government and the home government have either failed to see or have not yet been able to take full advantage of opportunities presented.

Categorization of government roles

The role of government in CSR has been studied much less than that of business. Fox et al. (2002) note that the contemporary CSR agenda is relatively immature and that, while current public sector agencies in developing countries face considerable capacity constraints, there are significant opportunities for them to take advantage of CSR initiatives to further their policy goals. Key mechanisms and activities that might help build capacity in these agencies include CSR awareness building, public bodies setting the terms of the CSR debate, building a stable and transparent CSR environment, engaging the private sector in public policy processes, and developing frameworks within which to consider local or national priorities in relation to CSR (Fox et al., 2002).

A general literature survey on CSR and conflicts in the mineral exploration and mining sector reveals few specific references to the role of governments. However, when a role for government is implied in existing literature, it frequently falls into one of two categories: governments can either 'accompany' or be directly involved in the CSR process. Accompaniment includes facilitation, capacity building, coordination, and conflict management strategies. Direct involvement covers a vision and goals for the role of business, human rights responsibilities, establishing a clear context, and boundaries for CSR and market rules. Figure 4.1 describes this categorization scheme.

Accompaniment

Possible *facilitation* roles of government authorities include participation in the development of codes of conduct, promoting innovation, guiding debates on the roles and responsibilities of the actors involved, and providing assistance on technical, legal, and human rights issues. They can sponsor relevant research and support dialogs, meetings, events, and promotion of the CSR concept in a variety of ways. For example, in Canada, the Department of Foreign Affairs and International Trade (DFAIT, 2008, 2010, 2011), Industry Canada (2006), and Natural Resources Canada (2007) have produced related documents and

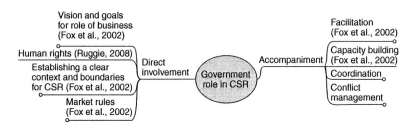

Figure 4.1 High-level categories for government roles

newsletters. The International Development Research Centre that is largely funded by the government of Canada undertook an extensive study of social-economic issues related to mining through its 1998–2003 Mineral Policy Research Initiative in Latin America, which achieved impressive results (International Development Research Centre, 2003b). *Capacity building* includes building expertise in developing compliance approaches, knowledge and application of best practices, and resourcing, developing, and institutionalizing capacity.

Examples of the government *coordination* role are the organization of roundtables on CSR in the mining industry in Canada (National Roundtables, 2007), the promotion of partnerships such as the Frente Regional en Defensa del Agua [Regional Front for the Defence of Water] in Piura, Peru (Todo Sobre Majaz, 2008), and the coordination of CSR activities across government departments.

Governments can also implement *conflict management* strategies. For example, the Peruvian Defensoría del Pueblo [Ombudsman's office] proposed that the government develop a transformative conflict management capability (Defensoría del Pueblo, 2007, p. 7). Elements of a conflict strategy that addresses underlying conditions for conflict (Arellano-Yanguas, 2008a, p. 24) could include righting past wrongs (Aste et al., 2003, p. 65), arbitration (Arellano-Yanguas, 2008a, p. 30), adding conflict management to Environmental Impact Assessments (Guttiérez, 2007, p. 85), and establishing conflict resolution institutions. The Ministry of Energy and Mines in Peru has established the Oficina General de Gestión Social, and part of its mandate deals with conflict management (Ministerio de Energía y Minas, 2007). The office is building the credibility it needs to become effective (Boon, 2009). However, many mining conflicts in Peru are related to imperfections in the implementation of the canón minero. The counselor's office that is part of the Government of Canada's CSR strategy, albeit weaker than an ombudsman function, also aims to help resolve conflicts (Department of Foreign Affairs and Trade, 2009). The UK's Department for International Development includes a Conflict Policy Team in its country programs (Department for International Development, 2010).

Direct involvement

CSR initiatives create opportunities for public sector agencies in developing countries (host governments) to further their development policy goals. They have also been used by politicians to obtain opportunistic advantage, when they take (undeserved) credit for company CSR initiatives. At the same time, there is a risk of the private sector filling policy space and reducing the legitimacy of government and its institutions. Unfortunately this happens all too often, as was pointed out by various respondents in Peru.

Extractive industry projects in developing nations are often associated with sometimes violent conflicts that cause regional and national crises. These opportunities and conflicts are set in the context of the governance ecosystem described elsewhere in this volume, of which government is an important part. The ecosystems view can highlight important inconsistencies between parts of the system such as a company's CSR department undertaking meaningful development initiatives while at the same time its Chief Executive Officer (CEO) may obtain barely defensible concessions from the national government using questionable tactics. Orihuela (2010) noted that in Peru, 'a revolving door was left open between the regulator and the regulated' (p. 22).

Governments of developed nations in which extractive industries have their headquarters also have a stake in CSR. For a mining and oil and gas nation such as Canada, the behavior of its extractive companies overseas has important foreign policy and local public opinion implications (Ottawa respondent). Whatever they do (or don't do), governments affect and are affected by CSR and it is to their advantage to influence the CSR agenda. Direct involvement offers increased control, but unfortunately many developing nations lack the capacity, resources, and stability required for it to be effective. While developed nations have the stability needed for a sustainable effort, many governments have been reluctant to invest heavily in this area, possibly because of lobbying by parts of the mining industry. This may have been the case in Canada when the government did not accept the recommendations of an industry–Non-governmental Organization (NGO)–academia advisory group (National Roundtables, 2007). Ways in which governments can become directly involved include setting a vision and goals for the expected role of business, paying attention to human rights issues, establishing a clear context and boundaries for CSR, and changing certain market rules.

Setting a vision and goals for the expected role of business includes adapting company law and corporate governance and joining efforts such as the Extractive Industries Transparency Initiative. For example in 2006, following a campaign by a CSR coalition of NGOs, faith-based groups, and parliamentarians, the UK passed a new Companies Act that strengthened requirements on social and environmental reporting. This forces directors of UK companies to consider the impacts of their operations on the community and the environment and it requires company reports to show how directors are performing on their corresponding duties (National Archives, 2006; Shah, 2006, p. 36).

Avoiding approaches that give companies political power is important, but the balance of power between the extractive industries and national governments may pose difficulties. For example the Peruvian government was reluctant to increase royalties to take advantage of high market prices for metals, as this might reduce foreign direct investment. As a compromise, it negotiated extractive industry contributions to a 'voluntary fund' for financing development projects. However, the expenditures of the fund are being managed by the companies and many consider its establishment to be a ridiculous abdication of responsibility (Boon, 2009). While leading by example (for instance, by transparency and consistency in government's own operations) is a powerful incentive, for many governments in developing nations this may be difficult to achieve. As one study respondent in Peru remarked, 'Mining companies are surrounded by a sea of societal irresponsibility, and the state is "playing dead"' (Boon, 2009, p. 89).

Governments have a duty to protect against *human rights* abuses by third parties. While states have full jurisdiction over human rights abuses committed within their jurisdiction, there is much debate about whether home governments can or are required to regulate against harm by 'their' firms caused overseas. However, many circumstances would allow home states to take such action (Ruggie, 2008, p. 7). The US Alien Torte Statute is increasingly being used to sue transnational extractive companies operating in developing nations for human rights violations through US courts (Cowman, 2010). However, the US Court of Appeals for the Second Circuit recently held that '"[f]or now, and for the foreseeable future, the Alien Tort Statute does not provide subject matter jurisdiction over claims against corporations." *Kiobel v. Royal Dutch Petroleum Co.*, Nos. 06-4800 & 06-4876 (2d Cir. Sept. 17, 2010).' This means that corporations can no longer be sued using this statute and claimants will now likely seek recourse by suing individuals within corporations (Cailteux and Grabill, 2010). Ruggie (2008) suggests that increasing coherence of states' domestic and international policies through approaches such as the incorporation of Human Rights requirements in the decisions of government Export Credit Agencies and their cooperation development agencies would help reduce human rights abuses.

Host governments should enforce their laws and regulations and provide punishment and redress in case of transgressions. Redress is very important, not only from a justice perspective but also from a conflict resolution perspective: unresolved past grievances are often at the root of ongoing conflict situations (Aste et al., 2003). Host governments should increase the coherence of their domestic policies, and Ruggie gives the example of investment protection agreements that negatively affect the ability of a government to implement needed social policies (Ruggie, 2008).

Governments have a role in *establishing a clear context and boundaries for CSR*, especially in establishing CSR policy and regulations that provide a level playing field for the actors. This includes establishing the boundaries of CSR (for example, where CSR stops and other actors' responsibilities start), providing a range of incentives (such as accelerated approvals for projects with a CSR strategy), developing instruments, exploring new roles (for example, making CSR agreements with communities compulsory), establishing a robust regulatory and enforcement context leaving room for voluntary CSR, and judicious application of fiscal policy (linking taxation levels to the presence or absence of acceptable CSR strategies). A European Union (2007) study found that many companies see CSR as a means to improve compliance with non-voluntary regulation – an interesting finding in view of usual private sector aversion to state regulation. It also noted that for CSR to contribute to EU sustainability goals, strongly binding social and environmental policies remain essential both to promote sustainability and to create a level playing field among companies.

Peru attempted to regulate CSR through a decree that, while not adopted by Congress, still has had a significant impact. The proposed decree, No. 042-2003-EM (Presidente de la República, 2003), would have required applicants of mineral concessions to meet basic commitments related to environmental excellence; respect for institutions, authorities, local culture, and customs; relations with the local population; dialog with and provision of relevant information; an institutionalized frame of reference for local development that extends beyond the life of a mine; local employment and related capacity building; and local purchasing of goods and services. Congress did not approve the decree ostensibly because it considered CSR to be a voluntary activity, but industry lobbying may have been an underlying driver (personal communication). However, according to staff at the Peruvian Ministry of Energy and Mines, the draft decree is in practice on its way to becoming a de facto standard.

Mechanisms for equitable mining revenue allocation and productive spending are another way in which governments can set a clear context. However, there may be pitfalls. For example, while Peru's Canón Minero directs a portion of tax revenues from mining back to the regions, imperfections in the allocation mechanism resulted in violent demonstrations pitting Moquegua and Tacna provinces against each other, and Arellano-Yanguas (2011) demonstrated a significant positive correlation between the size of canón minero transfers and the number of conflicts. A stable civil service with continuing staff is essential for managing the CSR interactions between actors throughout the mining and elections cycles. Many industry and community respondents in Peru commented that the 'CSR life cycle' is out of phase with the government election cycle and the staff changes that often accompany it. Also, many officials in Peru who received CSR training through development assistance programs subsequently found better-paid jobs in the extractive industry. Effective direct involvement of governments in developing nations presents considerable challenges.

Governments can also influence CSR through *market rules* such as guidelines for foreign direct investment, securities exchange regulations requiring disclosure of company activities and plans, fiscal policy (requiring pension funds to invest ethically), and labeling (provenance certification of diamonds, employment practices, and environmental management). In addition, government procurement rules, export credits, and credit guarantees can be made conditional on CSR performance. For example, the government credit agency Export Development Canada now makes adherence to International Finance Corporation Performance Standards (IFC, 2006) a condition for loans.

Governments' choices from the above options are political and are affected by ideology, local power structures, external influences, available resources, and other factors. Further, setting priorities is a complex undertaking. Most home governments focus their activities on accompaniment rather than direct involvement, and are reluctant to undertake actions that could be perceived as extraterritorial interference.

Stakeholder perspectives in Peru and Canada on the roles of government

The author obtained the perceptions of key stakeholders of the role of government in CSR in the mineral exploration and mining industry through a series of interviews (Boon, 2009) in Lima, Peru with host government, industry, NGO, and CSR consultant representatives, and in the areas of influence of an exploration project (Río Blanco, Piura Region, Peru) and a large mine (Antamina, Ancash Region, Peru) with campesinos, professionals, business/merchants, and service personnel. The latter groups were also asked for their perception of the actual performance of the various levels of government. Levels of government in Peru are, in descending order: national, regional, provincial, district, and municipal. In addition, home government representatives were interviewed in Ottawa, Canada. In total, 70 respondents took part in the study. A summary of the author's interpretations of their explicit and implicit suggestions regarding home and host government roles follows below.

Accompaniment

Facilitation

The home government and all levels of the host government should provide information to affected stakeholder communities relevant to the level of government and should support worthy industry initiatives. All levels of the host government have a role in facilitating communication between the company and the community; promoting early involvement; creating capacity and opportunities for dialog; being respectful, flexible, and creative in developing options; and tailoring approaches to local circumstances. They should also conduct public consultations and intervene with other levels of government when necessary. Respondents urged the national host government to be more present and impartial and to provide support rather than opposition, to conduct a real dialog with campesinos and, in general, to improve bottom-up dialog. All levels of host government could benefit from researching the use of local radio stations for information dissemination and capacity building, and the national host government could develop CSR models for informal, artisanal, and small- and medium-scale mining.

Capacity building

All levels of host government should build strategic and administrative capabilities and understand CSR and what it can do. District and higher levels of government should ensure communities understand what development is and have realistic expectations of what a mine could deliver. They also should provide cultural preparation for a money economy, pay attention to health, develop meeting and negotiation skills, and seek internal and external expertise. The regional and national host governments should support the development of local qualified personnel to work in the mine. Respondents recommended that the national host government and the home government establish a governance school and build preparedness for climate change and environmental challenges. The home government should provide scholarships.

Coordination

All levels of the host government should develop a system for coordination initiatives, establish associative partnerships, and improve linkages between CSR initiatives and the use of canón minero funds. They also should promote understanding between the various communities in the area of influence. The regional and national host governments should establish a government policy of coordination of social programs with each other and with company CSR initiatives, while the national host government and the home government were urged to establish mechanisms for mutual coordination (development of joint audit mechanisms, joint standards development, and adaptation of home country approaches to the Peruvian situation).

The home government was asked to better coordinate securities regulations, materiality considerations, and financing practices that have an impact on company CSR strategies.

Conflict management

All levels of host government were encouraged to avoid confrontations and to negotiate, and the four lowest levels of government to seek peace with the higher levels of government. Further, the provincial, regional, and national host governments and the home government should engage more in dispute resolution. Some respondents urged the regional and national host governments to play a mediating role, and to develop a capacity for transformative conflict resolution. The national host government needs to put in place a conflict management strategy (that could include a complaints office).

Direct involvement

Vision, strategy, and goals

All levels of host governments are urged to be strategic in their approach, with a leadership role falling to the national and regional governments. With the exception of the municipal level, they all should promote large development projects. The national and host governments should transfer strategic knowledge and technology, help provide markets, and change trade rules.

The national and regional governments should develop a *Government Social Responsibility* strategy and goals, which should be implemented by all levels of government, and they all should endeavor to attend more to people's needs, as well as to be transparent, recognize and assume their responsibilities, and eliminate corruption and bribery.

Human rights

All levels of host government and the home government should exercise some kind of control over companies. For the national host government this would include regulation and the application of sanctions and incentives. For the home government this could include supervision of 'its' companies, setting expectations related to investment support provided by government, and promoting consistency of company operations at home and abroad. A number of respondents suggested that the home government should institute an effective impartial arbiter function.

Context and boundaries for CSR

All levels of host government should ensure that CSR initiatives are aligned with development plans, and strive for infrastructure investments. Linking CSR to participatory budget processes could provide an opportunity for co-financing of development by the private sector. Respondents also urged stabilization of institutions to provide continuity, especially at the higher levels of government. There is a role for the provincial, regional, and national host government levels in providing legal frameworks for agreements, CSR initiatives, and community action and in legitimizing these. These same levels are urged to enforce existing land and resource use legislation and regulations and to promote increased investment in cattle farming, agriculture, and reforestation. It was suggested that both the national host government and the home government establish a mechanism to ensure that mines comply with agreements and promises. Another suggestion was the establishment of a comprehensive mining law that better integrates all mining-related activities (including CSR). Home government roles that were mentioned by respondents include clearly stating expectations regarding the social behavior of its companies abroad, setting expectations related to investment support provided by government, promoting consistency of company operations at home and abroad, and improving company reporting and disclosure practices. Both the home and the host government were urged to support the development of industry codes and standards, and to promote best practices. An additional suggestion for the national host government in this area was to establish CSR databases and performance-tracking mechanisms. It also was strongly advised to adjust the distribution mechanism for canón minero funds and regional, provincial, and district governments were strongly advised to put these funds to better use. A strong role for the home government was seen in cautious extrapolation and adaptation of 'home' CSR approaches to situations in Peru, as well as in establishing principles and conducting benchmarking studies.

These suggestions map well onto the categories summarized in Figure 4.1, while many provide more detail. Some touch on the core of the social responsibility concept, such as a suggestion that *Government* Social Responsibility is as important as *Corporate* Social Responsibility, if not more so. Others are more pragmatic, such as suggestions to invest in cattle farming, agriculture, and reforestation. While there was no consensus, a number of Canadian government officials saw a role for the Government of Canada in ensuring that companies fulfill their promises and that their CSR projects are sustainable. This would sound like music to the ears of the campesinos in the San Marcos area who made similar suggestions.

The breadth and number of suggestions indicates that respondents believe there to be an important role in CSR for both home and host governments. Also, all are 'enablers' in terms of the ecosystem model described in an earlier chapter of this volume. The author believes the suggestions in the 'Visions, strategy and goals' category to be among the most important, as they set a context for the success of other suggestions. The field is complex and in constant flux, and sustained leadership is essential for progress.

Having reviewed the range of suggested options for government action on CSR and corresponding stakeholder expectations in Peru and Canada, it is instructive to look at what their governments have actually done. The description of perceived government actions is preceded by a general discussion of the challenges faced by governments.

Governments' ability to act

Comaroff (1998) describes the state as simultaneously an illusion, a claim to authority, a cultural artifact, a present absence/absent presence, a principle of unity masking institutional disarticulation, and a potent construct that manifests itself, with tangible effects, in the quotidian activities of government and politics. In such context, the drivers of governments' choices are necessarily complex.

Many factors influence governments' ability to live up to expectations, not least of which is political disagreements about the meaning of 'good governance' and about the best route to achieve it. Many of those described below could be considered as 'disablers' or 'enablers' in terms of the ecosystem model described in Chapter 1. Important factors in developing countries include fiscal priorities inspired by free-market considerations and IMF Structural Adjustment Programs; the dependence on foreign direct investment to produce income and foreign exchange for government functioning; social demands created by foreign direct investment; lack of capacity and resources to inform decisions, and of political will to satisfy these demands; limits to their range of action imposed by the current economic international ordering; and lack of effectiveness and efficiency of government institutions. Many governments have become prisoners of circumstance, and their weakness has forced many to abdicate their responsibility and to rely on the private sector instead (after Szablowski, 2007, pp. 27–60). Be this as it may, governments play an important role in the functioning of society, whether through action or inaction. The context they set significantly influences the implementation of CSR initiatives in their countries.

The powers at the disposal of governments vary with circumstances and countries. The governments of developed nations helped set a neo-liberal agenda and they are now setting a context for CSR through their influence in international bodies. Some use the means at their disposal to influence the foreign activities of companies headquartered in their countries. They can (but often choose not to) influence 'home' activities of such companies through regulation or the threat of regulation.

The CSR movement could help governments of developing nations achieve their development goals, yet individual governments of developing countries have limited freedom of action. First, their need for foreign direct investment to create jobs and generate income puts them in a weak negotiating position. Also, state dependence upon loans from institutions such as the World Bank or the IMF, whose loan conditions require export-led industrialization, often has provided incentives to promote industrial expansion at the expense of social and environmental safeguards, which sometimes leads to communities becoming the victims of state aggression or discrimination (Garvey and Newell, 2005). Secondly, national and local governments often are poorly prepared for effective cooperation with both communities and corporations in development programs (International Development Research Centre, 2003a), and they frequently lack the resources to enforce regulations. Thirdly, they lack the skills and experience to avoid being taken advantage of by multinationals with deep pockets (for example, through complicated legal maneuvering). Fourthly, many governments face bribery and corruption challenges.

Governments of developing nations may perceive that a lack of regulation attracts foreign direct investment (Graham and Woods, 2006). The result can be a regulatory 'race to the bottom.' The pressure to deregulate exerted by international institutions such as the International Monetary Fund (IMF) has aided this phenomenon, and many countries have been left with a weak framework for ensuring the development of their societies (Graham and Woods, 2006). This is to the disadvantage of local communities, which can be left entirely unprotected by their own government.

Where the 'race to the bottom' and lack of resources and skills are combined with corruption, poor management and inappropriate use of resource rent, the result has been a 'resource curse' in countries like Nigeria, where extreme poverty continues to coexist with high resource revenues (Boele et al., 2001; Ite, 2005).

Blowfield and Frynas (2005) mention a shift toward a facilitation role for governments. A credible threat of government sanctions helps this role being realized in practice. States can provide both incentives and disincentives for CSR initiatives by establishing legal requirements and investment conditions for companies that operate in their jurisdiction. For example, the Government of India is drafting a mining code that if accepted, would make CSR mandatory (Canada-India Conclave on CSR, New Delhi, 30 April 2010). However, even where states are willing to use sanctions, they may be unable to implement them against more powerful transnational corporations.

Host and home government actions: Peru and Canada

The following sections describe perceptions of how the governments of Peru and Canada are currently approaching CSR. Unless indicated otherwise, the analysis draws on the observations made by participants in the study referred to earlier (Boon, 2009).

Peru

Most Lima respondents had a rather low opinion of the understanding and attitudes of authorities at most levels toward CSR and toward their own responsibilities. CSR allows the state to shirk its responsibilities ('it is playing dead'), thereby increasing the risk of unsustainable dependency on companies. Local authorities are using 'mining funds,' be they from CSR or the canón minero, to replace their regular budgets and diverting the latter to other areas. This undercuts the institutionalization of social benefit structures and risks making a community worse off after a mine closes than before it opened. As many mayors are elected with a fairly low percentage of the vote, their support base is weak and their political future uncertain, which understandably leads to political opportunism, such as using canón minero funds for political purposes.

The 4–5-year election cycle is out of phase with the 15-year 'CSR cycle,' and a government change is often accompanied by a change of officials in government institutions. This lack of institutional stability negatively affects the sustainability of CSR initiatives. The national government is mostly absent and lacks the will to solve problems. Where it does show some will, it is often incompetent, such as in the way it handled the Moquegua–Tacna dispute mentioned below. The political will to take action does not (yet?) seem to exist and the political agenda appears to be dominated by short-term considerations.

Arellano-Yanguas (2008a) argued that many government decisions in Peru are ad hoc and often not internally consistent. For example, while everyone knows that serious problems exist, action is only taken when an explosion occurs. Such was the case when violence erupted over a 2008 dispute about the distribution of canón minero funding between Tacna and Moquegua provinces in which 60 policemen were taken hostage by crowds. Prime Ministerial intervention was required for the situation to calm down, despite the fact that a report of the ombudsman's office almost a year earlier had anticipated this event (Púlsar, 2008). While the International Labour Organization Convention 169 on indigenous peoples has been ratified by Peru, actual Peruvian legislation is not consistent with this and leaves a vacuum in many areas (Salas Carreño, 2008, p. 207).

Leadership and coherent policies are lacking at many levels. Not all levels of government cooperate with companies, and coordination and cooperation between levels is often poor. All levels of government are affected by a lack of capacity and resources, and bureaucratic hurdles often add to the challenge.

Perspectives on the authorities' way of approaching CSR issues in the vicinity of the Río Blanco Copper S.A. exploration project differed much from those in the vicinity of the Antamina mine. Most of this difference is a consequence of the difference in mining cycle stage and in community characteristics. In the Río Blanco area, the camps are strongly polarized on the issue of whether there should or should not be a mine. The actions and attitudes of the authorities are judged in terms of their choice of camp. Within this framework, opinions on the authorities' approaches show a clear fault line: the municipal, district, and provincial authorities are 'good'; the regional and national authorities are 'bad.'

In the Antamina area the mine has power, there is an inflow of money and increased risk of corruption, and the community's social fabric is weaker. The actions and attitudes of the authorities are judged in terms of their development outcomes, and many respondents appeared to be concerned by what they observed.

Through its significant symbolic role the state is an important CSR actor, whether actually present or not. For example Damonte Valencia argues that the families of the comunidad campesina Angoraju de Carhuayoc have always supported the authority of the state (however weak) as a means of overcoming suppression by local elites, and in the process of their ongoing battles with the state they have developed a strong sense of identity. They are reproducing this same old 'bilateral' strategy in their dealings with Compañía Minera Antamina (Damonte Valencia, 2007).

While the UK and Switzerland were mentioned as having involved themselves in the local scene in some way, Canada was seen as an active player that is close to the problems faced by the mining sector, and various respondents commented that Canadian companies' competitive position would benefit from a strong Canadian government CSR framework that would provide recourse for communities negatively affected by the activities of Canadian companies. Canada's PERCAN (PERU–CANADA) program (aimed at building capacity in Peruvian government agencies) was mentioned repeatedly, as was the program's mining toolkit. While the PERCAN program aimed to strengthen government capacity, it may instead have created CSR capacity outside government: many of the trainees have left government and are now private consultants that provide services to the mining industry, or employees of international organizations or mining companies (private communication).

Canada

Canadian government officials believed that CSR is here to stay and that aspects of it may morph into regulation. They commented that the *role* of government is wider than its mandated *responsibility*; therefore the government supports and promotes CSR. They recognized both the upside of productive synergies with the CSR initiatives of Canadian companies in developing countries, and the downside risk of host governments abandoning their responsibilities. Canadian industry CSR performance was considered to be reasonably good, as the number of serious breaches reported over a ten-year period is low relative to the large number of projects in which Canadian companies are involved. Both Canada's reputation abroad and Canadian public opinion are policy drivers.

Public concern over alleged misdeeds by Canadian companies abroad led the Government of Canada to organize a series of public roundtables in Canada that were well attended and at which over 100 presentations were made. The Advisory Group of the National Roundtables on CSR in the Canadian Extractive Industry Abroad, composed of members from industry, the NGO sector, investors, and academia, produced a hard-won consensus report containing recommendations that include the establishment of a Canadian CSR framework with Canadian standards; reporting against the Canadian standards in an internationally recognized format; links to income tax incentives, stock exchange listing requirements, and institutional investor social, environmental, and governance disclosure requirements; and the establishment of an independent ombudsman and a tripartite compliance review committee (National Roundtables, 2007; Drohan, 2010).

The government, after two years of deliberation, announced a CSR policy that is considerable weaker than the recommendations of the Advisory Group (DFAIT, 2009). Drohan, in a policy brief prepared for the University of Ottawa Centre for Policy Studies commented:

> It rejected the idea of Canadian standards, an independent ombudsman, a tripartite review committee, or any threat of withdrawing of government support. Instead it set up a new counselor, answerable to the minister of international trade, to advise companies and to investigate complaints if all parties to the complaint agreed. The government accepted some of the consensus recommendations, including that it help host governments build institutional capacity to oversee their extractive sectors, and it joined the Extractive Industries Transparency Initiative, which puts the onus on host governments to be more transparent about resource revenues. However, by ignoring the main provisions of the consensus report, the government rekindled the decade-old debate and eroded support in the mining community for the compromise position. (Drohan, 2010, p. 2)

Not surprisingly, NGOs that had made significant concessions felt cheated and sought other ways to achieve their objectives. They helped develop and support private member's bill C-300 (House of Commons of Canada, 2009). Again quoting Drohan (2010):

> The bill calls on the ministers of foreign affairs and of trade to set out guidelines for economic, environmental and social performance of Canadian firms operating abroad, to accept and investigate complaints that firms have contravened these guidelines, and to withdraw consular support, funding by Export Development Canada and investment by the Canadian Pension Plan Investment Board when a company is found in non-compliance. Bill C-300 is not as comprehensive as the 2007 consensus report, nor does it command the same broad support. Yet critics of the current government policy have rallied to support it, perhaps because although it is a long shot, it is the only means on offer to tighten Canadian oversight of mining companies abroad. (Drohan, 2010, p. 3)

As bill C-300 was a private member's bill it could not commit the government to spending public money beyond the approved budget. This limited the scope of its drafters to incorporating only items that could reasonably be expected to be funded from existing Department budget. For example, the ombudsman's office that had been proposed by the Advisory Group could not be included in the bill (National Roundtables, 2007). It also forced the granting of discretionary power to the Minister (House of Commons of Canada, 2009). Thus the resulting bill suffered from a number of weaknesses and industry argued that it would leave them open to frivolous accusations that would unjustly endanger its members' reputations and destroy the companies thus targeted. Civil society organizations argued that the government's CSR strategy is weak and that it would be better to have weak legislation than to have none. After ferocious lobbying by both sides, the bill was narrowly defeated with 140 votes against and 134 in favor. C-300 is one of a few private member's bills that came so close to being passed. The debate drew great interest around the world, especially from communities affected by mining. While the CSR strategy that was mentioned earlier will remain the official Government of Canada approach, it appears that further challenges could be mounted.

The CSR strategy has four 'pillars': enhance the capacities of developing countries, promote widely recognized international CSR performance guidelines with Canadian extractive companies operating abroad, the Office of the Extractive Sector CSR Counsellor, and a CSR Centre of Excellence. The Department of Foreign Affairs and

International Trade is now intensively educating its trade commissioners on the subject of CSR and giving prominence to CSR at the international stage.

The Centre of Excellence is being hosted by the Canadian Institute of Mining (CIM). While the CIM is approaching the task in an unbiased way, the placement of the Centre in an organization related to mining has raised eyebrows (Hohnen, 2009). The CSR strategy is presumably promoted as 'building the Canadian advantage' because it maintains a level playing field between companies by not placing government expectations on Canadian companies that are more stringent than those faced by its foreign competitors. This reasoning may overlook community, societal, and NGO expectations that are playing an increasingly important role. There are many examples of the high cost of not meeting these expectations. It led Meridian Gold to lose its Esquel project in Argentina, worth hundreds of millions of dollars (Prospectors and Developers Association of Canada, 2007). Mass demonstrations forced Newmont/Minera Yanacocha S.A. to abandon its planned Cerro Quilish extension and Newmont was forced to reclassify the 3.9 million ounces of gold at Cerro Quilish to 'mineralized material not in reserves.' These would have yielded an approximate valuation of US$1.6 billion in 2004 and US$4.3 billion in 2009 (CERES, 2009, p. 76). Also, Peruvian respondents suggested that a strong Canadian CSR policy would provide a competitive advantage to Canadian companies.

On 20 October 2010, the Office of the Extractive Sector CSR Counsellor formally launched its review process that will be conducted in the context of the Government of Canada's endorsed performance standards: the IFC Performance Standards, the Voluntary Principles on Security and Human Rights, and the Global Reporting Initiative. The process applies to all Canadian oil, gas, and mining companies in their operations outside of Canada. Requests for review involving the Organisation for Economic Co-operation and Development (OECD) Guidelines for Multinational Enterprises will continue to be handled by Canada's OECD National Contacts Point. The most salient aspects of the process are its voluntary nature, its focus on constructive dialog and creative problem-solving, and the ability of individuals, groups, communities, and Canadian extractive companies to request a review. The process consists of an assessment of eligibility, followed by a two-phase informal mediation. The first phase is directed at building trust, while the second phase involves structured mediation. The parties can access formal mediation if they so desire. They can also withdraw from the process at any time (Foreign Affairs and International Trade Canada, 2010).

The existence of a separate process for reviews related to the OECD guidelines may create some confusion, and it will be interesting to see whether the Counsellor Office process will be used more often than the OECD National Contact Point has been so far.

As the recent comprehensive agreement between the Canadian forestry sector and environmental civil society organizations shows, paradigms in the resource industry are shifting. The Canadian Boreal Forest Agreement was signed in May 2010 between the 21 members of the Forest Products Association of Canada and 9 leading environmental organizations. FPAC members commit to the highest environmental standards of forest management and conservation, while environmental organizations commit to global recognition and support for FPAC members efforts (Canadian Boreal Forest Agreement, 2010). From this perspective, it is clear that the Government of Canada lost a rare strategic opportunity by not accepting the recommendations of the Advisory Group, an observation shared by various industry representatives (personal communication).

As mentioned earlier, there is growing international recognition that home governments can regulate against harm caused by 'their' companies overseas under many circumstances (Ruggie, 2008, p. 7), and the final report (targeted for 2011) of the UN special representative on business and human rights is expected to make corresponding recommendations. Drohan (2010) suggested that the Government of Canada should begin preparations now for the development of a more useful and comprehensive policy based on the recommendations in that report.

The Canadian mineral exploration and mining industry is a key sector of the Canadian economy and a codeterminant of Canada's place in the world. The industry is facing an environment of rapidly changing expectations and power relations and its ability to meet these challenges is being severely taxed. At the same time, the new circumstances present opportunities for industry and for Canada. The Government of Canada is a very important actor and as such it could benefit both the industry and the country by taking a more strategic and proactive view than it has taken so far. It can foster new types of alliances and partnerships and choose to set boundary conditions for steering the CSR ship into prosperous and much less troubled waters.

Summary and conclusions

The involvement of governments in CSR can take many forms, which can be grouped into 'accompaniment' and 'direct involvement' headings. Categories in the former group include facilitation, capacity building, coordination, and conflict resolution. The latter group includes provision of leadership and vision, human rights, establishing a clear context and boundaries for CSR, and market rules. The many factors that influence governments' choice of action include limitations imposed by the international economic ordering, capacity, resources, and political context. Whatever choices they make, both host and home governments are exceedingly important actors and even their inaction can have significant consequences. An analysis of the approaches taken by the governments of Peru and Canada indicates that lack of political will, and an ad hoc governance style prevent the former from taking a strategic approach that maximizes the benefits Peru can gain from CSR initiatives, while the latter's strategy may be based on a paradigm that may not remain in place much longer. This study's findings support the validity of the ecosystem model proposed in an earlier chapter. Future research applying Foucault's genealogical approach to CSR as an analytics of government may diagnose the limits and the possibilities of the present and identify suitable 'technologies of government' (Dean, 2010, pp. 58, 59).

5
Regulatory Frameworks, Issues of Legitimacy, Responsibility, and Accountability: Reflections Drawn from the PERCAN Initiative

Bonnie Campbell, Etienne Roy-Grégoire, and Myriam Laforce

Introduction

During the 1990s, the Peruvian mining sector was the object of a far-reaching process of liberalization which included the privatization of state enterprises and a series of measures aimed to attract foreign direct investment. Since 1992, the sector has received US$12 billion in national and foreign investment. In 2007, it represented 20 per cent of total foreign direct investment and generated 25 per cent of public revenue collected from corporate income taxes by the Peruvian government; 130,000 people are estimated to work in the mining sector. Since the 1990s, with the growth rate at 7.8 per cent as compared with the 4.1 per cent growth rate of the GDP, it is one of the fastest growing sectors. A large part of this activity is undertaken by foreign mining companies. Peru is at present the world's second largest producer of silver, the third largest of zinc, the fourth largest of lead, the fifth largest of copper, and the sixth of gold. In 2007, mining activities represented 62 per cent of the total of Peruvian exports and 10 per cent of GDP (PERCAN and MEM, 2009, p. 1). Canadian mining companies are key actors in the Peruvian mining sector, particularly with regard to the exploration and production of gold (DFAIT, 2010).

However, the Peruvian mining sector is characterized by a very high incidence of social conflicts (PERCAN and MEM, 2009, p. 1). Peru's Ombudsman reported that a significant proportion of the 197 reported social conflicts in 2008 were related to mining (DFAIT, 2010). At present, according to certain estimates, activities concerning the exploitation of natural resources represent by far the most important (50 per cent) source of conflicts in Peru (Defensoría del Pueblo, 2010, p. 6); their frequency inevitably has important implications for the governance and the viability of the sector.

In the context of increasing mining activities and of conflicts in this area, the Canadian International Development Agency (CIDA) introduced an ambitious program which had as its objective:

> [I]mproving the contribution of the mining sector to the sustainable development of the Peruvian mining regions, promoting the dynamic and firm integration of social issues within an integrated framework for socially responsible and efficient mining... through the implementation of activities aimed at strengthening the institutional capacity of the mining sector, favoring policies and practices that are appropriate for the Peruvian mining industry. (PERCAN, 2009)

The Peru–Canada Mineral Resources Reform Project (PERCAN) is the result of a bilateral agreement between CIDA and the Peruvian Ministry of Energy and Mines (MEM). The project has been implemented in several phases over the period 1998 to 2011, with a total budget CAN$17.7 million (PERCAN and MEM, 2009, pp. 6–7, 29). In view of the fact that the Canadian government has expressed its intention to reproduce this initiative in other Latin American countries (notably in Colombia), an examination of the context in which this project has been set up and the assumptions on which it rests seems pertinent in order to be in a position to evaluate to what extent the strategy pursued by the Canadian government is likely to resolve the problems raised by the widespread presence of conflicts related to mining in Peru.

This chapter adopts a historical perspective in order to explore the problems that PERCAN seeks to address, by examining the role that regulatory frameworks and the actors involved, whether public or private, play in shaping the environment in which mining activities take place. The first section draws attention to issues of legitimacy, responsibility, and accountability that are often at the origin of conflicts, and in doing so, attempts to cast light on the processes noted in the Introduction of this volume, notably, by addressing the question of *how* 'irresponsibility and lack of accountability are built into the system' (Sagebien and Lindsay, present volume, Introduction, p. 2).

The second section will analyze the strategies that PERCAN has implemented in order to determine the extent to which these address the problems which they set out to resolve. The example permits deepening understanding of the role of various actors, notably external, in current initiatives to address governance gaps.

Mining regulation in Peru: Legitimacy, responsibility, and accountability

The reforms ushered into the mining sector in the 1980s and 1990s under the leadership of the World Bank Group (World Bank, 1992, 1995a, 1996), whether in Latin America or Africa, had as their primary objective the opening of mineral-rich economies to foreign investment as a means to stimulate economic growth. In the new policy environment that resulted:

> The new agenda advocated comprehensive privatisation of state companies, an end to restrictions on foreign ownership and the repatriation of profits, lowering rates of taxation and royalties, restructuring labour laws to permit greater flexibility, and the termination of performance requirements such as those mandating local sourcing or local hiring. In addition, mining legislation had to be rationalised, administrative processes simplified, technical services to the industry (such as modernisation of the mining cadastre) improved, and 'subjective' elements of bureaucratic discretion removed from the permitting and approvals processes. (Szablowski, 2007, p. 34)

In Peru, the agenda that aimed to revitalize the mining sector beginning in 1991 under the first Fujimori government entailed a radical change in the country's political orientation away from statist economic policies and toward free market economics, ushering in a period of structural macroeconomic change and related legal reforms. The culmination of this process has been characterized as 'one of the most rigorous neoliberal strategies ever applied in Latin America' (Teivainen, 2002, p. 113), making the country a showcase among the international community (Boloña, 1996, p. 215).

In fact, while reducing inflation and restoring major macroeconomic balance were presented as the priorities of the new government's domestic policies, the objective of having Peru welcomed back into the international financial community and the reestablishment of relations with the multilateral financial organizations were central in shaping the reorientation of the country's economic policies according to many analysts (Aliaga and De Echave, 1994, p. 5; Boloña, 1996, p. 213; Gonzales de Olarte, 1996, p. 73; Iguíñiz, 1998, p. 26). According to Susan Stokes (1997, p. 219), when the new President took office, the stark contrast between the economic policies that had been part of the electoral program of Fujimori and the unequivocally neoliberal orientation of the measures that were in fact to be implemented by his government, can be explained by the conditionalities presented by the multilateral financial institutions as non-negotiable preconditions for the disbursement of new loans.

As a result, the Peruvian adjustment program benefited from the strong support, both financial and technical, of the World Bank, the Inter-American Development Bank and the International Monetary Fund (IMF). The IMF in particular was closely involved with the drawing up of the economic policy program (Teivainen, 2002, p. 117), through the Rights Accumulation Program and the Extended Facility Program, introduced respectively in 1991 and 1993 (Boloña, 1996, pp. 216, 222). In fact, in a single month in November 1991, Fujimori introduced 120 bills for new laws that aimed officially to reduce sate intervention in the economy. Of these, 78 per cent were in the end adopted (Cameron, 1994, p. 149). According to Philip Mauceri, '[t]ogether, these legislative decrees represented the most significant reordering of the Peruvian state since the Velasco era' (quoted by Cameron, 1997, p. 55).

The number of reforms and the rapidity with which they were adopted had both political and economic consequences. On the economic front, the development model that was introduced attributed overriding importance to the role that foreign direct investment was to play in the economy and called for the withdrawal of the state from a large number of its previous functions: 'In 1991, the administration redefined the state's role in production, that is, that the state should not be present at all in direct production but should return to its sole function as the promoter of private capital. This role was specifically expressed in the promotional laws' (Nuñez-Barriga, 1999, p. 153). In this regard, an advantageous package of fiscal incentives was introduced to attract foreign investments (UNCTAD, 1993; UNCTAD, 2000a, p. 20).

However, the stability of these reforms depended on the narrowing of political space and a realignment of political forces (Samford, 2010), which led Fujimori in 1992 to suspend the Constitution, dissolve Congress, and dismantle the judicial system. The new Constitution (1993) subsequently substantially strengthened the power of the Executive (Cameron, 1994; Crabtree, 2000). Moreover, the reforms significantly modified the state's mandate. To summarize, the Fujimori administration proceeded to introduce important modifications in the country's macroeconomic regulations and to construct a new legal order that was seen as 'more suitable for the needs of transnationalizing business transactions' (Teivainen, 2002, p. 132). According to UNCTAD, these measures 'place powerful constraints and obligations on the Government' (UNCTAD, 2000a, p. 26). The consequences of these policies with regard to the improvement of the level of well-being of Peruvian people were, at best, mixed. While in 1986, 37.9 per cent of the population lived below the poverty level, ten years later and after the far-reaching reforms, in 1996, this proportion had climbed to 49 per cent (Gonzales de Olarte, 1997, p. 80; see also Figueroa et al., 1996; Wise, 1997; Crabtree and Thomas, 1998; Machuca, 2002).

A further consequence of these economic measures was the increased importance of the place occupied by the primary sector in the Peruvian economy and in particular, by those sectors able to generate a relatively higher level of profitability such as the mining and fishing sectors (Iguíñiz, 1998; Jiménez, 2002). The mining sector, which is highly export oriented, has traditionally maintained a central role in generating foreign exchange for the country (IDEM, 1991), and was in fact essential in order for Peru to be in a position once again to begin repaying the country's debt (Aste Daffós, 1997). The World Bank played an active and determinant role in the reform of the mining sector, notably through the granting of loans for two programs over the period from 1993 to 1998: the Privatization Adjustment Loan (Mineral Resources and Mining) (for a sum of US$250 million) and the Energy and Mining Technical Assistance Loan (Proyecto EMTAL – US$11.8 million) (Mainhardt-Gibbs, 2003).

The process of reform introduced into Peru's mining sector at the time closely resembled that of Chile. Like Chile, Peru has a rich mining history and tradition, but its tradition had been one of administrative rather than adjudicative determination of mining rights. In 1991, Peru passed numerous amendments to the existing mining law of 1981. In 1992, the new legal framework for mining was promulgated as the General Mining Law. Like Chile, Peru gave stronger recognition and protection to private property rights in the new constitution enacted in 1993 after the amendment to the mining law. Also like Chile, Peru offers investors stabilization agreements. Unlike Chile, however, Peru carried out an aggressive program of privatization of nearly all state-owned mining enterprises and made environmental compliance an integral component of its new legal regime for mining (World Bank, 1996).

In conformity with the economic policy agenda that had been adopted and in order to establish a level playing field between public and private, and between foreign and national enterprises, the Peruvian General Mining Law of 1992 also redefined the role of the state in mining. In keeping with the World Bank recommendations which 'set out programs for the reform of the institutions to meet the needs of a private-sector-led mining industry' (World Bank, 1996, p. 30), the law also provided the legal framework for the reform of the public mining institutions which administer the mining law and simplified the administration of mineral rights. The Peruvian experience has in fact in many ways inspired the conceptualization by the World Bank of its Latin American Mining Law Model, which served as a barometer for the recasting of the mining legislation of dozens of countries around the world (Williams, 2005).

Summarized briefly, the reforms introduced into the sector resulted in five main changes. First was the conversion of *mineral rights* into *irrevocable property rights*, which are protected from being seized without just compensation and due process of law, are freely transferable, and can be held for an unlimited period of time (Williams, 2005, p. 746). Failure to comply with a requirement on the part of the investor no longer entailed the cancellation of the mining title but only a simple penalty or 'administrative sanction,' as it has been called (Sánchez Albavera et al., 1999, pp. 19–20). Secondly, the *procedures for access to these rights* take place in a non-discriminatory, transparent, and efficient manner that provides security of the right to proceed from exploration to exploitation, due to the introduction of the unified concession for exploration and exploitation in the Peruvian Mining Investment Law of 1991 (Chaparro, 2002; Williams, 2005). Thirdly, the *security of tenure* reduces the possibility of discretionary practices on the part of the administration. Maintenance fees are fixed at an annual rate of US$2 per hectare (Campodónico Sánchez, 1999), a figure which was to be increased slightly at the very end of the 1990s. Fourthly, with regard to *operating rights and obligations*, most observers recognize that the former were extended and the latter reduced and simplified (Chaparro, 2002). Fifthly and finally, the *competitiveness and stability of the investment parameters* led to the introduction of a fiscal regime based on revenues rather than on inputs or production, as well as the possibility for certain companies to sign fiscal stability agreements with the government. These stability agreements, which are signed for terms of ten to fifteen years, represent a clear incentive that is considered to have played a major role in attracting new investment (World Bank, 1995a). Between 1991 and 1998, 22 mining companies took advantage of this right (Campodónico Sánchez, 1999). Contrary to the guarantees granted through such stability agreements to investors more generally, those offered to mining companies apply to all taxes normally collected by the government (UNCTAD, 2000b).

These reforms were well received by international mining companies, resulting in the spectacular growth in exploration investment in Peru from about US$10 million in 1989 to about US$200 million in 1995 (World Bank, 1996). As noted at the time, however, the inadequacy of public monitoring of investments, notably with regard to environmental issues (World Bank, 1996, p. 6; Nuñez-Barriga, 1999), as well as the soaring number of projects that were simultaneously to increase considerably the land area devoted to mining activities (Bedoya, 2001), represented over the years that followed (among other factors) fertile ground for the emergence of tensions and conflicts between companies and communities affected by mining operations.

In order to better understand the nature of the difficulties that have persisted to the present and certain factors at the root of such conflicts, three implications of the manner in which the process of liberalization was introduced into the Peruvian mining sector in the 1990s merit closer attention.

First, underlining the approaches of the 1990s which aimed to create a suitable investment environment for the private sector, the one past trend that stands out above all others concerns the radical redefinition of the role and functions of the state and the new delineation between public and private spheres of authority that have accompanied this redefinition. As summarized in a World Bank document entitled *Strategy for Mining in Africa* (World Bank, 1992), which systematized the approach to mining regime reform in Africa adopted by the multilateral financial institutions, this required: 'A clearly articulated mining sector policy that emphasizes the role of the private sector as owner and operator and of government as regulator and promoter' (World Bank, 1992, p. 53). As we have seen, the primary role of government in the 1990s was that of facilitator of private investment. What resulted was a sectoral approach that focused on mining rather than one which sought to articulate the contribution of the mining sector to macroeconomic objectives involving inter-sectoral linkages, with a view of seeing to what extent the sector could play a transformative role and contribute to broader developmental goals. In Peru, for example, little provision was made to build eventual backward and forward linkages in the industry, such as the possibility of value-added processing of minerals (UNCTAD, 2000a), which in a resource extraction economy would normally be considered important development objectives.

Secondly, and as elsewhere, as states were caught between the competing imperatives of attracting foreign investment and at the same time responding to national and local concerns, their latitude for action was often circumscribed by the legal and practical conditions set out in reformed regulatory frameworks. The lack of room for maneuver led some states to respond to, if not reconcile, competing internal and external pressures involved by awarding 'rights to the investor accompanied by an informal delegation of local regulatory responsibilities' (Szablowski, 2007, p. 27). It led as well to states effectively 'transferring legal authority to mineral enterprises to manage social mediation' (Szablowski, 2007, p. 27). In this context, the coping strategy adopted by governments to deal with new mining regimes can be described as one of the 'retreat of the state' (Strange, 1996) or of 'selective absence.' According to the latter formulation, the state basically 'absented itself from substantial parts' of the legal regimes intended to help 'mediate between investors and community interests' (Szablowski, 2007, pp. 28, 45).

A third consequence of the liberalization of the mining sector has been the way in which past public functions of the state have increasingly been delegated to private operators. These include service delivery and also rule-setting and implementation. The tendency has been for 'an increased (and often reluctant) assumption of state-like responsibilities by transnational mining enterprises at the discreet behest of weak governments' (Szablowski, 2007, p. 59). While past and current trends may allow governments to shift the locus of responsibility for what were previously considered state functions (such as clinics, roads, infrastructure, and security) to the private operators of large-scale mining projects and Non-governmental Organizations (NGOs), and help explain the pressures on companies to engage in Corporate Social Responsibility (CSR), such transfers not only silence the legitimate and indeed necessary right of governments to offer services to their populations, a precondition to their being held publicly accountable, they also contribute to obscuring the issue of government responsibility itself. The current sidestepping of the state, by suggesting companies can gain better legitimacy for their operations by offering social services, runs the risk of undermining a precondition for building responsible governments and the basis of democratic practice: the need for governments to be accountable to their populations.

In a context of the weakened institutional and political capacities of states resulting from the reform process and, consequently, of their weakened capacity to pursue developmental objectives, to enforce regulations in areas of key importance to communities, and to meet national economic objectives, along with the trend of transferring public responsibilities to unaccountable private actors, these trends are likely to raise issues of legitimacy for the operations of mining companies themselves.

There have been various interpretations and responses to the problems and risks resulting from the 'legitimacy gaps' with which companies have been faced and which may well be increasing. For authors such as Szablowski (2007) who are interested in responses at a multilateral level, 'the transnational dimension of legitimation brought on by globalization has prompted the development of global policy arenas and has sparked the need for transnational law-making, with far-reaching consequences' (p. 60). What is being debated through the resulting 'global *legal* politics' concerning the rights and obligations of enterprises 'is the regulatory terms on which different audiences are willing to find that the entitlements of transnational enterprises will be deemed legitimate' (p. 65).

For other analysts who adopt a different point of entry and analytical framework, the focus is placed on processes internal to the countries in which companies operate, which are seen as dysfunctional and described as the 'governance gaps.' According to this perspective, 'CSR provides firms with a strategic response to the risks that systemic dynamics present, by addressing governance gaps that can, in turn, increase the potential for obtaining a "social license to operate"' (Sagebien and Lindsay, present volume, Introduction, p. 2). As with approaches formulated in terms of 'weak governance' this type of perspective raises a series of questions, for it appears to be postulated that these problems concern above all poor (or corrupt) management issues which can be resolved by the introduction of the right set of administrative practices and procedural measures, and monitored by using 'governance indicators.' However, such approaches often lead to the introduction of parameters that seek to quantify the performance of historically constructed, country-specific, highly complex institutional relations, using notions that are variously defined and the object of debate, and that are often highly subjective, such as 'government effectiveness,' 'regulatory quality,' and 'voice and accountability.' The increasing technicization of decision-making processes runs the risk of sidelining important substantive debates and, notably, of depoliticizing issues such as resource distribution, which may be treated as technical questions even though they are clearly political. Consequently, these issues are difficult to track, monitor, and measure because they often involve political choices and not only technical decisions.

In this regard, it seems difficult to reconcile how CSR might provide firms with a strategic response to the risks that systemic dynamics present by addressing governance gaps around key issues where institutional and technical capacity are lacking, such as in contract negotiation, revenue collection, or even environmental monitoring, *all areas in which they would be in conflict of interest*. Moreover, with technical and administrative aspects of 'governance' given overriding emphasis, current proposals to contribute to 'capacity building for resource governance' in developing countries unfortunately miss the key point that past reform measures, which have sought to open the extractive sector to investment, have done so in a manner that severely weakened the political and institutional capacity of local governments. Consequently, it becomes a circular argument to call for the reinforcing of local capacity if the nature of past and ongoing reforms, which weaken local capacity, is not questioned.

Our analysis suggests that issues of legitimacy with which companies are increasingly faced can be seen as a consequence of evolving structural relations that result in part from the manner in which the mining sector has been reformed. These structural relations have significantly modified and sometimes obscured the demarcation of spheres of responsibilities, whether public or private, and have frequently also blurred distinctions between the political and the technical domain.

On a conceptual level and as an alternative to approaches formulated in terms of 'weak governance' or 'governance gaps,' this chapter proposes the usefulness of the notion of 'modes of governance.' This term has been defined in many ways. Applied to mining, Belem (2008) defines it as the sum of the forms of regulation that determine the conditions of exploitation of mineral resources for any particular project. This definition puts emphasis on the actors responsible for the forms of regulation, as well as the evolution of these forms resulting from the evolving positioning of these actors. 'The notion of modes of governance permits identifying the implications for social relations of emerging institutional arrangements, as well as the role of actors who represent alternative values' (Belem, 2008, p. x [our translation]). Hence, the mode of governance in the mining sector 'represents the sum of the forms of regulation for each of the related dimensions (economic, social and environmental), which determines, in any given period, the conditions of exploitation of mining resources' (Belem, 2008, p. 232 [our translation]).

It is against this background stressing the importance of taking into account the reforms at the origin of the reshaping of institutional arrangements, the structural relations of influence and authority which characterize these reforms, and the role and responsibilities of the various actors involved, that PERCAN will be examined as a specific program introduced as a response to the challenges raised by the emergence of conflicts around mining activities. In a manner complementary to debates that have arisen concerning responses at a multilateral level to issues of legitimacy as analyzed by Szablowski, as well as those that look to CSR as a tool and suggest possible avenues of responses at the corporate level, this chapter focuses on a bilateral initiative to address these issues as they have become manifest in Peru.

The PERCAN initiative

In 2007, the Canadian government announced its intention to make Latin America a region of priority in its foreign policy. A key measure underlining this re-engagement in the Americas is the establishment of free trade agreements with countries and zones considered strategic allies (such as Peru, Colombia, and Central America). These agreements are also seen as instruments of democratic consolidation (Cameron and Hecht, 2008).

In a parallel sense, Latin America has become a privileged destination over the last decade for Canadian investment in the extractive sector. According to the Canadian Ministry of Natural Resources, between 2002 and 2008, the distribution of assets of Canadian mining companies located in Latin America, as compared with those invested abroad in general, rose from 24 per cent to 51 per cent. In absolute figures, this represents an increase from CAN$8.5 billion in 2002 to nearly CAN$57 billion in 2008 (NRCAN, 2008a, 2009). The presence of Canadian companies is of such importance that these have been described as 'quasi-diplomatic envoys of the Canadian state' (Sagebien et al., 2008, p. 120). In Peru, the book value of the assets of Canadian mining companies was CAN$2.3 billion in 2008 (NRCAN, 2009).

In view of the fact that the relations between companies and local communities give rise to an increasing number of conflicts, this area of concern has become more and more important in the strategies adopted by Canada concerning the mining sector in the hemisphere; in this regard, 'problems of governance' in host countries are often raised in order to explain the disappointing results of mining investments.[1] Thus, the PERCAN project has as its central objective responding to 'problems of governance.' In the description drafted by the team responsible for the project, mining activities are described as the cornerstone of the Peruvian economy; the problem, it is suggested, is that the latter is characterized by a generalized absence of confidence. In spite of the reforms, it is noted that companies find administrative processes and the legal framework cumbersome. Further, the Peruvian Ministry of Energy and Mines (MEM) considers that companies could and should devote more resources to improving the welfare of local communities and that communities 'suffer from unfulfilled expectations as well as the scars of environmental degradation, which fuel the perception that new mining operations will leave behind additional environmental legacies, in addition to the hundreds of hazards from the past that dot the country and have still not been addressed' (PERCAN and MEM, 2009, p. 1[2]).

Given its objective to reinforce the capacity of the MEM as the 'facilitator and promoter of better company-community relations,' and its specific objectives to 'increase the level of acceptance of mining operations,' 'diminish the number of violent crisis in the mining sector,' and 'increase the number of mechanisms and multi-stakeholder participatory, consultation and dialogue processes' (PERCAN and MEM, 2009, p. 17), PERCAN has produced a number of tools to help improve the management of social issues. These tools are to be used by the MEM in order to 'provide guidance to companies and local population...so that they exercise their rights and obligation adequately' (PERCAN MEM, 2009, p. 72). It is suggested that the MEM is to undertake a systematic monitoring of the commitments made by the companies in order to ensure that their activities are articulated within the sustainable development framework (PERCAN and MEM, 2009, p. 63). Other tools should allow the MEM to develop a strategic approach to managing crises and conflicts in the mining sector (PERCAN and MEM, 2009, p. 66), while keeping in mind that although these will not be eliminated, acts of violence will be reduced (PERCAN and MEM, 2009, p. 67).

One of the tools created to reach these objectives is the *Guide on the management and communication in situations of crisis* (Cabrera, 2007). Written in order to reinforce the capacity of mining operators to manage social conflicts, the guide presents a specific conception of the origin of crises and the solutions to be envisaged: 'Non-violent crises...should be considered as opportunities to promote a better distribution of the wealth created by mining activities and to prevent...the occurrence of violence, destruction and aggression whether individual or collective' (Cabrera, 2007, p. 3 [our translation]). In order to do this, the guide recommends forms of relations that are proactive rather than reactive (Cabrera, 2007, p. 7).

The guide analyzes the 'social risks' with which the mining operator is faced depending on the degree of acceptance of its activities on the part of the affected population and the degree of confidence and credibility with which its actions are perceived. In this regard, the analysis and course of action proposed for social risk management aim to mitigate the risks that result from inadequate modes of communication and which give rise to a loss of confidence and a lack of credibility and legitimacy that call into question the 'social license to operate' (Cabrera, 2007, p. 90). The guide warns against perceptions that put emphasis on the asymmetry of relations of power or a presentation of actors in terms of a victim/aggressor dichotomy. This type of perception is unfortunately favored by the media, according to the guide, which projects simplistic interpretations concerning complex problems and privileges attitudes that are antagonistic among actors (Cabrera, 2007, p. 97). It also warns against the existence of '"pseudo-crises" which are not based on real facts but fabricated to promote certain interests' (Cabrera, 2007, p. 10). In summary, according to the guide adequate communication of the risks associated with mining activities should determine the communities' acceptance or rejection of a given mining project (Cabrera, 2007, p. 106; see also PERCAN and MEM, 2009, p. 2).

PERCAN also supports the dissemination of the regulations and norms that apply to the participation of citizens (PERCAN and MEM, 2009, p. 85). This normative framework (MEM, 2008, 2010) commits mining operators to a certain number of activities in the area of information and consultation. However, although it encourages conciliation and the signing of agreements between companies and communities, it does not question the security of a mining title. In contrast to the principles set out in the application of Convention 169 of the International Labour Organization, for example, these norms clearly stipulate that indigenous communities do not have the right to a veto concerning mining activities or over the decisions of the authorities responsible for them (MEM, 2008, p. 4).

In a similar manner, the activities that aim to ensure a better technical understanding of environmental impacts of mining activities do not imply that the procedures which lead to the granting of mineral rights be questioned or that there could be modifications made in the measures of control and sanctioning of environmental damages. In this sense, the confusion surrounding the role of the Ministry of the Environment created in 2008, to which the MEM did not transfer responsibilities linked to the approval of environmental impact assessments of mining projects, is in itself telling (Sagebien et al., 2008; PERCAN and MEM, 2009).

The technical assistance supplied by PERCAN needs to be reset in a particular political context. In this regard, the conflicts between companies and communities over the last 15 years have become a 'highly contested arena of global conflict' (Szablowski, 2007, p. 3). As well as involving states and companies, many NGOs and advocacy networks have established contacts with affected communities in an attempt to bring about changes in the modes of governance of the sector (Szablowski, 2007). Accompanying the informal transfer of responsibilities of the state to private operators noted above, this area of debate has encouraged the development of regimes of negotiated justice and direct engagement regimes between companies and communities (Szablowski, 2010).[3] Consequently, different regimes set out different roles for the state and the other actors, as well as different conceptions of the rules that should govern the relations between companies and local communities and the manner in which the asymmetries of power among actors should be taken into account (Szablowski, 2010).

These regimes, which are in fact in competition with one another, result in competing political identities for the actors, that is: 'the kind and degree of political recognition that is conferred by a legal order on those who are subject to it' (Szablowski, 2007, p. 303). For in this area, two basic tendencies can be identified: the promotion of political identity defined in terms of 'stakeholders' and that in terms of 'rightholders.' Hence,

> [The] participation of non indigenous NGOs in the mining struggles of indigenous peoples takes place in a context in which local and international indigenous organizations have developed rights-based positions with respect to their lands, territories and natural resources and with respect to the right to determine the kind of development that takes place on their land. (Coumans, 2008a, p. 42)

These forms of collaboration give rise to demands based on human rights and to a perspective that is antagonistic in terms of company/community relations. In such a context, the access that communities have to rights-based support is fundamental in order to ensure that the options open to them are not reduced *a priori*, including the option of refusing a mining project.

For two decades, the government of Canada and the mining industry have attempted to define the engagement between companies and communities in terms of bilateral relations that concern above all technical questions (Coumans, 2008a, p. 62), and in which the interventions of a third party are to be seen as interference or, at times, instrumentalization.[4] The foundations of this approach were made explicit within the framework of a process of 'multiparty dialogue' initiated in 1992 by the mining industry and called the Whitehorse Mining Initiative (WMI). This framework was to serve to encourage the emergence of a broad consensus concerning the best way for the mining industry to contribute to sustainable development in Canada. At the time, the process had already brought to the fore a certain tension between the status of Indigenous communities as *stakeholders* and that of Indigenous peoples entitled to collective rights (*rightholders*) (Weitzner, 2010). The WMI is referred to explicitly with regard to Canadian financial and technical support of CSR in South Africa, Peru, Brazil, and several other Latin American countries (Weitzner, 2010, pp. 88, 99).[5]

In keeping with the strategies proposed by PERCAN, the Canadian government developed a number of partnerships with industry in order to make CSR 'tools' available to companies operating abroad and to develop specific tools for indigenous communities, including a Mining Information Kit for Aboriginal Communities, developed in partnership with PDAC (Prospectors and Developers Association of Canada), MAC (Mining Association of Canada), and CABA (Canadian Aboriginal Mining Association) (Coumans, 2008a, p. 43). The Canadian embassy in Peru financed an adaptation of this document for use abroad and ensures its distribution in collaboration with PERCAN (PERCAN and MEM, 2009, p. 13).

When it created the position of Extractive Sector CSR Counsellor, the Government of Canada again defined its responsibility concerning the activities of Canadian mining companies abroad in keeping with this perspective of the relations between companies and communities. In this regard it stated:

> Unresolved disputes directly affect businesses through expensive project delays, damaged reputations, high conflict management costs, investor uncertainty, and, in some cases, the loss of investment capital... [T]here [is] strong support for a mechanism to enable the sector to resolve CSR disputes related to the Canadian extractive

sector active abroad in a timely and transparent manner. (DFAIT, 2009)[6]

As mentioned above, multi-stakeholder dialogue tools thus imply a particular conception of the nature and sources of conflicts and of the appropriate means to resolve them. For PERCAN '[The] existence of conflicts, latent or potential, is intrinsic to social relationships, making imperative the creation of a favorable climate for the development of mining operations and the training of stakeholders in conflict management. Thus, for the PERCAN project, conflict management is an important element in the strengthening of the industry as a whole' (PERCAN, 2010).

To summarize, the measures proposed by PERCAN are based on a notion of political identity that is expressed in terms of 'stakeholder,' corresponding to a regime that has as its object the mitigation of 'violent crises,' and in which the desirability of carrying out mining activity and its priority over the other uses of the soil and of its resources has been determined *a priori*.

Obviously, the proposals set out in PERCAN in theory do not prevent the Peruvian state from introducing legislative and regulatory measures based on a political identity expressed in terms of rights ('rightholders'). Such measures would in principle give precedence to the respect of the rights of the people who were affected by mining operations, and it would be in this perspective that the planning of mining activities and of their regulation would be determined. Several Peruvian political actors in fact advocate for this type of approach. However, when compared with the voice of other actors in the Peruvian mining sector, particularly those of civil society and the communities affected by mining, PERCAN has a relatively privileged access to and capacity to be heard by the Peruvian authorities, notably when Canadian mining interests are involved. For example, an activity proposed by PERCAN in 2009 had as its objective that the MEM should adopt a guide on the environmental management for the mining of uranium, and justified it in these terms:

> Peru has no experience in mining uranium. However, recently a junior Canadian company undertook uranium exploration...which has been the cause of concern on the part of the Peruvian authorities and communities...It is therefore necessary for the Peruvian government to have access to the information and appropriate framework

in order to regulate this activity in an appropriate manner... Canada has developed vast experience in the management of the extraction of uranium... The objective is to develop a technical guide... and to advise the MEM in the development of the necessary legal norms. (PERCAN and MEM, 2009, p. 110)

Conclusion

As PERCAN appropriately suggests, 'During the last decade of growth, in spite of the considerable effort put forward... policies and institutions have not been developed sufficiently in order to deal with environmental impacts and social conflicts arising from mining activity' (PERCAN and MEM, 2009, p. 2). However, the objective of promoting and consolidating a process of social management led by the MEM risks being constrained by the consequences of the process of liberalization and the reform of the mining sector described in the first part of this chapter. By significantly reducing the capacity of state intervention, these reforms have simultaneously reduced the range of policies that Peru could introduce in order to better address the imminently conflict-ridden nature of mining activity in the country.

Although the documents produced by PERCAN recognize the historical origins of conflicts and the fact that social conflicts are exacerbated by contemporary mining activity, the solutions that are put forward in order to solve them appear limited in terms of the real impact they are likely to have. Consequently, the measures introduced by PERCAN are likely to prove disappointing. As laid out in the risk assessment of the PERCAN project, this may provoke 'the adverse reactions of mining title holders who consider that these proposals entail a certain control over the commitments of companies with regard to social issues, when in fact the MEM no longer has jurisdiction in this area; [and] due to the fact that relations with communities depend on voluntary initiatives which should not be supervised' (PERCAN and MEM, 2009, p. 72).

On the basis of the analysis of a bilateral initiative, this chapter has attempted to suggest that much closer attention needs to be paid to the links between issues of legitimacy, responsibility and accountability, and the manner in which regulatory frameworks condition these links and issues. Such an analysis needs to be reset in its specific historical settings if further conflicts are to be avoided in the mining sector in countries such as Peru. Moreover, the analysis also suggests that the manner in which the PERCAN project defines the issues at stake in the Peruvian mining sector and the solutions proposed cast doubt on the neutrality of the interventions of external actors concerning the role they play in shaping the modes of governance of the sector.

Notes

1. In its answer concerning the corporate responsibility of Canadian companies to the Standing Committee on Foreign Affairs (House of Commons, 2005), the Government of Canada stated that 'Canadian investment abroad can provide a much needed infusion of capital for developing countries' and that host governments bear 'the primary responsibility for monitoring company compliance with local laws' (DFAIT, 2005).
2. Citations from the document (PERCAN and MEM, 2009) are originally in Spanish and have been translated by the authors into English.
3. Negotiated justice implies '[a move] toward new governance models in which normative authority is pluralized and multiple actors are engaged in horizontal relationships... [I]ncreasing conflict and calls for the social mediation of mining investment at a time when states have made legal and ideological commitments to limit the formal regulatory burden placed on extractive firms [have contributed to the development of] regimes to promote and constrain engagement between extractive firms and affected communities in order to delegate (often informally) the responsibility of social mediation onto extractive projects themselves' (Szablowski, 2010, p. 113).
4. Canadian officials have sometimes played a role in delegitimizing the critics of the impact of Canadian extractive investment, as in Guatemala where the Canadian ambassador has been condemned to pay damages for the defamation of a Canadian journalist (Ontario Superior Court of Justice, 2009; Schnoor and Murray Klippenstein, 2010). They have also been accused of 'promoting conflict between indigenous communities and NGOs that are critical of mining' (Coumans, 2008a, p. 48).
5. For Weitzner, '[i]t is unclear to what extent officials involved in these strategies are learning from... the lessons from the Canadian experience, especially in Indigenous participation. Without targeted attention to increasing the voice of marginalised peoples in these forums... the process that Canada is encouraging could be seen as "greenwashes" that enable Canadian industry to continue to operate overseas, under the pretext that something is being done, when in fact business continues as usual' (Weitzner, 2010, p. 100).
6. As a result, the proposed mechanisms only permit making inquiries into this kind of allegation if the explicit endorsement of all parties – including that of the companies involved – is confirmed. However, several authors suggest that Canada has not adopted extraterritorial legislation in keeping with its international obligations, nor adequate mechanisms required to deal with eventual cases of human rights violations that might arise abroad (Webber, 2008, pp. 35, 36; see, as well, Belem et al., 2008, p. 60).

6
Conflict Diamonds: The Kimberley Process and the South American Challenge

Ian Smillie

Diamonds, the purest form of carbon, were produced by great heat and pressure 75 to 125 miles below the earth's surface a billion years ago. Most of the diamonds that subsequently came to the surface did so through small volcanoes that produced carrot-shaped 'pipes' of gray–green rock called 'kimberlite.' Some of these pipes can be mined by digging down through their core. The initial capital cost of this kind of mining can be enormous, running in some cases to a billion dollars or more. The surfaces of some pipes, however, have been eroded over millions of years, washing diamonds down rivers and across vast expanses of land. These 'alluvial' diamonds are often close to the surface and the technology required to mine them is basic.

Known since ancient times, diamonds were first mined in India and later, Brazil. The modern diamond era began in South Africa in 1867 with the discovery of the appropriately named 21-carat 'Eureka Diamond.' Subsequent finds were made on a farm owned by two brothers named De Beer. They sold their land and their name to a consortium, and by 1872 there were 50,000 diggers in the area. By 1890, British imperialist Cecil Rhodes had concluded a series of buyouts under the De Beers name, gaining control of 95 per cent of the world's diamond production. Rhodes' aim in controlling the supply was to control the price of an increasingly common commodity.

Between the 1930s and the 1980s, De Beers wielded absolute control over all aspects of the rough diamond industry, handling directly more than 80 per cent of world production. Today the company controls a little less than half of the global market in rough diamonds but its influence remains proportionally much greater. Other major players include the mining giants BHP Billiton and Rio Tinto.

Corporate social responsibility in the diamond industry

Corporate social responsibility and the diamond industry are two concepts that for a century were related to one another only tangentially. This is, in part, because despite the overwhelming presence of De Beers, 'the diamond industry' itself is little more than a concept. At one end of the scale are the giant mining firms; at the other, the quiet ambience of the Cartier and Tiffany showrooms where the diamond is advertised as a symbol of love, purity, wealth, and eternity – where diamonds are *forever*. In between there is a raucous free-for-all where nothing is forever: there, diamonds come and go with lightning speed. In Africa, where almost 60 per cent of the world's diamonds, by value, are mined, the industry is characterized by a few gigantic, well-fenced holes in the ground, and in contrast, by hundreds of thousands of alluvial diamond diggers, known variously as *garimpeiros* in Angola and Brazil, *creuseurs* in the Congo and Central African Republic, *diggers* in Sierra Leone and Liberia.

The most productive and profitable diamond mines in the world are those in Botswana, where De Beers, in 50–50 partnership with the government, digs straight down into volcanic kimberlite pipes and pulls up huge volumes of diamonds – 34.9 million carats in 2007, representing sales of US$2.9 billion. This is a capital-intensive, high-technology operation that employs 6000 people, about three per cent of the formal labor force.

But where Africa's kimberlite pipes have been eroded by millions of rainy seasons, the result is alluvial diamonds. The same is true in South America, where diamonds are mined in Brazil, Venezuela, and Guyana. Scattered over hundreds of square miles – along riverbeds, in valleys where rivers once flowed – alluvial diamonds are close to the surface. They are often available to individual diggers with little more than shovels, sieves, and a source of water for straining gravel. This is where hundreds of thousands of *garimpeiros* and *creuseurs* dig, often illegally, always under unhealthy, unsafe, and usually subsistence circumstances. Here the industry is essentially unregulated, unwatched, and nameless. Here the concept of corporate social responsibility exists only in its absence. Corporations as conceived in industrialized countries – even governments – barely exist in these diamond fields. Middlemen pass the diamonds on to other middlemen and then still others. If and when the diamonds are noticed by government, they may be taxed, but few of the benefits filter back to those who dig, or to those on whose land the diamonds are found. Here, corporate social responsibility is nonexistent.

Conflict diamonds and the Kimberley Process

Because of their high weight-to-value ratio and because of the industry's historically secretive nature, diamonds have always attracted criminals and have been used as a medium of exchange in a variety of illicit endeavors: tax evasion, as a substitute for weak currencies, as a medium for money laundering, for the clandestine purchase of weapons, drugs and other illicit goods. During the 1990s, rebel armies in Angola, Sierra Leone, and the Democratic Republic of the Congo occupied alluvial diamond fields, reaping gems that could be sold for weapons. Known as 'conflict diamonds' or 'blood diamonds,' these stones represented as much as 15 per cent of global supply during the mid 1990s. Taken together with diamonds used for money laundering and other nefarious purposes, it is estimated that 25 per cent of the world's rough diamonds during the 1990s were in some way illicit (Smillie, 2002, pp. 8–10).

Exposed by both Non-governmental Organizations (NGOs) and the United Nations, conflict diamonds became the subject of international attention in 1999. Between 2000 and 2003, international negotiations involving governments, industry, and NGOs led to the creation of the Kimberley Process Certification Scheme (KPCS) for rough diamonds. Since 2003, when the KPCS came on stream, all international shipments of rough diamonds have required a government certificate of origin, backed by a system of prescribed internal controls in each participating country.

Supported by diamond-specific legislation in each of 50 nations plus all those represented by the European Union, the KPCS comprises all major diamond-producing, -trading, and -processing countries. In its early days, the Kimberley Process accomplished a great deal. The very presence of the negotiations helped choke diamond supplies to rebel movements in Angola and Sierra Leone and contributed to the end of hostilities. The KP has the best diamond database in the world,

---

Box 6.1 Basic elements of the KPCS

- Each participant (i.e., each participating government) undertakes to maintain internal controls over rough diamonds. For producers this means establishing an audit trail between mines and the point of export; for others, it means maintaining a chain of warranties between the point of import and either the cutting factory or the point of re-export.

- Each participant agrees that a Kimberley Certificate will accompany each export; certificates, to be designed and issued by each participating country, have certain common features and must also contain adequate security features. All international shipments of diamonds must be made in tamper-proof packages.
- No participant will permit the import of rough diamonds unless accompanied by a KP certificate from another participant. Penalties are provided in the case of breaches.
- Each parcel received will be acknowledged by the importing authority to the exporting authority.
- Each participant must submit quarterly trade statistics and semi-annual production statistics within 60 days of the reference period. A centrally maintained statistical website allows participants and observers to compare and verify exports from one country with imports to another.

Source: Kimberley Process, 2003

and it is credited by several countries with the growth in legitimate diamond exports and thus of tax revenue. The KP is discussed as a model for other extractive industries, and as a model of participation and communication between governments, industry, and civil society, all of which play an active and meaningful role in its management. Refer to Box 6.1 for an overview of the basic elements of the Kimberley Process Certification Scheme.

Weaknesses revealed: Kimberley Process in South America

But while the formal diamond industry – from the giant mining firms at one end to the retail showrooms at the other – relied on the KP to buff a tarnished image, things in the middle began to fall apart. Some of the countries most damaged by the 'diamond wars' were unable to create meaningful controls in the informal alluvial diamond fields. The most important part of the KPCS – the ability of a certifying government to know where its exported diamonds were mined – has failed badly in Angola and the Democratic Republic of the Congo (DRC). As much as 40 per cent of the diamonds being exported from the DRC cannot, with any accuracy, be tracked back to the place where they are mined (Partnership Africa Canada, 2009). But the KP, being an intergovernmental body, preferred chastisement to suspension, recommendations to action, and amnesia to follow-up.

Although the three South American diamond producers represent a small part of the global diamond industry, each in turn soon demonstrated the pathologies of the industry and the weakness of the KP. Brazil was first. Spurred by rumors of corruption in the management of Brazil's certification system, Partnership Africa Canada (PAC) carried out an investigation (Blore, 2005). PAC had been one of two leading NGOs in the campaign to create the KPCS, and was a member of KP plenaries and working groups. What it found in Brazil, where diamonds are mined by small companies and large numbers of independent *garimpeiros*, was a weak system of Kimberley controls that had been subverted by illicit diamond traders and willing government accomplices. In 2006, the Brazilian government denounced PAC and its report at the Moscow Kimberley plenary, but within a few months another PAC report and an investigation by Brazil's Federal Police led to the arrest of several officials and diamond traders and a vindication of what PAC had uncovered. Brazil halted all diamond exports for six months while it tightened its internal regulations and controls. Throughout, the KP had remained silent.

Meanwhile, in Venezuela – a charter member of the KP – something else was going on. Following a reorganization of Venezuela's ministry of mines in 2005, government stopped issuing Kimberley Certificates. No reasons were given, although it was probably at first a bureaucratic bungle. Whatever the reason, Venezuelan diamond miners continued to mine, buyers continued to buy, and exporters continued to export, albeit illegally. The Venezuelan government was thus actively condoning the smuggling of 100 per cent of its diamond production out of the country.

The KP did little until PAC produced an investigative report on the Venezuelan diamond industry in 2006, exposing the smuggling (Blore, 2006a). The KP procrastinated until October 2008, when a KP team was finally allowed to visit Venezuela. The team, which did not travel to the diamond-mining areas, reported essentially what the Venezuelan government told it: mistakes had been made but they would be corrected. Finally, after much debate, Venezuela 'self-suspended' itself from the KPCS, saying it would halt all diamond production and exports for at least two years while reorganizing its diamond sector. This face-saving measure seemed to solve the problem. The KP, which had faced vociferous demands that it suspend Venezuela, breathed a sigh of relief.

In Venezuela, however, diamonds continued to be smuggled out of the country on a daily basis. In 2009, the mineral leases of five diamond-mining cooperatives held by a state-owned mining concern were renewed. Visits by reporters and investigative NGOs to the mining areas and the border towns where diamonds are sold revealed that diamond traders remained actively engaged in buying and selling into the illegal cross-border traffic. The KP, wedded to the polite fiction of a Venezuelan 'self-suspension,' did nothing. Effectively, the body created to halt illicit diamond flows was now condoning them.

Many of Venezuela's diamonds were being smuggled to Brazil, but the bulk appeared to be going to Guyana, another KP member and a country with its own alluvial diamond production. On paper, Guyana's system for tracking alluvial diamond production is good, but in practice, lax enforcement and pervasive corruption among government officials has made Guyana the laundry of choice for tens of millions of dollars' worth of Venezuelan diamonds every year (Blore, 2006b). As in Brazil and Venezuela, the KP has turned a blind eye to detailed and widespread reports of Guyanese non-compliance.

These problems, however, seemed minor in comparison with human rights abuse in the diamond fields of Zimbabwe. There, in an area called Marange, Zimbabwean armed forces opened fire on illicit diamond diggers late in 2008, killing almost 200 of them, and setting on others with truncheons and dogs (Human Rights Watch, 2009). Here the KP had a new challenge. It had been designed to halt conflict diamonds, but 'conflict diamonds' had been narrowly defined as diamonds used by rebel armies to fuel wars against legitimate governments. While human rights had been mentioned tangentially in the KPCS agreement, there was no specific requirement that a government should abide by, say, the UN Charter of Human Rights in its enforcement of KPCS minimum standards.

Zimbabwe was not the first instance of the problem. The Venezuelan army had killed several *garimpeiros* in 2006, in an attempt to clear an illegal alluvial diamond mining site. Angola had been using officially sanctioned violence, robbery, rape, and forced marches to expel tens of thousands of illicit Congolese diamond-diggers from its alluvial diamond fields between 2006 and 2009.

Although the Zimbabwe massacre became a media *cause célèbre* through 2009 and into 2010, Robert Mugabe's traditional southern African neighbors rallied to Zimbabwe's defense. In a KP where consensus means unanimity, strident calls by large numbers of NGOs, industry representatives, and governments for Zimbabwe's expulsion were doomed from the outset.

Kimberley Process and Corporate Social Responsibility in the diamond industry

What does all of this say about corporate social responsibility in the diamond industry? The 'industry,' like a number of the major campaigning NGOs, had been party to KP negotiations from the outset. A body called the World Diamond Council (WDC) had been formed by a range of companies – in fact all of the major players, from mining through to retail. A strength of the KP is that it is not strictly governmental. The WDC and NGOs were part of the architectural design process and they are part of the ongoing management of the KPCS. This helped to make the KPCS both practical and credible.

But there was no provision in the KP to do what all regulators *must* do: there was no provision to plug holes, tighten loose bolts, and fix the parts that were not working. A fundamental part of any regulatory system is the need to keep one step ahead of those who seek to evade it, as they figure out new ways around rules and regulations. However, in the KP, there has from the beginning been a prohibition against 'opening the document' and an odd 'consensus' decision-making process that requires unanimity on all questions. In practical terms, this means that any change to which one or two participants object can be blocked, sometimes even by a single veto.

There were three fundamental flaws in the agreement from the outset: a weak monitoring system, the KP's unworkable 'consensus' decision making, and few penalties for noncompliance beyond suspension. These flaws almost guaranteed that it would run into intractable problems sooner rather than later.

Because the KPCS is rooted in national laws rather than an international treaty, it is in many ways stronger than other international regulatory systems. Canada, like South Africa, China, Russia, and all other members of the KP, has passed laws requiring those who handle rough diamonds to comply with KPCS minimum standards. NGOs thought this gave it strength to which voluntary codes and guidelines could never aspire. Industry thought the same. What they and many governments failed to foresee, however, was that several governments would be either unable or unwilling to enforce KPCS regulations. And in 2003, when the agreement was finalized, they could not foresee that some governments would resort to the kind of brutality that the KPCS had been designed to halt. For the formal industry, much of the exercise had been about governments getting a grip on the informal industry – those alluvial diamond fields that had been so attractive to rebel armies, and the 1.3 million artisanal diamond diggers who operated largely beyond the purview of permissions and licenses and taxation. For industry, the KP was also about protecting an expensive product whose reputation was based on appeals to love and fidelity, and therefore a product vulnerable to public opinion and discretionary spending. As early as 2006, only three years into implementation, it was clear that the design flaws in the KPCS were going to limit its ability to clean up a tarnished industry.

There was a further issue where the reputation of the industry was concerned. An initial strength of the KP had been its fixation on one thing: ending conflict diamonds. That single goal made it possible to reach an agreement relatively quickly. The debate and the final agreement were not hobbled by arguments about the environment, child labor, working conditions, or fair prices. But these, of course, *were* prominent issues in the alluvial diamond fields where large companies were absent and where conditions had been so ripe for rebel armies. As it turned out, eliminating the armies while ignoring the social ills was not a formula for complete or long-term success.

When NGOs began to argue that the KP's regulatory mandate had to be complemented with a development approach, some governments objected, and the need for consensus made the idea a non-starter. But industry listened. The formal industry understood without hesitation that this underbelly of the diamond world had the capacity to be as damaging as conflict diamonds. Engagement on the issue was swift, and over the period 2005–07, representatives of industry, interested governments, and NGOs created a new organization called the Diamond Development Initiative International (DDII).

DDII parallels and complements the KP, but it is an NGO and it draws its strength from willing supporters rather than a captive audience of cranky and unbending governments. DDII is still new. Its objective is to tackle development and economic problems with development and economic solutions. It has a research function, and its on-the-ground projects are mainly about learning what works and what does not in a complex political economy that has remained largely unchanged in 75 years. It has received financial support from the diamond industry and industry representatives sit on its board but it is not an industry body, and care has been taken to ensure DDII's independence. It has also received funding from the governments of Sweden, Belgium, and the UK, where it is seen as a novel approach to deep-seated problems in countries where diamonds may no longer fuel war, but where they still contribute much to dissention and little to development.

There is a second initiative worth discussing in the context of corporate social responsibility in the diamond industry: the Responsible Jewellery Council (RJC). Founded in 2005 to add a unique industry perspective to the debate, the RJC is where issues more common to all extractive industries are discussed: land rights, the environment, labor standards, basic mining practice. The RJC is 'an international, not-for-profit organization established to reinforce consumer confidence in the jewellery industry by advancing responsible business practices throughout the diamond and gold jewellery supply chain.' Its website says that 'the Council aims to build a community of confidence across every step of the diamond and gold jewellery supply chain in all geographies, and among businesses large and small. It seeks to work with a wide range of stakeholders in defining and implementing responsible jewellery practices through the RJC's certification system' (Responsible Jewellery Council, 2010).

The RJC is interesting on three counts. First, it covers the entire supply chain: corporate members come from mining, trading, smelting, processing, retail, and banking services. Secondly, it has taken the best from a wide range of industry codes of conduct and standards.[1] Thirdly, the hoary old idea that standards and codes must be voluntary and self-regulating has been rejected. Membership in the RJC is voluntary, but it comes with a *requirement* that all members meet rigorous RJC standards through a process that includes independent third party audits. Members in good standing at 31 December 2009 must undergo independent verification by December 2011 – or December 2012 if they have mining facilities. New members must undergo independent verification within two years of becoming members.

This is one of the toughest and clearest extractive industry certification systems to date, and as such it is highly commendable. Its weakness is that there is, as yet, no chain of custody tracking system. This means, for example that AngloGold Ashanti and De Beers on the mining end will ensure that there is no child labor in their mines, and at the retail end, Piaget and Cartier will ensure that there is no child labor in their shops. But Piaget and Cartier are not limited to RJC members for their gold and diamonds, either directly or through the trading, refining, and manufacturing companies that are also members of the RJC. So child labor might still be a feature in products sold by Piaget and Cartier. The flaw in the system is its stovepipe nature and the absence of a chain of custody. However, the weakness is acknowledged, and in 2010 the RJC began to investigate the applicability of such tracking systems to jewellery products. 'It is not the intention of the RJC to develop its own system of product tracking,' the organization says, 'but to investigate the possibility of certifying the effectiveness, validity and claims made by proprietary tracking systems' (Responsible Jewellery Council, 2009).

A further weakness is that RJC members, by and large, do not operate in the alluvial diamond fields of Africa and South America. There, diggers, traders, and exporters are left to their own devices. But their product will – as diamonds always have – enter the pipeline at one stage or another, if not openly, then through covert means. Left out of the ethical standards created by RJC, they will likely suffer from depressed prices offered by the ever-present bottom-feeders eager to find an opportunity or an excuse to cheat the most vulnerable. Because an estimated 15 per cent of the world's gem diamonds are produced artisanally, the RJC will not cover more than 85 per cent of the supply. And while it will be able to give some assurances, it could actually make life more difficult for those who are left out.

To deal with this problem in the artisanal sector, DDII too is developing a set of standards. These are not 'fair trade' standards that might have to rely on an NGO connection of some kind. These 'development diamond standards' (DDS), which are still under development, aim to be widely accessible to small producers, dealing with basic issues of health and safety, fair prices and wages, environmental remediation, and community engagement. DDS diamonds, like those under the RJC banner, will be subject to third party verification, and it is hoped that they may achieve an advantage with jewelers and consumers looking for ethical products produced both artisanally and in developmentally sound ways.

Conclusion

It was unclear at the beginning of 2010 how the turbulence in the diamond industry would evolve. The RJC and DDII standards were still aspirational, and proof of their viability lay in the future. There were loud calls from industry, NGOs, and some governments for KP reform. But in the absence of a democratic decision-making making process, it was hard to see how this would happen, when several key governments objected to the kinds of change required: scrupulous third party monitoring and meaningful penalties for non-compliance. The tragedy of the KP is that while it has such a strong foundation – supported by dedicated legal instruments in 50 countries and across the European Union – it has been unable to enforce even the most basic chains of custody in the countries worst affected by conflict diamonds.

One emerging possibility is the creation of a 'Kimberley Process-Plus' – a supplementary set of tougher standards – including many of the elements currently missing from the KP – that could be bolted onto the existing agreement. These new standards, to be negotiated among like-minded governments, industry, and civil society, would be *voluntary*, but would be inaugurated with fanfare and acceptance by industry, civil society, and several key governments. The idea is that there would be a rush by other governments to 'join' rather than be left behind in a camp with a set of second-class performers. This could be done without 'opening the document' and without unanimity. 'Voluntary' would gradually, or even quickly, become 'compulsory' in the same way as participation in the KP itself is voluntary – and yet compulsory for any government wanting to participate in the world's legitimate diamond trade.

Some suggest that with the wars now over, perhaps there is no more need of a regulatory system for rough diamonds. Nothing could be farther from the truth. The combined 2009–10 cost of UN peacekeeping efforts in Liberia, Côte d'Ivoire, and the DRC was US$2.3 billion, and in some areas the effort was barely keeping a lid on the problem. A collapse of the KPCS could mean a quick return to the criminal diamond free-for-all of the 1990s and its accompanying bloodshed and destruction.

If there is a lesson in the diamond experience for other extractive industries, it is that certification systems – whether they aim to end and prevent conflict or to ensure good corporate citizenship – must be clear in their purpose, and must have a vision that extends beyond the immediate. The early KP vision was understandably narrow, but it had no room to develop and evolve. Participants failed to think about what was fuelling the conflict – poverty, bad governance, human rights abuse, and illicit behavior – and they saddled themselves with a mechanism that could not adapt to these challenges when they became more apparent.

The diamond industry, which once understood corporate responsibility in terms of building schools and orphanages, has changed dramatically in recent years. New ideas about Corporate Social Responsibility have emerged from a variety of places – civil society campaigning, governmental pressures, market-driven forces, and from within the industry itself. Reforms have obscured the clear distinction between 'voluntary' and 'compulsory' that may once have seemed inevitable, and this helped to make real change possible. Where the governmental KP has failed the test of compulsion, the industry has accepted on its own that tough standards are required to enforce responsible corporate behavior and that these require independent third party verification. Where the KP found itself unable to deal with the developmental problems that characterize the alluvial diamond sector, civil society and industry were able to come together around the creation of the DDI. The industry has accepted that it has a responsibility to deal with issues that once seemed tangential if not alien to its core business: criminality, war, poverty, and underdevelopment. Whether the recognition and the change will result in meaningful long-term change in an industry where 'forever' has many connotations, remains to be seen.

Note

1. Its standards, for example, would meet all of the requirements of Canada's proposed Bill C-300 on corporate accountability, debated through much of 2009 and 2010.

7
Whose Development? Mining, Local Resistance, and Development Agendas

Catherine Coumans

Introduction: Local resistance to mining's negative impacts on development

In the 1960s, villagers on the small island of Marinduque in the Philippines lived mainly on subsistence fishing and farming with copra, bananas, and marine products as their main sources of trade and income, until 1969 when the opening of a world-class copper mining operation co-owned and managed by Canada's Placer Dome[1] raised new expectations for economic prosperity in the region. However, nearly 30 years of large-scale open-pit mining in the mountains of Marinduque instead became a lesson learned by communities throughout the Philippines of how mining can erode the very basis of economic sustainability and development.

Negative development impacts of the mine resulted from dislocation, economic distortions, punishing labor practices for mine employees, lengthy and ad hoc tax holidays and exemptions that deprived local governments of mining revenues, political corruption, and the severe degradation of three major ecosystems. Disposal of 200 million tonnes of metal-leaching mine waste in Calancan Bay between 1975 and 1991 severely affected the subsistence and livelihood of 12 remote fishing communities along the bay, as well as raising concerns about lead contamination following medical testing of Calancan Bay children. In 1993, a dam holding back mine waste in the mountains near the mine burst, inundating the Mogpog River and nearby villages with acidic and metal-leaching sludge. Two children drowned in the waste and the river remains severely contaminated to this day (Tingay, 2004). Three years later in 1996, a second waste containment failure filled the 26 kilometre-long Boac River with tailing waste from the mountains to the sea. While this latest catastrophic failure resulted in the suspension of the mine, the Boac River, the Mogpog River, and Calancan Bay have never been rehabilitated, resulting in ongoing economic impacts for local communities.

Recognizing the adverse long-term development effects of unremediated ecosystem damage related to the mine, in 2005 the Marinuque provincial government filed a lawsuit against Placer Dome in the US to seek compensation for damages and funds to rehabilitate the effected ecosystems.[2] The provincial government also passed a resolution declaring a 50-year moratorium on large-scale mining in the region.

The Boac River spill occurred in 1996, just one year after the Philippines had adopted a new mining Act with provisions aimed at making the Philippines a hospitable jurisdiction for global mining companies. Since then, community conflict over proposed mining projects has become widespread throughout the Philippines and is a key reason for the country's low ranking by global mining companies in spite of favorable investment conditions for mining put in place by the national government (McMahon and Cervantes, 2010, p. 55). In many of these community struggles, the negative impacts that long-term mining had on fertile ecosystems, livelihood and health in Marinduque are raised as a cautionary tale. Since the Boac River disaster in Marinduque there have been at least 10 provincial and 32 municipal moratoria on large-scale mining[3] and in 2010 a resolution was tabled in the House of Representatives, calling for a moratorium on large-scale mining in the Philippines and noting that local government units who have declared moratoria 'are not convinced of the claimed development benefits of mining, and that their local development plans have not identified mining as a development option' (House Resolution No. 528).

The evolution of resistance to large-scale mining in the Philippines reflects increased local-level opposition to mining in other regions of the world (Szablowski, 2007, p. 3). Fifteen years ago communities were more likely to be receptive, or at least resigned, to the prospect of hosting a large-scale mine, only starting to protest when unanticipated impacts became overwhelming. Now, potentially affected community members, even in remote locations, are increasingly opposing mining before it starts. Even as the head of Manhattan Minerals in Peru, Americo Villafuerte, argued: 'Our company has a concrete proposal for development in Tambogrande. It's a proposal with three concrete social and economic aspects that will resolve many of the problems affecting thousands of children and adults in Tambogrande' (Hennesey, 2003), more than 90 per cent of voters rejected the mine in an informal referendum in 2002. Altemira Hidalgo from Tambogrande said 'We were born here. We grew up here. Our children and our homes were formed here... our ancestors were also here. We want these men to go. We don't want mining. We want agriculture' (ibid.).

However, even as community members and locally elected officials urgently articulate reasons for opposing mining that include protection of local-level security and future economic development, mining industry associations, home and host country governments of multinational mining companies, and international financial institutions increasingly assert positive local development outcomes associated with mining, while emphasizing the need for increased local-level 'benefit sharing.'

This chapter discusses the evolution of development discourses, analysis, and strategies in relation to mining projects as projected by the global and Canadian mining industry. It subsequently examines the roles that the Canadian government and Canadian development NGOs have adopted as 'development partners' of mining companies. Finally, it further explores the development-related reasons community members give for opposing mining and considers the potentially negative implications of government–industry–NGO alliances on the agency of local actors who oppose, or seek to modify, mining projects that they consider harmful to values of importance to them and to the futures they envisage.

The 'social license to operate' and development agendas

The Boac River catastrophe in the Philippines was but one of a number of high-profile environmental disasters linked to mining in the 1990s.[4] If the 1990s were characterized by concern about the environmental impacts of mining, the 2000s have been marked by greater focus on a wide range of social impacts, including human rights abuses related to mining in conflict zones forcing evictions; abuses by security guards; indigenous rights violations; and impacts on women, minorities, and workers, among others. The 2000s have also seen a marked increase in social conflict associated with mining projects and local-level opposition to mining. This reality has led to a number of high-profile international and national reviews of the mining sector.[5] The increase in global high-profile mining conflicts is recognized by mining companies and industry associations when they refer to the need to obtain a 'social license to operate' if they are to avoid further reputational risk, costly delays, the potential loss of a project, as well as increased pressure for more effective regulation of the industry.[6] The question of how mining companies set out to obtain a 'social license to operate' is at the core of the issues explored in this chapter.

As local conflicts have intensified, mining companies and their associations have more urgently emphasized the need for local-level 'benefit sharing' and community-level development projects at their mine sites within a context of voluntary corporate social responsibility (CSR). The global industry is actively promoting a positive association between mining and development, by emphasizing the provision of jobs, taxes, and royalties, even as it seeks ways to assure greater dispersal of funds and programs at the local level. The push to link 'mining' and 'development' in public discourse and perception is reminiscent of the earlier effort to counter reports of high-profile environmental disasters by asserting positive associations between mining and sustainability. But the messaging effort is only part of a more substantive effort to counter local opposition and achieve a social licence to operate by channeling 'benefits' to the local level.

The following sections provide an overview of the 'mining and development' theme as it is reasoned and articulated by various interactive stakeholder groups at the international level and in Canada.

Industry discourse, analysis, and strategy on mining and development

At the international level, the International Council on Minerals and Metals (ICMM), established in 2001, has led industry efforts to respond to the findings of academic research that show a link between natural resource extraction and increased national poverty, and to address rising local-level conflict associated with mining projects. The so-called 'resource curse' literature sets out a number of interrelated explanations and conditions for national economic decline associated with resource extraction, particularly in resource-dependent poor counties.[7] In 2004, ICMM established a multi-year program called the Resource Endowment Initiative aimed in part to push back on the findings of resource curse literature (ICMM, 2010f). In partnership with United Nations Conference on Trade and Development (UNCTAD) and the World Bank Group, ICMM began by analyzing data from 33 mineral-dependent countries (where mining contributed 20 per cent or more of exports from 1965 to 2003) with respect to socioeconomic data (ICMM, 2006a), and subsequently prepared a Community Development Toolkit that would provide a methodology to assess the socioeconomic impacts of a particular mine project. This toolkit was ultimately field-tested at a mining project in Tanzania, Ghana, Chile, and Peru (ICMM, 2010f).

The major findings that have come out of this work firmly shift responsibility for negative national-level economic outcomes and for local-level conflict away from the activity of mining or mining companies and toward host country governance. ICMM argues that weak governments do not manage revenues from mining well enough to ensure national development and do not ensure sufficient social investment or benefit sharing, particularly at the local level where mines frequently have to deal with social conflict, asserting that 'governance weaknesses are the basis of many of the most heated local criticisms against (or issues faced by) the companies' (McPhail, 2008, p. 8).

ICMM's main strategy for tackling these national and local mining-related problems lies in 'multi-stakeholder partnerships [which] are key to enhancing mining's economic contribution' (ICMM, 2010e). In comments to the International Finance Corporation (IFC) with respect to the Performance Standards under review, ICMM suggested that the IFC 'emphasize the importance of fostering collaborative engagement between governments, development agencies and civil society' (ICMM 2010a, p. 5).

ICMM has identified a number of priority areas – among these, poverty reduction, revenue management, social investment, and dispute resolution – and has identified partners including 'governments, NGOs, donors and international organizations' (ICMM, 2010d). By 'governments' ICMM refers to home country governments, where its multinational members are headquartered, and host country governments, mainly of developing nations, where its members operate.

Most importantly, however, ICMM's analysis of the problem places it outside its members' sphere of influence. ICMM's solutions draw in a range of actors, making them co-responsible for outcomes and in the process generating far-reaching consequences, particularly for host governments. For example, the ICCM suggests that the mining sector should be integrated into poverty reduction strategies, pointing out that:

> This, in turn, will require government departments (for example, treasury and mining) to collaborate more closely, and will require chambers of mines and companies to participate in national development dialogues... In addition, social funds and donor agencies should connect funds better to the special needs of communities affected by mining. (McPhail, 2008, p. 9)

In the Paris Declaration of 2005, donor countries agreed on the need for '[i]ncreasing alignment of aid with partner countries' priorities, systems and procedures' (OECD, 2005, p. 1). Partner countries are those that receive aid. Under 'Partnership Commitments,' donor countries agree to 'base their overall support on partner countries' national development strategies, institutions and procedures' (ibid., p. 3).

By integrating the mining sector into a host developing country's poverty reduction strategies it becomes possible for the host government, as well as the multinational mining companies operating there, to call on the home countries of those mining companies to direct official development assistance toward the host country government to directly support its efforts to 'manage the revenues' it receives from mining. It also opens the door for the home country to direct development aid to development projects at the mines of its own mining companies. Additionally, this strategy, which was developed by ICMM in partnership with the World Bank Group, further justifies involvement in mining in developing countries by the World Bank Group, including its private sector lending arm, the IFC, and its Multilateral Investment Guarantee Agency.

Before moving on to a discussion of how these themes have been playing out in the Canadian context, it is worth noting that neither the analysis of the problem nor the suggested solutions coming out of the collaboration of ICMM and the World Bank are undisputed (see Pegg, 2003; Campbell et al., 2007; Columban Fathers, 2007). The role of individual mining companies and mining projects in creating serious development deficits through unacceptable environmental impacts with long-lasting implications, such as those highlighted in the Marinduque case above, and their role in local-level conflict associated with abuses of a range of human and indigenous rights, cannot simply be put down to bad governance on the part of host countries. It can just as easily be argued, and has been, that multinationals have for decades taken advantage of governance weaknesses and lack of regulatory capacity in host countries to avoid costs associated with best environmental practices and meaningful community engagement. The ICMM analysis also falls short in addressing the fact that even when faced with possible jobs or other benefits from a potential mining project, some communities (or groups within communities) are choosing to reject mining in favor of existing or alternative economic activities. This puts these communities into direct conflict with the interests of the company, elements of the state, and possibly other community members.

Furthermore, ICMM and other industry associations and companies emphasize the financial contributions of mining operations to development through taxes and royalties. They express concern that these revenues are not well managed as a result of weak and corrupt governance: '[t]here are a number of well-documented challenges associated with translating mineral wealth into national wealth. Too often, for example, mining revenues have been squandered by governing elites or exacerbated corruption' (ICMM, 2010b). ICMM and other industry associations support the implementation of the Extractive Industry Transparency Initiative in host countries as a way to address loss of revenues through corruption.[8] However, far less is said by the industry and by the international financial institutions about the various mechanisms – including confidential contracts and stability agreements, mining Acts, and trade agreements – through which mining companies secure lengthy tax holidays, keep tax and royalty levels to a minimum, and secure protection from potential costs associated with future environmental or social legislation aimed at protecting communities from negative impacts from mining (Akabzaa, 2009; Belem, 2009; Campbell, 2009a, 2009b, 2009c; Sarrasin, 2009). Additionally, little is said by the industry about the various means by which taxes are avoided and revenues related to resource extraction are removed from host countries through accounting mechanisms such as transfer pricing and the use of tax havens such as the Cayman Islands.[9]

Finally, the various development solutions that ICMM and the World Bank propose not only involve spreading the responsibility for positive development outcomes to a wide array of 'partners,' but also involve externalizing costs to these partners, in particular to home and host governments – which means the taxpayers and citizens of those countries. An interesting example of this is a solution that ICMM proposed for dealing with the fact that few resources for local development projects are available in the early and final phases of a mine. The ICMM suggested that host governments take out a loan to cover this period, rather than that mining companies themselves cover these costs. The ICMM suggests that at the beginning and the end of mine project development when tax revenue is lower:

> [T]here is an opportunity for governments to monetize the assets and bring the cash flows forward by, for example, using the long-term loans and guarantees offered by the World Bank to transform future project revenue flows into current capital funding for social and infrastructure investments in local communities, while reserving

sufficient funds to deal with subsequent mine closure. (McPhail, 2008, pp. 9–10)

In Canada, high-level policy debates about mining and development have been triggered by civil society and parliamentary concern about the environmental and social impacts of Canadian mining companies operating in developing countries and the lack of accountability mechanisms at the disposal of the Canadian government for assuring that public financing does not support mines that fail to adhere to international human rights or environmental standards. Following recommendations made to the Government of Canada in a parliamentary report, *Mining in Developing Countries* (SCFAIT, 2005), the Government of Canada spearheaded a multi-stakeholder process called the National Roundtables on Corporate Social Responsibility and the Canadian Extractive Industry in Developing Countries (hereafter, the CSR Roundtables) in 2006.[10] The CSR Roundtables focused on themes that were drawn from the 2005 parliamentary report. These themes were discussed by members of eight Government of Canada departments, 60 invited global experts, 156 members of the public representing a range of stakeholder groups, and the Government of Canada's Advisory Group, which was made up of members from civil society, academia, labor, and the resource industry.[11]

One of the themes discussed during the CSR Roundtables was 'resource governance.' Under this theme the issue of the relationship between resource extraction and development was thoroughly discussed. The analysis and findings of the resource curse literature and the counterarguments provided by ICMM's Resource Endowment Initiative found their way into the Final Report of the CSR Roundtables, which is a consensus document authored by the members of the Advisory Group (DFAIT, 2007, pp. 50, 51).

In closed-door discussions concerning the role of Canada's bilateral and multilateral aid, industry representatives of the Advisory Group appealed for the Government of Canada to engage more actively with host country governments where Canadian extractive companies operate to strengthen their capacity to 'manage' resource revenues. Industry also appealed to the Canadian International Development Agency (CIDA) to become more active in supporting development projects at Canadian mine sites and 'play a role in helping companies to engage effectively with key stakeholders, including local communities' (DFAIT, 2007, p. 53).

Perspectives of the Canadian mining industry regarding its achievements, as well as its needs, with respect to development were also reflected in its responses to a private member's bill, Bill C-300, introduced in the Canadian House of Commons in February of 2009. Bill C-300 was aimed at making Canadian government financial and political support for extractive companies contingent on those companies meeting international human rights and environmental standards. The Bill was strongly opposed by Canadian mining companies and their associations. In a letter to Canadian Members of Parliament, the Prospectors and Developers Association of Canada (PDAC), the largest mining industry association in the world, asserted that 'many tangible benefits are being delivered to local communities and host governments in the form of employment, infrastructure, health, nutrition, education, infrastructure [sic], technology and economy' (PDAC, 2009b). PDAC also called on the government of Canada to do more to address what it called 'challenges arising from weak host-country governance' through programs 'designed to assist host countries to strengthen institutional capacity,' noting that 'these programs are currently given only limited support' (ibid.).

Even as the Final Report of the CSR Roundtables was presented to the Government of Canada in March of 2007, a series of closed-door meetings started at the Munk Centre for International Relations at the University of Toronto.[12] The meetings brought together major Canadian mining companies, mining associations such as PDAC, and major Canadian development organizations such as World Vision, Care Canada, and Plan Canada. These development organizations had not participated in the CSR Roundtable process. The aim of the meetings was to establish the potential for long-term institutionalized collaboration around development projects at Canadian mining projects overseas. The new organization was named the Devonshire Initiative (DI), after the street address of the Munk Centre.[13] While some of the development Non-governmental Organizations (NGOs) that engaged in these meetings had already been partnering with mining companies through individual contracts, the DI offered an opportunity to seek external funding for these collaborations. Sources that were discussed included the Government of Canada and the Clinton–Giustra Fund.[14]

Shortly after the establishment of the DI, Inmet Mining provided insight into industry's assessment of the accountability-focused CSR Roundtable process, versus the potential of the 'solutions-oriented' DI:

> The Devonshire Initiative...grew out of the multistakeholder national Corporate Social Responsibility Roundtables convened by the Canadian government to address corporate social responsibility

by the Canadian extractive industry in developing counties. The Roundtables highlighted the polarized and unproductive nature of the relationship between the extractive sector and a small segment of non-governmental organizations (NGO). To overcome this polarization and move towards productive, on-the-ground solutions to issues of mutual concern, development NGOs and some industry members have come together in a new dialogue that could see both sides working together in developing countries to achieve mutual objectives. This is an exciting opportunity and we are fully committed to the DI process. (Inmet Mining Corporation, n.d.)[15]

A recent statement by the new director of the Devonshire Initiative notes that 'the mission of the DI is to position itself as the leading multi-stakeholder mechanism for direct dialogue, capacity building and coordination between the Canadian extractives industry, development NGOs and Government' (Thomas, 2010). At the most recent gathering of the DI on 27 September 2010, participants included civil servants from five Canadian government departments, as well as from the World Bank, the IFC, United States Agency for International Development (USAID), and the Inter-American Development Bank.

The DI can be seen as a coming to fruition in Canada of a number of key elements of the vision set out in ICMM's Resource Endowment Initiative: a focus on partnerships between industry, development NGOs, and government, and on community development projects at mine sites – sharing responsibility for outcomes and externalizing significant costs for CSR projects to governments and foundations.

The Canadian government, mining, and the development agenda

The Canadian government has traditionally supported the Canadian mining industry operating overseas politically through embassies, trade missions, and bilateral and multilateral relations involving government departments such as CIDA, Natural Resources Canada, and DFAIT, financially through the Canadian Pension Plan (CPP), Export Development Canada (EDC), and CIDA Inc.,[16] as well as indirectly through support to multilateral financial institutions such as the World Bank Group and regional development banks. As noted above, during the CSR Roundtables mining company representatives pushed for greater involvement by the Government of Canada, and in particular by CIDA, both at the national level and at the mining project level in host countries to better ensure local-level benefit sharing and development outcomes. Government participants in the CSR Roundtables responded by claiming that the agency 'does not have a major presence in the mining, oil and gas sectors' because 'to date such issues have not been identified as a priority area for official development assistance by developing countries' (DFAIT, 2007, p. 52).

Nonetheless, the CSR Roundtable process identified areas in which CIDA had been active on mining in developing countries and provided important feedback from civil society groups in those countries: 'several invited experts from developing countries expressed concern about CIDA's work in this area' especially with regard to 'changes made to mining legislation that, with the support of CIDA, undermined a number of existing legal provisions, including those that were protective of indigenous rights' (DFAIT, 2007, p. 52). Participants also noted that in the past 'CIDA has partnered with civil society organizations that provided support to communities affected by mining operations' particularly to 'strengthen communities' capacity to effectively engage in decision-making processes regarding mining investments' but that 'CIDA no longer supports this type of work' (DFAIT, 2007, p. 52). Participants in the CSR Roundtables also noted that CIDA was engaged in exchanges between indigenous peoples affected by mining through the Indigenous Peoples Partnership Programme but that 'organisations in developing countries' had noted a 'lack of transparency around the criteria being used to determine how Canadian individuals and organizations are selected for exchanges and whether these individuals and organizations represent a diversity of indigenous perspectives on extractive industry development' (DFAIT, 2007, p. 53).

Two years after the Final Report of the CSR Roundtables was provided to the Government of Canada, the government issued its official response, *Building the Canadian Advantage: A Corporate Social Responsibility (CSR) Strategy for the Canadian International Extractive Sector* (March, 2009). This report acknowledges the role CIDA has been playing at the national level in developing countries in 'building and modernizing governance regimes' related to natural resources and CIDA's ongoing support to 'bilateral and multilateral initiatives that enhance the capacities of developing countries to manage natural resource development,' including developing and promoting regulatory requirements and 'legal and judicial reform' (Government of Canada, 2009). The government's report particularly notes the role CIDA has been playing in Latin America 'to develop and promote regulatory requirements for social and environmental management' (ibid.).[17] The government's new strategy supports and expands CIDA's role in these national level activities with a goal of enhancing 'the capacities of developing countries to manage the development of minerals and oil and gas, and to benefit from these resources to reduce poverty' (ibid.).

The consensus recommendations coming out of the CSR Roundtables did not recommend that CIDA fund CSR projects at Canadian mine sites, nor did they recommend that CIDA partner with Canadian extractive companies and Canadian development NGOs to deliver CSR projects at Canadian mining operations in developing countries. The Government of Canada's new strategy in response to the Final Report from the CSR Roundtables also does not explicitly indicate a role for CIDA in funding or partnering on CSR projects at Canadian mine sites. Nonetheless, development NGOs and industry members of the DI have jointly and separately petitioned CIDA to provide funds for CSR projects at Canadian mine sites in developing counties. CIDA is also reportedly considering a possible role for the agency as a partner in tripartite partnerships between CIDA, Canadian mining companies, and development NGOs.

There is little transparency from CIDA regarding its funding for CSR projects at mine sites. An order paper request to CIDA from a Member of Parliament received confirmation that CIDA has received CSR project proposals in which 'NGOs and firms have described arrangements of shared leadership and responsibility.'[18] CIDA asserted that the projects 'being considered by CIDA would support the strengthening of alliances and partnerships that increase the impact, quality and sustainability of development activities' and that they involve 'strong engagement in project design, implementation, and evaluation on the part of local communities.' CIDA also asserted that 'CIDA does not directly fund development projects that benefit the commercial interests of private sector companies.' It is unclear how CIDA defines development projects that 'directly benefit the commercial interests of private sector companies.' CIDA did not respond to questions about the level of public funding for CSR projects at mine sites that it has spent, or plans to spend, other than to refer the Member of Parliament to CIDA's proactive disclosure website, which is not designed to provide this level of information.

In response to a letter requesting more information, CIDA divulged that the agency is funding an initiative in Peru involving a community where Canadian mining company Barrick Gold has been operating.[19] The reforestation and agroforestry project is being managed by Canadian development NGO SOCODEVI. In a letter from CIDA to MiningWatch Canada dated 14 January 2011, CIDA indicated that 'Barrick Gold is contributing approximately $150,000 directly to the

Peruvian non-governmental organization... The total amount budgeted for this sub-project is $499,445, and to date in the 2010–2011 fiscal year a total of $158,241 has been disbursed.'

Barrick has been less reticent about discussing the project, noting in a 2010 publication that 'Barrick and the Canadian International Development Agency (CIDA) have teamed up to launch a reforestation pilot project in remote communities near the company's Lagunas Norte mine in Peru' (Barrick Gold Corporation, 2010). In the January 2011 letter to MiningWatch Canada, CIDA also indicated that it has approved CAN$500,000 over three years to a second project, in Ghana, 'whose objective is to maximize the development impact of mining operations' for mining communities around Rio Tinto Alcan's mine. The mining company will contribute CAN$268,000 in that case.

It is unclear why CIDA would fund CSR projects, which have traditionally, and by definition, been funded by mining companies themselves, particularly for the world's most well-endowed mining companies. However, Canadian development NGOs already have a history of partnering directly with large Canadian mining companies to carry out projects at their mine sites. CIDA's new involvement supporting CSR projects at mine sites raises a number of concerns. Given local-level conflict and community divisions that accompany many mine projects, sometimes escalating to violence and allegations of human rights abuses, Canadian government involvement in a local CSR project could be perceived by community members as overt support by the Canadian government for the mine project or the mining company, possibly in the face of their own opposition to the project. This raises a number of questions. Aside from evaluating the particulars of a given development project, how does CIDA evaluate which mine projects to engage in? What role does community consent for a mine, or lack thereof, play in this decision making? How well placed is CIDA to assess community support for a mine project, or even for a particular CSR project?

There appears to be reason to question CIDA's assertion that the CSR projects it funds always have strong community engagement. A recent study of mining and local development in Peru, conducted by Emilie Lemieux (2010b), discusses the CIDA-funded CSR project at Barrick Gold's Lagunos Norte mine described above.[20] Lemieux writes:

> This project seems to fulfill the basic social needs the company is looking to address, as well as the Canadian embassy's interest to work in CSR, rather than the needs of the local population. The project

was implemented in a very top-to-down way, based on the mine's desire to re-grow the forests, and CIDA's need to work in CSR and promote the good initiatives of Canadian mines. Consultation with local actors is non-existent and the project is led exclusively by external actors. (Lemieux, 2010b, p. 25)

Canadian development NGOs, mining, and the development agenda

In the last decade, some Canadian development NGOs have entered into contracts with Canadian mining companies to carry out development projects in communities around mines in developing countries. These contracts are typically worth hundreds of thousands to well over a million dollars and may run for multiple years. Development NGOs have been reticent to discuss their contracts with mining companies and there has not been much public dialog about these arrangements (Coumans, 2011). It is therefore perhaps not surprising that the DI members started to meet behind closed doors in March of 2007 and deliberately kept the development of the initiative 'under wraps' until it was rolled out to the public in October of 2007.[21]

World Vision stands out for the degree of transparency it displays on its website regarding these arrangements. World Vision actively recruits corporate partners, noting on its website that 'there are many reasons why your corporation may like to engage with World Vision' (World Vision Canada, 2008c). With regard to community programs in particular, World Vision notes that a 'philanthropic program based on your business interests or geographical focus enhances your corporate profile in local and overseas markets.' World Vision provides brief sections on its partnership programs at Barrick Gold's mines in Peru and at Centerra Gold's mine in Mongolia, including the value of the contracts (World Vision Canada, 2008b). However, these brief sections do not provide any insight into why World Vision has decided to partner with mining companies to do its work, nor why it chose these particular companies or projects.

Similarly, the DI website offers little insight into the reasons development NGOs would partner with mining companies to carry out CSR projects at mine sites, other than the rather anemic explanation that development NGOs might welcome the 'opportunity to steer private sector development in a more socially sensitive and equitable way' (Devonshire Initiative, 2010). In that light, it is noteworthy that the NGOs involved in the DI have been largely absent in the remarkable series of efforts to secure greater accountability in Canada for the activities of Canadian mining companies operating overseas, starting with the parliamentary hearings and report of 2005, through the CSR Roundtables of 2006 and the 2009–10 hearings and discussions around Bill C-300, each of which did engage large segments of Canadian civil society, including development NGOs that are not engaged in the DI.

The lack of publicly available elaboration on the rationale for NGO partnerships with mining companies by the development NGOs engaged in these partnerships stands in stark contrast to the large volume of analysis and reflection on this same topic available on the ICMM website.

Mining-affected communities, resistance, and development

In the discourse of industry, government and partnering development NGOs 'communities' and 'development' feature prominently. Communities associated with mine sites are primarily characterized as the recipients of programming created for them and 'development' is described as a good that is delivered to them. This discourse stands in stark contrast to the language that can be found in joint letters, petitions, and mining moratoria of communities who are struggling, often at great costs to themselves, to stop a mine from starting on their land, continuing to operate, or expanding. In these statements of community intent, mining is commonly described as incompatible both with community members' *own* aspirations for their futures, and with their understanding of themselves as agents of their current lives and of their futures.

Industry associations such as ICMM maintain that community opposition to mining may be characterized as a consequence of weak governance (see McPhail, 2008). The Government of Canada (2009) is focused on providing assistance to governments overseas in 'managing their natural resources' in part to increase local-level benefit sharing. However, these approaches rest on a rationale focused on the provision of goods to mining-affected communities and fail to adequately recognize the desire of communities to preserve existing natural resources and cultural and economic practices in relation to those resources. So to close this section an example from the US, presumably not a 'weak governance state.' In Alaska, the massive gold–copper Pebble Mine is proposed to be located at the headwaters of the Bristol Bay Watershed, famous for being the world's single largest wild salmon spawning habitat. Every year some 30–40 million sockeye salmon come to this area to spawn, providing a livelihood worth more than 100 million dollars annually to commercial fishermen and sport-fishing guides and also providing subsistence livelihoods to indigenous Yup'ik. In spite of development promises made by the mine's proponents, the project is opposed by a majority of indigenous and non-indigenous local people. Bobby Andrew is Yup'ik and a lifelong resident of the Bristol Bay Area. He says that 'I don't think we should have a mine. There's too much at risk: our water, our air, our fish, our subsistence way of life' (Sherwonit, 2005, p. 20). Brian Kraft, a businessman and local lodge owner, also opposes the project: 'This is a very, very sensitive place, one that should be left alone. Salmon are the lifeblood here' (ibid., 2005, p. 58).[22]

Conclusions

This paper contends that local resistance to mining, whether in strong or weak governance jurisdictions, is often rooted in efforts by community members to protect economic, environmental, social, and cultural values of individual and collective importance to them and essential to maintaining control over their lives and to realizing the futures they envision. Communities actively opposing mines openly articulate these values and needs in clear and coherent ways.

The mining industry, through its associations, also articulates its needs clearly. As opposition to mining has expanded and intensified and is becoming more organized globally, achieving a social license to operate has become a critical focus for the industry. Industry associations acknowledge these facts openly and emphasize the need for partnerships, particularly with development NGOs and with governments as essential to minimizing conflict, risk, and cost and assuring continued access to land and resources.

Of the actors that have been discussed here, Canadian development NGOs provide the least in the way of publicly available explanation for their interest in partnering with mining companies. On the part of most home and host governments, the rationale provided for collaboration with industry is sparse and largely aligned with, and responsive to, the rationale and stated needs set out by industry. There is little evidence that governments question the assertion that mining is a vehicle for sustainable local-level development.[23] The Canadian government's development agency, CIDA, has also done little to articulate a clear public policy position and unambiguous rationale to explain why it is now funding development NGOs to carry out CSR projects at mine sites in partnerships with mining companies.

In response to an inquiry from a Member of Parliament, CIDA referred to its role under the Government of Canada's new CSR Strategy with respect to the extractive sector: 'close cooperation with stakeholders is, therefore, required for the successful implementation of the *Building the Canadian Advantage Strategy*' (Inquiry of Ministry, 2010). The Government of Canada's CSR Strategy, as interpreted by CIDA and by Canadian mining companies, mandates CIDA to provide public funds for CSR projects. CIDA says that it will entertain 'CSR project proposals' in which NGOs and firms share 'leadership and responsibility' (Inquiry of Ministry, 2010).

This facilitating role of the state has been described as a shift in which '[d]evelopment is no longer the responsibility of the state; rather, the state sets the wider framework, the market must be its motor, and civil society would give it direction' (Rist, 1997, in Blaser et al., 2004, p. 11). This shift is seen to be based on neoliberal economics and liberal political theory that:

> assigns to the state the role of a legislator and guarantor of the rules that allow the market to operate unhindered on a transnational and global scale. The assigned role for the market is to generate the wealth with which development can be built. The task of making development 'human'... that is to input other values than economic efficiency – has been increasingly assigned to organizations from civil society, or NGOs. (Blaser et al., 2004, p. 12)

The questions this chapter poses are whether the Government of Canada's new CSR Strategy, which in its very title speaks of building a *Canadian* advantage, whether CIDA's new mandate to fund corporate social responsibility projects at mine sites in developing countries, and whether Canadian development NGOs' partnerships with mining companies at their mines in developing countries are in fact making development 'more human' for the communities affected by Canadian mining companies. Do these policies and partnerships respond to the stated aspirations of the communities they target? Do they respect community decisions not to host a mine? How do they respond in cases of mining-related violence, human rights abuses, or severe environmental degradation? How do development NGOs position themselves in these conflicted terrains?

In order to answer these questions there needs to be far greater transparency regarding social and environmental conditions at mine sites, as well as regarding the nature of the CSR projects CIDA is funding and in which Canadian development NGOs are participating. CIDA insists that its projects ensure 'strong engagement in project design, implementation, and evaluation on the part of local communities.' The project described by Lemieux (2010b, p. 25) above raises serious questions about the level of community consultation in that particular CIDA-funded CSR project. These questions become particularly acute in the case of communities in which members are actively opposing a mine. These communities may be united or divided on the issue, but for community members opposing a mine, it is likely that their definition of development is not the same as that of the Government of Canada.

While public information is sparse, both on the side of CIDA and on the side of the development NGOs, there is nothing in the information that is available that would indicate that CIDA, or Canadian development NGOs, are actively supporting the agency of communities that have articulated a vision of their own future, of 'development' that does not include a mine.

Notes

1. For more on the history of mining by Placer Dome in Marinduque, see Coumans (1994, 1995, 2000, 2002a, 2002b).
2. In 2005 the Provincial Government of Marinduque filed a lawsuit against Placer Dome in the State of Nevada in the US. This suit now includes Barrick Gold since the takeover, in 2006 of Placer Dome by Barrick Gold. For more on this suit, see Diamond McCarthy (n.d.).
3. The following provinces have declared mining moratoria in the Philippines: Capiz; Iloilo; Oriental Mindoro; Occidental Mindoro; Marinduque; Palawan; Northern Samar; Eastern Samar; Western Samar; Romblon. In addition to provinces and towns that have declared mining moratoria, many smaller government units known as barangays have also declared and provided rationales for such moratoria.
4. Other high-profile mining-related environmental disasters were the Cambior cyanide spill in Guyana in 1995 and the Boliden spill in Spain in 1997.
5. See, for example, World Bank (2003), CFAIT (2005), and DFAIT (2007).
6. See, for example, the recently defeated Private Member's bill tabled by John McKay MP in 1999, Bill C-300 (McKay, 2009). For more on Bill C-300 see Coumans (2010a, 2010b).
7. In brief, reasons for national economic decline associated with high levels of resource extraction include exchange rate overvaluation leading to loss of competitiveness and development of other economic sectors (so-called Dutch Disease); volatility in mineral values; overconsumption based on a non-renewable resource; and corruption and unequal distribution of benefits associated with mineral wealth. For a literature review regarding the resource curse, see Kuyek and Coumans (2003, pp. 1–5).
8. ICMM members have been strong public supporters of the EITI since its inception. 'We believe that improving transparency should be part of broader governance improvement programs in order to deliver long-term economic growth and poverty alleviation.' (ICMM, 2010c).
9. 'Transfer pricing happens when affiliated companies – say, a parent and a subsidiary, or two subsidiaries – can set their own artificial (non-market)

transaction price' (Emmons, 2010) and through this mechanism avoid paying taxes on profit. For more information on this topic, see Baker (2005) and Christian Aid (2009).
10. For a more in-depth discussion of the National Roundtables on Corporate Social Responsibility (CSR) and the Canadian Extractive Industry in Developing Countries, see Coumans (2010a).
11. Members of the Advisory Group sat as individual experts, not as representatives of their organizations or institutions. The members were Tony Andrews, Andrea Botto, Diana Bronson, Jim Cooney, Craig Forcese, Dennis Jones, David MacKenzie, Reg Manhas, Robert Walker, Gerry Barr, Henry Brehaut, Bonnie Campbell, Catherine Coumans, Pierre Gratton, Karyn Keenan, Audrey Macklin, and Gordon Peeling.
12. The Munk Center is named after its benefactor, Peter Munk, founder and chairman of the world's largest gold mining company, Barrick Gold Corporation.
13. The DI meetings were headed up by Marketa Evans, then-director of the Munk Centre. Evans is now the government appointed CSR Counsellor for the Extractive Industries.
14. The Government of Canada has granted the Clinton Foundation charitable status in Canada. One of the foundation's largest donors is Vancouver mining executive Frank Giustra, who has pledged CAN$100-million to the foundation.
15. This text has since been modified on Inmet's website.
16. CIDA Industrial Cooperation provides funds to Canadian private sector companies to carry out 'commercial development projects.' See CIDA (2007).
17. For an in-depth analysis of the long-running Peru–Canada Mineral Resources Reform Project (PERCAN) project that resulted from a bilateral agreement between CIDA and the Peruvian Ministry of Energy and Mines (MEM) see (Campbell et al., present volume, Chapter 5).
18. The Order Paper Request was submitted on 9 June 2010 and received a response from CIDA on 20 September 2010.
19. A letter from MiningWatch Canada to CIDA dated 30 August 2010 received a response on 14 January 2011.
20. The SOCODEVI website page that Lemieux references in her publication mentioned Barrick as a partner in the project (personal communication with Lemieux, 18 February 2011). This page has been changed on the SOCODEVI website and no longer mentions Barrick. See SOCODEVI (2011).
21. Personal communication with a DI participant on condition of anonymity.
22. Other examples of community opposition to mining from North America include the cases of Kitchenuhmaykoosib Inninuwug in Ontario and the Tsilhqot'in Nation in British Columbia. For more on these cases and others see www.miningwatch.ca, accessed 10 June 2011.
23. There are exceptions. The governments of Costa Rica and El Salvador are attempting to curtail large-scale open-pit mining in those countries.

8
Mining Industry Associations and CSR Discourse: Mapping the Terrain of Sustainable Development Strategies

Nicole Marie Lindsay

> Discourse is not the expression of thought; it is a practice, with conditions, rules, and historical transformations.
> (Escobar, 1995, p. 216)

Introduction

The core concern of this chapter is in seeking to understand the discursive dynamics of industry-led Corporate Social Responsibility (CSR) initiatives in the mining industry. In this chapter, I consider CSR as an emerging discourse that is both shaped by and, in turn, shapes the political economic context in which it operates to inform the practices, conditions, rules, and historical transformations of the business–society relationship in the mining industry.

Taking this broad, political economic view, I consider the context in which CSR discourse has emerged in the mining industry, particularly as it has been taken up and institutionalized through standards and codes developed by mining industry associations. In doing so, I hope to illustrate the dialectical relationship between discourse and institutions as two closely related elements of the practice of rule-making and policy-setting. Such an approach requires an acknowledgment of the power dynamics inherent in discursive and institutional social relations – in other words: who makes the rules and who has to follow them? Further, in whose interests are rules and policies formed, and who has recourse or protection from or by those rules?

In bringing together insights from the diverse fields of discourse and semiotic analysis, political economy, and institutional theory,

I explore why CSR discourse as an institutionalized practice of rule-setting through industry codes and standards offers a limited opportunity for genuine transformation in the mining industry, particularly in terms of offering protection and recourse for those lacking institutional power – namely, vulnerable populations in the communities most directly affected by mining operations. Drawing from the findings of research demonstrating those failures (King and Lenox, 2000; Sethi and Emelianova, 2006; Frynas, 2009), through the lens of discourse analysis I argue that a critical understanding of the limitations and opportunities represented by CSR is a crucial first step in improving real world outcomes in social responsibility in the mining industry. In what follows, I explore in more detail both these limitations and opportunities.

CSR as discourse

The discipline of critical discourse analysis is based on the view that semiosis – or, the ways in which we think, talk about, and represent the world – is 'an irreducible element of all material social processes' (Chiapello and Fairclough, 2002, p. 193). Further, since social practices are inflected with power dynamics, a central concern of critical discourse analysis is to examine the role of discourse in the production, reproduction, and challenge of relationships of dominance (van Dijk, 1993; Chiapello and Fairclough, 2002).

Chiapello and Fairclough (2002) explain that discourse figures in social practices in three broad ways. First, discourse figures as a part of the social activity within a practice. For example, in the case of CSR and mining, the *genre* of CSR discourse constitutes the activities undertaken by mining executives and corporate social responsibility practitioners in their everyday management: they analyze the impact of their operations, engage in stakeholder discussions, interact with other professionals and government agencies on CSR-related issues, and so on. In all of these interactions, executives, CSR practitioners, stakeholders, and government actors employ the genre of CSR discourse – they might discuss 'stakeholder interests,' 'win–win solutions,' 'environmental management,' 'community development,' and so on. In doing so, they draw from a reserve of commonly understood ideas, expectations, and language regarding the roles and practices of mining companies, communities, government, and other groups considered stakeholders.

Secondly, discourse figures in *representations*. In the course of any social practice, actors interpret and understand their own and other roles, practices, and relationships in representative terms – that is, terms that can be redescribed in contexts other than those in which they take place. For example, in the discursive field of CSR and mining, mining company executives and NGO actors formulate different representations of the interests, goals, and practices of various groups of actors, and they each represent the lives of poor and disadvantaged communities in different ways (that may or may not be accurate representations of the lived reality of those communities). However, accurate or not, these discursive representations bear directly on how actors interact with each other. If an NGO actor represents a mining executive as ruthlessly self-interested and instrumental, and the executive represents the NGO actor as irrational and confrontational, these discursive representations have very real material implications on their potential for genuine communication.

Thirdly, discourse figures in ways of being, or in the constitution of identities. Again, the discursive construction of the identity of *mining company executive* carries strong connotations that can be subjectively internalized in very material ways of acting and being. This is not to say that identities are constituted simply and directly through semiosis. Rather, social relationships and institutions also contribute to the discursive constitutions of identities – consider the impact of an MBA education versus an education in political theory or community development, not to mention all of the social, economic, and subjective factors that influence a decision to obtain a business degree or to engage in social activism.

I want to make three points about these insights. The first is that the semiotic functioning of genres, representations, and identities described above largely takes place *under the surface* of what is easily observable as reality for social actors (Fairclough, 2001). Secondly, these processes take place within a social order (or order of discourse) structured by relations of *dominance*. As Chiapello and Fairclough (2002) point out, 'some ways of making meaning are dominant or mainstream in a particular order of discourse, others are marginal, or oppositional, or "alternative" ' (p. 194). Thirdly, like any social order or process, orders of discourse are subject to change over time, and thus, they are frequently the sites of social and political struggles over meaning.

Understanding CSR as a discursive order sheds light on aspects of the business–society relationship that are frequently under-theorized in CSR-oriented management approaches – namely, questions regarding power, hegemony, and political struggles for meaning and legitimacy (one notable exception to this is Banerjee, 2008). Indeed, given that discursive orders are no less significant in academic disciplines than in business and the rest of society, it is little surprise that this is the case.

Thus, a central concern of this chapter is to explore the notion of CSR as a site of discursive struggle over the legitimate *meaning* of the notoriously slippery concepts of 'social responsibility' and 'sustainability' as practiced in the mining industry. Further, keeping in mind that discourse is not simply the expression of thought or ideas, but that it informs and shapes material social practices, institutions, policies, and the like (Escobar, 1995), I argue that it is important to examine the ways in which social practices and institutional policies in the mining industry shape and are in turn shaped by CSR discourse. This chapter makes some initial steps toward such an inquiry.

On this point, the concept of semiotic selection (Jessop, 2004) usefully shows how discourse works to either reinforce, or in moments of crisis, transform dominant 'economic imaginaries,' or discursively ordered ideas about the economic world. Jessop argues that maintaining coherence in an otherwise chaotic political economic context requires that economic imaginaries are continually reinforced, repaired, and reproduced. He identifies four mechanisms that ensure the dialectical reinforcement of economic imaginaries (or orders of discourse) and political economic structures. These are as follows:

(1) selection of particular discourses for interpreting, legitimizing, or representing social phenomena;
(2) retention of certain resonant discourses through organizational routines, rules, and strategies, and state projects;
(3) reinforcement of discourses and their associated practices through elaboration, selective strengthening of appropriate genres, and elimination of less appropriate alternatives; and
(4) selective recruitment, inculcation, and retention of social agents 'whose predispositions fit maximally with...the preceding requirements' (p. 164).

In the absence of disruption or crisis, this process of semiotic selection will ensure the smooth reproduction of economic imaginaries; in other words, discourses not resonant with the existing social order will be weeded out in favor of a set of ideas more consistent with the dominant economic imaginary. However, in moments of rupture or crisis, space opens for the contestation of previously dominant economic imaginaries and the introduction of alternatives.

In the next section, I explore the contours of neoliberalism as representing both the dominant economic imaginary and the source of crisis shaping the political economic context for the emergence and institutionalization of CSR discourse in the mining industry, using Latin America as a case study to illustrate what is truly a global phenomenon.

The double movement: CSR and two phases of neoliberalism in Latin America

Karl Polanyi's (1944) seminal study of the social, political, and economic transformations accompanying the early industrial capitalist period holds particular relevance for a study of the emergence of CSR toward the end of the twentieth century. In *The Great Transformation*, Polanyi argues that the eighteenth-century development of the market economy as an idealized self-regulating, unfettered, *laissez faire* economic space was accompanied by a transformation of society into *market society*. This represented 'a complete transformation in the structure of society' in which 'human society became an accessory of the economic system' (Polanyi, 1944, pp. 74, 79). This transformation, and the social dislocation and environmental devastation it brought about, gave rise to what Polanyi called the 'double movement':

> While on the one hand markets spread all over the face of the globe and the amount of goods involved grew to unbelievable dimensions, on the other hand a network of measures and policies was integrated into powerful institutions designed to check the action of the market relative to labor, land, and money... Society protected itself against the perils inherent in a self-regulating market system – this was the one comprehensive feature in the history of the age. (Polanyi, 1944, pp. 79–80)

The cyclical nature of the double movement – free-market economic reform, followed by state regulation aimed to offset its more harmful social and environmental effects – can roughly be traced up to the present moment, and is particularly salient in recent debates about neoliberalism in Latin America and other regions in the developing world.

Following Polanyi's theory of the double movement, scholars such as Peck and Tickell (2002) and Porter and Craig (2004) have observed that neoliberal policy reforms beginning in the latter half of the twentieth century in most parts of the world involved two phases. The first phase was marked by a 'disembedding' or 'roll-back' of existing economic and social policies seen as a barrier to the circulation of capital. This phase generally involved wide-scale deregulation, market liberalization, and an apparent rolling back or retreat of the state from economic and social governance. The second phase of neoliberal reform involved the development of 'roll-out' policies focused on governance building, re-regulation, and 'purposeful construction and consolidation of neoliberalized state forms, modes of governance, and regulatory relations' (Peck and Tickell, 2002, p. 384). It is in this later, roll-out, or 'inclusive' phase of neoliberal economic development, with its re-embedding of social and environmental concerns, that CSR and sustainability emerged as politically salient concepts featured in debates about the role of business in society, as well as in corporate strategy and government policy-making in the mining industry.

Following the insights of discourse analysis as discussed above, the spread of neoliberal policy reforms throughout Latin America can be understood as a process of semiotic selection of policy preferences – a very specific set of responses to the economic crises of the 1970s and 1980s based on the emerging neoliberal economic imaginary of the era. As Peck and Tickell (2002) and many others have pointed out, the discourses influencing policy decisions during the roll-out phase of neoliberalism emerged from a specific locus of neoclassical economic thought originating from the Chicago School of Economics, and eventually becoming imbedded in Washington-based Bretton Woods Institutions such as the World Bank (Veltmeyer et al., 1997). Neoliberal policymaking in this era reinforced and reproduced the central tenets of what came to be known as the Washington Consensus through a process of selection, retention, reinforcement, and recruitment – the latter, largely in the form of technocrats educated in neoclassical economics who cycled through key policy-making positions in a variety of institutions (see Teichman, 2004, for a discussion on the 'epistemic communities' of technocrats and policy-makers influencing policy reform in Mexico and Argentina during the 1980s and 1990s).

Latin America's experience with the early phase of neoliberalism has been well documented, as have the devastating social and environmental effects of the roll-back neoliberal policies of this era (Veltmeyer et al., 1997; Harris and Nef, 2008). A central feature of the period of economic restructuring and reform of the 1980s and 1990s, however, was the opening (and in some cases, reopening) of the region's mineral reserves to foreign investment and development. In response to recurring debt crises and poor economic performance, nation after nation in Latin America received assistance from the World Bank to undergo dramatic structural reforms in the interests of creating an investment-friendly climate aimed to attract foreign capital into the mining sector (Bridge, 2004). In just over ten years, the region's share of global mining investment increased by 20 percentage points, from 18 per cent in 1990 to 39 per cent in 2001 (ibid.). Between 2000 and 2005, global mining exports from Peru roughly quadrupled (US$911.4–$44267.6 million), tripled from Chile (US$2868.8–$9387.9 million), and almost doubled for the already large producer, Mexico (US$15,453.5–$29,550.7 million).[1]

However, despite the World Bank's heavy emphasis on mineral extraction as a preferred route to economic stabilization and the obvious economic benefit that the dramatic expansion of Latin America's mining industry has brought to many nations in the region, these benefits have not translated into a significant reduction in poverty in Latin America, and poor rural communities located nearest mine sites are particularly vulnerable to the social, health, and environmental risks associated with the mining (Pegg, 2006; Bebbington et al., 2008). These risks – both real and perceived – have resulted in a high level of local opposition to large-scale mining projects in Latin America, often accompanied by violent conflict in confrontations between military, police, and mine security forces on one hand, and on the other, civil society groups defending the rights and demands of mining-affected communities, many of which consist largely of historically marginalized indigenous groups.

It is in this context of crisis and conflict – one marked both by the persistent historical inequalities in Latin American society and the conflicts arising from a rapid expansion of large-scale mining projects operated by foreign multinationals in a context of neoliberal regulatory reforms – that CSR has emerged as an important management and governance strategy for the mining industry, advocated by industry and governments alike. As Himley (2008) suggests, CSR can be seen in this sense as a 'hallmark' of neoliberal governance (p. 446) insofar as it remains resonant with the central features of the first phase of neoliberalism (for example, in its emphasis on voluntary self-regulation by corporations as opposed to binding state regulation and sanctions). Indeed, many critical commentators have pointed out that it is precisely the voluntary, self-regulatory aspect of CSR initiatives that undermines their legitimacy, leading some to argue that CSR might actually *deepen* neoliberalism in the sense that it institutionalizes the emergence of toothless governance mechanisms in which 'decision-making powers regarding the social and environmental consequences of mining are, to a significant extent, ceded from the state to the corporation' (Himley, 2008, p. 446; see also Banerjee, 2008).

Industry associations and CSR codes: Reinforcing an order of discourse

In the preceding sections, I have argued that both CSR and neoliberalism operate as *discursive orders* insofar as they dialectically interact with social and economic practice to shape the rules, conditions, and historical transformations of the business–society relationship. Further, and from a broader, political economic perspective, one might usefully consider that CSR and neoliberalism could be seen as *nested* discursive orders. That is to say, CSR discourse has been semiotically shaped (or 'selected,' to use Jessop's language) from within a broader discursive order – that of neoliberalism. Following the same logic, then, one might argue that neoliberalism has also been semiotically shaped by a still broader discursive order unique to global capitalism. I will return to this thought at the conclusion of the chapter, but for now I turn to a short analysis of industry associations as one locus for the institutional selection, retention, reinforcement, and inculcation of CSR discourse in the mining industry.

Previous research on private standards initiatives and organizations has pointed out that the proliferation of voluntary standards and codes involving new constellations of private and non-governmental organizations has changed the 'rules of the game' for governance of the business–society relationship (Tallontire, 2007). Although the specific contours of these changes are up for debate, few would disagree that globalization has brought about not only increasing integration of national economies into transnational networks based on global circulations of capital, goods, and services that are far more difficult to govern, but it has also brought about increasing involvement of private, non-governmental interests in attempts to address the complex and persistent problems that emerge as a result of global governance gaps (see also Ruggie, 2008, and Chapter 1 in this volume). In the global mining industry, as in many other globally oriented industries, industry associations have begun to develop ways to protect the interests of their members from the political, social, and economic risks associated with governance gaps.

Voluntary codes and standards of practice have emerged as perhaps the primary strategy for industry groups to address governance gaps that widen the potential for corporate abuse of human rights and/or the environment in many parts of the world. That corporations and industry associations prefer voluntary self-regulation to state regulation is virtually an undisputed given, and is frequently cited as one central reason for the proliferation of industry efforts on this front (Levy and Prakash, 2003; Dahan et al., 2006; Cutler, 2008).

However, avoidance of state regulation is not the only reason that industry associations and corporations have devoted significant resources toward the development of private CSR standards and codes. Although public pressure in the form of targeted consumer campaigns, NGO naming and shaming, negative media exposure, along with socially responsible investing and shareholder and employee campaigns also play a role in the adoption of civil society codes or industry-based codes, these pressures also only partially explain the phenomenon. As Vogel (2010) suggests, the pressures listed above are relatively weak – intergovernmental regulation of business activity on a global level is hardly a valid threat at this point, consumer and NGO campaigns rarely hurt the bottom line of most multinationals, and the finance sector remains largely indifferent to firm CSR performance.

Rather, or perhaps, in addition to such external pressures, some researchers have pointed to the institutional and subjective internalization of changing global norms as having a significant influence on the adoption of CSR codes and standards (Haufler, 2001; Dashwood, 2007; Kollman, 2008). Thus, as Vogel (2010) points out:

> For many highly visible global firms, engaging in various forms of global CSR, including having a CSR office, issuing a CSR report, cooperating with NGOs, and agreeing to be bound by one or more voluntary industry codes, has become an accepted part of managing a global firm in a more politicized and transparent global economy. (p. 11)

My argument here is that the changing norms associated with the 'more politicized and transparent' global economy can be seen as a response to the negative externalities resulting from neoliberal market reforms – or in other words, as a contemporary example of Polanyi's double movement. In this way, the rise of CSR discourse, changing global norms regarding the responsibility of business toward society, civil society campaigns for corporate accountability, and so on, all might be seen as competing (or complementary, as the case may be) responses to a genuine crisis or rupture in the much-contested semiotic order of global capitalism. That is to say, old ideas about economic and social relations within a (global) capitalist system are losing legitimacy, and in the space they leave behind, new ideas are jockeying for dominance.

Industry associations are key players in these semiotic struggles. Indeed, as Dashwood (2007; see also Chapter 2 in this volume) has shown, industry actors in the mining industry (executives, corporations, and industry associations) are not only influenced by emerging CSR norms and standards but, in some cases, they have taken a proactive role in shaping and defining them. Nowhere is this more evident than in the codes and standards that have been developed and promoted by mining industry associations.

Before turning to a brief overview and analysis of three such codes, it is worth reviewing some recent literature regarding the dynamics of industry-led voluntary CSR standards. An overwhelming focus of much of this literature is on the efficacy of voluntary standards in improving companies' behavior. King and Lenox (2000) point out that industry self-regulation, seen as a 'middle way' between state regulation and *laissez-faire* market-based solutions, has become the preferred route to improving corporate environmental performance, despite a marked lack of evidence that such approaches actually work. King and Lenox set out to analyze the effectiveness of the Chemical Manufacturers Association's (CMA) Responsible Care program, finding that despite the promise represented by Responsible Care, participating companies were no more likely (and in some cases, even less likely) to improve their environmental performance than non-participating companies (King and Lenox, 2000).

King and Lenox explore several possible explanations for the disappointing outcomes of the Responsible Care program, concluding that the difficulty in establishing and maintaining industry self-regulation has much to do with an inability to counteract opportunism among poorly performing members in the absence of enforceable sanctions. Similarly, Prakash and Potoski (2007) suggest that the institutional design of what they term 'voluntary clubs' must address the issue of shirking through three crucial mechanisms: third-party monitoring, public disclosure of audit information, and sanction of non-performing members. Following Hobbes' assertion that 'covenants without swords are but words' (Hobbes, 1651 in Prakash and Potoski, 2007, p. 780), Prakash and Potoski differentiate between standards that contain all three monitoring and enforcement mechanisms ('strong sword' clubs), those that use only third-party audits and public disclosure without sanction ('medium sword' clubs), and those with only third-party audits ('weak sword' clubs). As I discuss in more detail below, no 'strong sword' clubs currently exist in the mining industry.

Writing at very nearly the same time, Campbell (2006, 2007) and Sethi and Emelianova (2006) both present a series of institutional conditions which they hypothesize would lead to improved CSR performance, summarized in Table 8.1.

Table 8.1 Preconditions for CSR performance and existing conditions for Canadian mining companies with operations in Latin America

Institutional preconditions that facilitate CSR (Campbell, 2006, 2007)	
Healthy economy and corporate financial performance	Variable, depending on company
Moderate levels of competition in industry	Increasing competition in many sectors as resource scarcity increases
Strong and well-enforced state regulations to ensure responsible behaviour	Weak in most jurisdictions
System of well-organized and effective industrial self-regulation	Emerging
Oversight and monitoring by private, independent organizations (NGOs, investors, media, etc.)	Variable
Institutionalization of CSR norms in business publications, business school curricula, etc.	Emerging
Membership in trade associations that support socially responsible behaviour	Variable
Institutionalized dialog with stakeholders	Dialog rarely institutionalized

Institutional preconditions for effective development, monitoring, and enforcement of industry CSR codes (Sethi and Emelianova, 2006; Prakash and Potoski, 2007)	
Selective membership during initial stages of code-creation	Variable
Code addresses issues of concern to communities	MAC 'community of interest' input, ICMM and PDAC stakeholder consultations
Broad principles amplified into objective, quantifiable, and outcome-oriented standards	Variable, needs improvement
Governance structure includes balanced representation and external input	Governance primarily by member companies
Performance monitoring and compliance verification by independent experts	MAC only, ICMM and PDAC in development
External audit results made public without prior censorship	MAC limited, ICMM at company's discretion
Enforceable sanction for non-performing members	None

Note: MAC (Mining Association of Canada); ICMM (International Council on Mining and Metals); PDAC (Prospectors and Developers Association of Canada).

Campbell's (2006, 2007) analysis is focused more on the macro-political economic context in which firms operate, and accordingly, he presents a series of contextual preconditions that will likely lead to more responsible firm behavior. These include a healthy economic context and strong corporate financial performance, a moderate level of inter-firm competition (too much or too little competition discourages responsibility performance), strong state regulation, collective industrial self-regulation, monitoring by NGOs and other independent organizations, a normative institutional environment that encourages socially responsible behavior, membership in industry associations, and institutionalized dialog with stakeholders.

Sethi and Emelianova (2006) focus on the institutional preconditions specific to industry association codes and standards that must be met 'if industry-based codes are to succeed in narrowing the performance-expectations gap between the industry and large segments of society' (p. 230). These preconditions include limited membership in the early stages of code development (to avoid the free-rider and adverse selection problems); focus on issues of importance to the community as opposed to just the industry; a governance structure that includes balanced representation, external input so that codes are not developed and monitored by those whose performance is being monitored, and membership fees to defray the costs of code implementation; executive accountability for code compliance; independent external monitoring and compliance verification; and finally, publicization of independent external audit results.

Taken together, the above recommendations for effective industry self-regulation show a significant departure from the 'weak sword' approaches that currently exist in the mining industry. Indeed, an analysis of the existing codes developed by the Mining Association of Canada (MAC), the International Council of Mining and Minerals (ICMM) and Prospectors and Developers Association of Canada (PDAC) shows as of April 2011 that only one – MAC's Toward Sustainable Mining (TSM) – subjects participating company reports to external verification (however, both ICMM and PDAC are planning to implement external verification). In all three, disclosure is limited, and currently none of the three have any sanctioning mechanism for non-performing members (see summary of codes in Table 8.2).

Further, uptake of industry association codes appears to be quite limited among mining corporations. While a little over half (17 out of 30) of MAC's members reported on the TSM initiative principles, of the six major mining companies with membership in all three industry organizations: Anglo-American, Barrick Gold, Newmont, Teck,

Table 8.2 Mining Industry Association CSR initiative comparison chart

	Mining Association of Canada (MAC)	International Council on Mining and Metals (ICMM)	Prospectors and Developers Association of Canada (PDAC)
CSR/Sustainability initiative name	Toward Sustainable Mining (TSM)	Sustainable Development Framework	e3 Plus: A Framework for Responsible Exploration
Date launched	2004	2003	2003 (original e3 framework) 2009 (e3 Plus)
Elements of initiative	Guiding principles: broad statements Performance Indicators and measurement – management system focus Reporting – based on self-assessment Verification – third-party verification	Principles – 10 broad statements Reporting – in accordance with Global Reporting Initiative (GRI) standards; based on self-assessment against performance criteria (outlined in 2005 ICMM Reporting Resource Guide) Assurance – third-party verification (implemented for 2009–10 reporting) Good practice – information-sharing online resource	Principles & guidance – broad statements with guidance on implementation Toolkits – outline core areas of performance Reporting (Phase 2 – not launched as of April 2011) Verification (Phase 2 – April 2011)
Content focus	Guiding principles: Involvement of communities of interest (COI)	Ethical business practices and sound corporate governance	Principles: Adopt responsible governance and management

Table 8.2 (Continued)

Mining Association of Canada (MAC)	International Council on Mining and Metals (ICMM)	Prospectors and Developers Association of Canada (PDAC)
Encouraging and supporting dialog about operations	Integrate sustainable development principles within decision-making process	Apply ethical business practices
Fostering leadership		Respect human rights
Transparent and accountable business operation		Commit to project due diligence and risk assessment
Protecting stakeholder health and safety	Uphold human rights and respect cultures, customs, and values of stakeholders	Engage host communities and other affected and interested parties
Promoting responsible production, use and recycling of minerals	Implement risk management strategies based on valid data and sound science	Contribute to community development and social well-being
Minimize impact of operations on environment and biodiversity	Continual improvement of health and safety performance	Protect the environment
Working with COIs to address orphaned and abandoned mine issues	Continual improvement of our environmental performance	Safeguard the health and safety of workers and the local population
Application of best practice and innovation	Conservation of biodiversity and integrated approaches to land use planning	*Toolkits:*
Respect human rights and cultures		Social responsibility – framework for how explorers can contribute to a more sustainable society. Topic areas include:
Recognize and respect concerns of aboriginal communities	Responsible product design, use, re-use, recycling, and disposal of our products	Responsible governance
Ethical business operations		Ethical conduct
Comply with laws and regulations		Human rights
		Due diligence

	Support community capacity to participate in mining opportunities Be responsive to community needs and concerns Provide lasting benefits to communities through self-sustaining programs Performance indicators: Tailings management Energy and greenhouse gas emissions External outreach Crisis management planning In development: Aboriginal relations and biodiversity	Contribute to the social, economic, and institutional development of the communities in which we operate Effective and transparent engagement, communication and independently verified reporting arrangements with our stakeholders	Stakeholder engagement Community development Environment Environmental stewardship – topic areas include: Archeological site management, land disturbance, site management, air management, fish & wildlife, water use, hazardous material, spill management, waste management, reclamation and closure, radiation protection Health and safety – focus on employee safety
Assessment	Self-assessment of performance against performance indicators and annual reporting by company	Self-assessment of performance and public reporting in accordance with GRI	Self-assessment and reporting planned in Phase 2 – not implemented as of October, 2009
Verification	External verification of company reports every three years by MAC-approved and certified verification service provider	Third-party verification to be implemented by 2009–10 reporting period. External assurance providers selected by company based on	Verification planned in Phase 2 – not implemented as of October, 2009

Table 8.2 (Continued)

	Mining Association of Canada (MAC)	International Council on Mining and Metals (ICMM)	Prospectors and Developers Association of Canada (PDAC)
	Letter of assurance from a CEO or authorized officer confirming the verified results Annual post-verification review of two or three member companies' performance by the COI Advisory Panel Public reporting and external verification of results will be phased in beginning in 2012	criteria of independence, individual competencies, and organizational competencies.	
Member participation	Voluntary participation – 17 company reports in 2008 (out of 30 member companies)	Participation requisite for ICMM membership.	Voluntary participation. All member companies expected to adhere to e3 principles, but no formal requirements for participation.
Governance	MAC governed by executive board and board of directors (member companies).	ICMM governed by CEO council (CEO of every member company)	PDAC governed by officer board and board of directors (member companies)

Consultation	TSM standards created in consultation with Community of Interest Advisory Panel (aboriginal, trade union, finance, and civil society representatives)	Sustainable Development Framework based on Mining, Minerals and Sustainable Development (MMSD) project, which involved two years of stakeholder consultation.	E3 Plus developed with input from exploration industry representative and in consultation with NGO, academic, government, and aboriginal representatives.
Communication of results	Annual TSM report posted on MAC website	Results of assurance (third-party verification) to be published in company annual reports (assured public corporate report)	No reporting online as of April 2011
Sanctions for poor performance	None	None	None

Sources: MAC (2010); ICMM (2010g); PDAC (2011).

Vale, and Xstrata, only Barrick and Teck listed memberships in all three associations on their websites, and none appeared to put much effort into publicizing their participation in the initiatives – despite the fact that the corporate websites for all six companies feature detailed sections devoted to sustainability and CSR, and all six publish annual sustainability reports.

Despite these observations, it is worth noting that semiotic change is a slow and contested process and CSR is still an emerging order of discourse. Although the effectiveness of the three industry codes analyzed here may be quite limited in practice, all three are still in active development. Research into the evolution and effectiveness of these codes should be ongoing and aimed to seek out both the limitations and the potential of these industry-specific codes. The sheer volume of information (both within the codes themselves and in terms of company reporting), combined with the difficulty in accessing the information required to verify self-reporting results, makes this a difficult and involved research task. However, without accurate information that can compare standards and self-reporting against actual performance, valid analysis of the effectiveness of industry-driven CSR codes and standards over time is impossible.

Comparing discourse communities: CSR and mining opposition

Perhaps nowhere else is the limitation of industry-led CSR discourse more evident than in a comparison between the language used by industry associations and that of oppositional movements engaged in active resistance against mining development. Using the lens of discourse communities, which considers the social construction of knowledge and 'fact' within and between groups of people, Hutchkins et al. (2007) identified six distinct discourse communities surrounding a proposed high-grade underground nickel mining project in Michigan and discussed three of them in detail. They found significant differences in the modes of expressing and understanding the risks and benefits associated with the proposed project between the discourse communities. The company, Kennecott Minerals Corporation, emphasized the value of the project to the local economy and the assurance of safety through 'the application of rigorous science and technology coupled with adherence to regulations' (Hutchkins et al., 2007, p. 23). Save the Wild UP (SWUP), a community group formed from concerned local residents, highlighted the environmental risks associated with the proposed project using more emotional language, framing an opposition between the long-term interests of the community and the economic interests of 'outsiders.' The discourse of the Keweenaw Bay Indian Community centred on the theme of spirituality linked to heritage and way of life based on honoring the environment.

Similar differences can be seen in language and emphasis through a comparison of the three industry associate codes discussed above and a declaration released by a number of Latin American community organizations and NGOs at the third meeting of the Latin American Observatory of Mining Conflicts (OCMAL) in Quito, Ecuador, in July 2009 (OCMAL, 2009). Where the codes of MAC, ICMM, and PDAC focus on key themes of public trust, social and environmental management, social license to operate, firm reputation and performance, alignment with communities of interest, continuous improvement, informed decision-making, excellence and implementation of good practice, the key themes emerging from the OCMAL declaration include transformation of capitalism; justice (including social justice, justice for the earth, and construction of a more just world); solidarity; and resistance to a number of negative trends including the subjugation of peoples, stripping the earth of natural wealth, persecution of indigenous leaders, restriction of democratic liberties, criminalization of dissent, and global warming (see Table 8.3).

As this relatively extreme example shows, key actors in negotiations both over the practical considerations and material conditions of mine development and in the definition of a mining company's social and environmental responsibilities can frequently be seen to be 'speaking

Table 8.3 Comparison of key themes – industry association CSR Codes/Guidelines and OCMAL declaration

MAC, ICMM, PDAC	OCMAL
public trust	transformation of capitalism
social and environmental management	social justice
social license to operate	justice for the earth
reputation	construction of a more just world
performance	solidarity
alignment with communities of interest	subjugation of peoples
continuous improvement	stripping of natural wealth
informed decision-making	persecution of indigenous leaders
excellence	restriction of democratic liberties
implementation of good practice	criminalization processes
	global warming

Sources: OCMAL (2009); MAC (2010); ICMM (2010g); PDAC (2011).

different languages.' Not only are there significant cultural and linguistic differences in an obvious, literal sense between the executives of foreign mining companies who may be involved in determining CSR policy at the firm or industry level and the communities with which they come into contact, but there are also important differences in the way each group of actors understands and conceptualizes the key issues and concerns related to mining. An industry-led CSR policy may intend to align the company's interests with the interests of the community, but what can CSR do if a significant group of community members is concerned more fundamentally with deep social and economic justice? This discursive divide is particularly problematic in political environments already inscribed by the histories of conflict and contradiction related to neoliberal capitalism described above.

Conclusion: On words and swords

I have argued that, as an emerging order of discourse, CSR policy and practice has been shaped by a political economic context in which neoliberalism forms the dominant economic imaginary. Through a process of selection, retention, reinforcement, and inculcation, emerging global norms regarding the business–society relationship have been taken up and institutionalized by industry associations and corporations. However, as research has shown, these efforts remain very limited and are unlikely, in their present form, to address the core conflicts and contradictions that they have ostensibly been developed to ameliorate. While industry actors have been more successful in applying technical solutions to address environmental problems in the mining industry, they have had considerably less success in community development and governance. As Frynas (2009) points out, 'the positive impact of social investments is severely constrained by the companies' own motives for community development,' particularly when projects are 'driven by short-term expediencies rather than the long-term development needs of a community' (p. 12).

The limited capacity – both material and semiotic – for industry actors to effectively address social and governance issues needs to be taken seriously by policy makers and analysts alike. Although semiotic change is inevitable and ongoing, it is also limited by strong forces of tradition, habit, and institutional path-dependency. While we may well be entering an era where social and environmental concerns gain ground on purely economic thinking, we are not there yet, and it may be decades before such semiotic shifts are internalized and institutionalized. In the absence of a devastating economic, political, and social crisis in the mining industry (which is not beyond imagination), the trajectory of CSR discursive development appears to be set toward a distinct policy preference for 'weak sword' types of regulation without disclosure or sanction. Any modification to this trajectory will only be the result of some considerable semiotic struggle.

However, given the deep social pressures that continue to build in the absence of any decisive movement toward improved governance of the global mining industry, 'weak sword' solutions hold little long-term promise for anyone. Here, the onus falls on state regulatory capacity as a primary alternative for 'stronger sword' approaches. Indeed, as Vogel (2010) points out: 'Until the world's developed countries are willing to more closely integrate the norms of civil regulations into their domestic laws and international relations, the global regulatory failures private social regulation was intended to redress will persist' (p. 83).

Note

1. Figures from CEPAL (2006), *Annuario Estadístico de América Latina y el Caribe*.

9
Drivers of Conflict around Large-scale Mining Activity in Latin America: The Case of the Carajás Iron Ore Complex in the Brazilian Amazon

Ana Carolina Alvares da Silva, Silvana Costa, and Marcello M. Veiga

Introduction

Social conflict is a constant in the Brazilian Amazon and can be largely associated with access, use, and ownership of land. In southern Pará – where the Carajás Iron Ore Complex is located – land conflicts stem largely from the removal of aboriginal communities, small-scale farmers, and resource extractors from their lands. Removal and the odd resettlement were a result of 1960s' and 1970s' corporatist development policies, reflected in large-scale projects and rapid expansion of cattle ranching. A major outcome of these policies was the establishment of new transportation corridors – mostly penetration roads and railroads – which allowed migrants to enter previously isolated areas of Pará. Other conflicts are a result of competing interests of artisanal miners (*garimpeiros*), the widespread landless people movement (*MST*, which advocates and takes considerable and controversial action for more equitable distribution of land and resources in Brazil), and legal and illegal logging industry operators. The latter often take advantage of strong markets for Amazon wood, weak institutional capacity, and aboriginal communities' inability to protect their lands from illegal extractive activities.

This chapter draws from research carried out on the Carajás Iron Ore Complex case that identified key drivers of conflict associated with several aspects of company practice, taking into account the political and social environment in which the company Companhia Vale do Rio Doce (CVRD) operates. An extensive literature review was comprised of scholarly articles, as well as government reports, civil society publications, news articles, and interviews with consultants and company representatives. Conflict drivers that enable government dynamics were identified empirically. Analysis revealed on-the-ground challenges in focusing on engagement with all necessary issues and groups due to pre-existing and complex political and social conditions. The conclusion provides lessons that can be applied by companies, local communities, and policy makers for the successful implementation of firm-level Corporate Social Responsibility (CSR) policies on the ground.

As noted by Sagebien and Lindsay (2009), increasing the efficacy of socially responsible practices 'requires a deeper understanding and awareness of the system-wide dynamics that either disable or enable multi-lateral efforts to preserve and conserve the commons – or, in other words, to contribute toward social and environmental value-creation.' With this in mind, this chapter takes a systemic approach to understanding the challenges inherent in managing a mining operation in an ecologically sensitive area where many political and social complexities exist. In the particular case of the Carajás Iron Ore Complex in Brazil, many of most challenging issues stem from lack of governance coupled with a long history of military oppression. This context presents serious challenges to a company in carrying out effective social practices and leaves the company vulnerable to criticism from a variety of national and international groups. This chapter also aims at shedding light on how a large corporation can respond to regulatory, social, and political pressures to contribute to inclusive development in remote areas characterized by a wide diversity of players and conflicts.

Large-scale mining in the Brazilian Amazon

Natural resources have been exploited in the Brazilian Amazon for centuries. The first resource subject to large-scale exploitation and exportation was rubber, which generated a remarkable boom in the early 1900s and culminated in a similarly spectacular bust in the twentieth century. However, gold mining has long been a significant and steady enterprise in the Amazon (Hecht and Cockburn, 1990). The dream of fast and great wealth typified in the legend of *El Dorado* has, since the seventeenth century, fueled mineral exploration in the Amazon. The search for gold spurred exploration expeditions and drove the history of settlements, migration patterns, and conflict dynamics in the Brazilian Amazon (Hecht and Cockburn, 1990).

Only in the twentieth century would other mineral products play a significant role in the mineral exploration and mining in the Amazon. For about two decades, the exploitation of manganese was the only large-scale mining activity in the eastern Brazilian Amazon. This was fundamentally changed with the 1964 military coup. The military government established a policy of occupation based on ideals of sovereignty protection, supported by a clear articulation of private capital interests, financial and fiscal incentives, and significant infrastructure development to support industrial interests (Monteiro, 2003).

In the early 1970s, the large-scale exploitation of alumina, aluminum, and bauxite was a catalyst for significant infrastructure development in the Brazilian Amazon, and provided a model for joint ventures of national and international capital investment in mineral resources. In 1970, a 49 per cent international and 51 per cent national ownership structure was applied to the joint venture between a subsidiary of US Steel and the Brazilian Companhia Vale do Rio Doce (referred to in this chapter as CVRD, but known as Vale since a rebranding campaign in 2008) for the development of the Carajás Mineral Province, one of the world's largest iron ore reserves. Mineral rights were granted to CVRD, as the Brazilian Constitution of the time disallowed foreign exploitation of national mineral resources (Monteiro, 2003).

The Great Carajás Project (GCP)

In investigating and analyzing the social conflicts associated with the Carajás Iron Ore Complex, it is imperative to understand this specific operation from within its larger context of rapid change and marginalization of aboriginal communities, local inhabitants, and local elites resultant from the implementation of the immense agro-mineral Great Carajás Project (GCP). In 1980, the GCP was established as a means to rapidly plan for and implement existing mineral projects, as well as to focus efforts toward industrial activity development in the Brazilian Amazon (Monteiro, 2003). The Carajás Iron Ore Complex is an important part, albeit not the only significant project of the GCP, established by the military regime that governed Brazil from the early 1960s to the mid-1980s to facilitate the planned and integrated development of the Brazilian Amazon.

As with other large-scale projects supported or implemented by the Brazilian military regime of the era, the GCP was not discussed publicly;

decisions were made by few and behind closed doors. The marketing surrounding these projects tended to highlight isolated benefits focused mostly on employment and wealth creation (Instituto Brasileiro de Análises Sociais e Econômicas, 1983). Given its large scale and complete absence of public consultation (Almeida, 1986), the GCP resulted in severe and unrelenting land conflicts in Pará that continue to generate controversy and debate nationally to this day.

The mineral discovery in Carajás redefined the social, economic, and political context in the south and southeast region of Pará. The estimated reserves in the Carajás Mineral Province at the time of discovery were of 18 billion tonnes of iron ore, 1 billion tonnes of copper, 40 million tonnes of bauxite, 60 million tonnes of manganese, and 124 million tonnes of nickel (Chaves, 2004). The Brazilian government became the main driver of the economy and new towns emerged as a result of a massive investment in infrastructure. Road-building efforts were accelerated, opening up access to the region by a multitude of migrants seeking economic opportunities. These incentives contributed to the substantial growth of large cattle ranches and soybean plantations (Ondetti, 2008). In the 1980s, a new gold rush, particularly along the Tapajós River, attracted thousands of artisanal miners – the *garimpeiros* – who in the late 1980s constituted 85 per cent of the gold mining occurring along the Tapajós (Hecth and Cockburn, 1990).

The Carajás Iron Ore Mining Complex

The Carajás Iron Ore Complex, located in the municipality of Parauapebas, southern Pará, is one of the largest mining operations in the world. The Complex is an integrated mail–railroad–port system, built and operated by CVRD. It includes five open pits in over 3.4 hectares. The life of the mine is speculated to last up to four centuries (Chadwick, 2005).

In 1997, control of CVRD transferred from the Brazilian federal government to the private sector when the Consórcio Brasil (Brazil Consortium), led by the National Steel Company (Companhia Siderúrgica Nacional) acquired 41.73 per cent of the federal government's common stock. The decision to privatize CVRD caused much controversy (Velasco, 1999) because it was seen as the crown jewel of Brazilian state enterprises and as an important player in the social and economic development of large rural areas. As a result, the Brazilian government had to appease many nationalist interests, including several sectors of civil society and politicians who feared that privatization would result in a decrease of jobs in remote areas. To address these concerns, the Brazilian government retained a 'golden share' and therefore the right of veto. It also announced that one half of the proceeds of the sale of CVRD would be put into the Economic Reconstruction Fund (FRE). This new fund would help fund infrastructure projects at low interest rates and support regional development. The other half of the sale proceeds was to be used to reduce public sector debt (Kirk, 1997).

Contextual drivers in southern Pará

Contextual drivers are the features of the political, social, and environmental setting over which the company had little control, but which play a significant role in shaping the outcomes and the risk environment for a project (Odell and Silva, 2006). Examples of contextual drivers in southern Pará include cattle ranching, logging, inequality in land distribution and the Brazilian Landless Workers' Movement (*MST, Movimento Sem Terra*), aboriginal groups, and artisanal miners. While cattle ranching and logging have resulted in the serious conflicts in the context of the mining operation, the *MST*, artisanal miners, and aboriginal groups have sustained direct conflicts with the company and in varying degrees have affected CVRD mining and logistic operations in southern Pará.

Cattle ranching and logging

The large-scale conversion of forest to cattle ranching was the single most controversial aspect of fiscal policies in the Amazon during the 1960s and 1970s. Fiscal incentives accelerated pasture development, particularly in southern Pará and northern Mato Grosso until the early 1990s, when the incentives were removed (Smith et al., 1995). Cattle ranching continues to be the preferred occupation of small- to large-scale operators in southern Pará. In the vicinity of Parauapebas, cattle pasture accounted for about 90 per cent of the cleared land after ten years of its settlement.

From a social standpoint, cattle ranching provides minimal employment and has often led to conflicts over land rights (Schmink and Wood, 1992). Pastures have also significantly affected the environment, resulting in large-scale clearing of forests and frequent burns (*queimadas*) that deplete the soil and are often uncontrolled and hazardous for neighboring areas of natural forest.

Conflicts around logging activities in southern Pará have been fueled by the high demand for quality woods in national and international markets, as well as by the inadequacy of Brazilian regulatory and enforcement agencies, high levels of corruption, and unsettled aboriginal land claims. Illegal loggers access vulnerable aboriginal lands to harvest high-value trees, often using corrupt legal authorities and/or the manipulation of aboriginal peoples who often need the money to pay off debts originating from in local commercial establishments (Beltrão and Domingues-Lopes, 2003).

Inequality in land distribution and the MST (The Brazilian Landless Workers Movement)

The Brazilian Landless Workers' Movement (*Movimento Sem Terra – MST* in Portuguese) is considered the largest social movement in Latin America. Brazilian peasants have organized themselves to carry out land reform in a country of dramatic inequalities, historically based on an exceptionally skewed land distribution pattern.[1] *MST* activities have been controversial and often resulted in acts of violence by different parties involved in conflicts. These conflicts directly impacted CVRD operations in the region.

Even though large *MST* settlements have been established in the periphery of Parauapebas and therefore in close proximity to the mining concession, *MST* continues to be a driver of conflict in the Carajás region. In recent years *MST* groups have repeatedly demonstrated against CVRD by attacking company trains transporting employees and the general public. These actions have been part of a declared strategy to demand significant changes in national mining policies (for example, elimination of export incentives and strengthening of the government agency responsible for mineral policies in Brazil). *MST* has also made some very specific demands of CVRD, including new mineral taxes in the states where CVRD operates, increased royalties, and more stringent environmental assessment of and control for the Carajás Iron Ore Complex.

MST has also advocated for the return of the company to state ownership, the immediate resolution of several labor-related legal claims against CVRD, increased salaries for CVRD employees, and the creation of a multi-stakeholder committee (including company representatives, the State of Pará, and civil society) to discuss mining projects and the sustainable use of natural resources in the region.

CVRD has not followed through on any of the *MST* requests. In response to MST attacks on company property and illegal confinement of company personnel, CVRD seems to have avoided direct negotiations and dialog. The company used legal tools to address the crisis situations while the Brazilian media reported that CVRD expected that the state and even federal police would need to act to resolve these conflicts without company involvement (Tereza, 2007; Jardim 2008).

Aboriginal peoples

Before the establishment of the Carajás Iron Ore Complex, around 1000 aboriginals lived in 92 villages in the areas surrounding the mine site (Companhia Vale do Rio Doce, 2006). Currently, these aboriginals live in various villages in proximity to the mine operation, the railway, and port facilities built to support the mining activity (Costa and Scoble, 2006).

Under its environmental conservation project, CVRD committed to give priority to community development efforts in the communities identified most affected by its operations (World Bank, 1995b). According to Rodrigues (2003), such an approach has been criticized by both independent observers and CVRD employees who shared the opinion that the assistance provided by the company to directly impacted communities (including aboriginal groups) affected by its operation during the 1980s and early 1990s had been, almost from the start, unsatisfactory. On several occasions since the mid-1980s, aboriginal groups have succeeded in disrupting company operations by invading the Carajás company town, interrupting railroad traffic, and burning power-line towers located within their territories.

In response to these disruptions, CVRD publicly stated that it would not negotiate with groups 'using illegal acts to impose their requests' and that it would enter negotiations under the direction of *Fundação Nacional do Indio*[2] (FUNAI), which, in the company's view, would be the sole group responsible for negotiations with aboriginal groups. According to CVRD, the company contributes over R$9 million (roughly US$4.5 million) annually to the Xicrin group, which receives the lion's share of CVRD's support to aboriginal groups because of its proximity to the Carajás operations (Potiguara, 2006). However, there is little evidence that such large investments have significantly contributed to improving quality of life and local capacity of the aboriginal groups directly impacted by CVRD operation.

Artisanal miners (*garimpeiros*)

The modern gold rush in Brazil started in 1979, when artisanal miners found gold in Serra Pelada, about 54 km from the Carajás Iron Ore Complex. The area had been a CVRD prospecting claim since 1974. Severe conflicts between mining companies and *garimpeiros* followed, as Serra

Pelada was considered an invaded area. In 1983, the Brazilian Institute of Mining (IBRAM) and CVRD started a ruthless campaign to remove all *garimpeiros* from Serra Pelada and stop the 'illegal' mining activities. In October 1983, Brazilian President João Figueiredo demanded the closure of Serra Pelada. This act created a high level of instability in the area, resulting in demonstrations and road blocks organized by the frustrated miners. In 1984, to bring peace to the region, Figueiredo re-opened Serra Pelada as an artisanal mining reserve, but without any technical assistance to the *garimpeiros*. To increase its popularity with the *garimpeiros*, the federal government created other areas designated for artisanal mining activity, some of these in existing aboriginal reserves (Cleary, 1990).

The presence of CVRD – practically controlling all mineral rights in the area – has been constantly disputed by *garimpeiro* associations. The existence of Serra Pelada was also a paradox for the technocentric military regime, which focused on creating top-down, large-scale and often ineffective projects in the area, such as the notorious Trans-Amazon Highway. The military government's vision of modernity stood out against the millions of landless workers and *garimpeiros* who migrated to the Amazon from the poorest regions of Brazil.

Recently, the Brazilian Ministry of Energy and Mines has worked toward the legalization of sites where the *garimpeiros* are active. Unfortunately, however, little support has been provided to organize the miners and improve their labor and living conditions. The main challenge for *garimpeiros* in the region is the lack of government presence in the mining sites, either to provide technical assistance or to give orientation about the legalization procedures.

Company practice drivers

According to Odell and Silva (2007), company practice drivers are actions that a company took or failed to take which contributed to or mitigated a potential conflict situation, such as environmental practice, engagement, and human resources policy.

In recent years, several international standards and principles with respect to social responsibility and human rights have been designed to guide best practice for internationally financed projects. For instance, the International Finance Corporation Performance Standards and the Equator Principles are rapidly becoming *de facto* international reporting systems for companies who wish to raise project financing from international banks.

Starting in the 1980s, the state-owned CVRD invested heavily in its image as a socially and environmentally responsible company, in part to address public criticism of its environmental record but also to address the increasing levels of global environmental awareness required by important international credit sources, such as the World Bank and the European Union (Rodrigues, 2003).

Following privatization, CVRD started to prioritize socio-environmental policies on which it stood to gain in economic and operational terms. A number of initiatives were implemented, such as the establishment of environmental management systems as a necessary step to obtain an ISO 114001 certification (Rodrigues, 2003). Other initiatives included enhancement of engagement practices with rural and aboriginal communities, advancement of development practices, definition of Human Resources policies, and continuing the preservation of Carajás National Park. As the company expanded its activities internationally, it also intensified focus on socio-environmental practices based on internationally accepted CSR principles and practices, such as the Global Reporting Initiative, a network-based organization which sets out the principles and indicators that organizations can use to measure and report their economic, environmental, and social performance.

Environmental practice

After privatization in 1997, CVRD implemented a formal international standard and reporting mechanism for environmental monitoring. The company also adopted improved technology for the rehabilitation of degraded areas around mining sites and along the Carajás railroad (thus avoiding mudslides that could compromise production and transportation). It improved effluent treatment, with direct benefits to water and soil quality (thus minimizing future recuperation efforts). It also assumed the primary responsibility for the preservation of 1.2 million hectares of tropical forests in and around its mining concession (Da Silva Enriquez, 2007).

Following the start of operations at the top of the Carajás Mountain Range, the Carajás National Park was established in 1988 to 'promote scientific research, sustainable exploitation of natural resources, the study and conservation of biodiversity and bring social benefits to the local communities through the rational exploitation of its resources as well as tourism' (Instituto Brasileiro do Meio Ambiente e Recursos Naturais Renováveis, 1998).

CVRD entered in an agreement with the Government of Brazil to assist the *Instituto Brasileiro do Meio Ambiente e Recursos Naturais*

Renováveis (Brazilian Institute for Renewable Natural Resources – abbreviated as IBAMA), to protect and maintain the park. Since then, CVRD and IBAMA have shared the responsibility for managing over 1 million hectares of forests in the Park (Costa, 2008). Company responsibilities include the provision of resources for the surveillance and protection of conservation areas, funding research, and managing the sustainable exploitation of forest resources. The conservation areas are located within the company's concession (the Carajás National Park) or immediately outside its boundaries (the Tapirape-Aquiri National Forest and the Igarapé Gelado's Area of Environmental Protection), thus securing the effectiveness of buffer zones in preventing invasions by squatters (Rodrigues, 2003).

Da Silva Enriquez (2007) explains that these reserves have acted as 'green belts' that insulate mining areas from costly land struggles with other actors, such as loggers, *garimpeiros*, cattle ranchers, and *MST* peasants. As a result, extensive forested areas were preserved right in the middle of an extensive regional swath of deforestation, nicknamed the *arco do desmatamento* – the deforested belt.

Notwithstanding CVRD efforts to improve its environmental performance in Carajás, it is clear that public perception of good environmental practice goes beyond improved reporting and adoption of ISO Environmental Management Systems. Experience in other mining projects in Latin America demonstrates that local communities and organizations often need to be actively involved in the environmental monitoring process for the company to be perceived as a good environmental steward. Companies can therefore reduce the negative perception of environmental impacts to the extent that they can familiarize people and organizations with the expected impacts of technology and environmental management systems. Such an approach, if adopted by CVRD, would have contributed to moving relations with stakeholders to the next level of trust and understanding and helped consolidate the company's engagement strategy.

Engagement practices

It is apparent that limited public consultation took place during the implementation of the project by the state-owned company, and the small number of public interactions focused on a few groups in the immediate area of influence.

Since its privatization, however, CVRD has advanced its engagement with aboriginal groups, particularly the Xicrim and Gavião groups, located in close proximity to the mining concession area. The company assists the Xicrim with the surveillance and protection of their land, provides health care assistance, and co-finances a project of sustainable forestry management at an estimated cost of US$1.2 million per year (Rodrigues, 2003). The project, conceived by the NGO Instituto Socio-Ambiental in partnership with the Xicrim association Bep-Noi, counts on the institutional and financial support of several entities, including the FUNAI and the Brazilian Ministry of Environment. For the Gavião, CVRD has maintained an assistance fund of approximately US$150,000 per year to be used according to priorities set by the community (Rodrigues, 2003).

Since 1998, CVRD has consolidated its Community Relations Department and created the Vale Foundation. The Foundation has developed various long-term sustainable projects in partnership with the civic authorities, NGOs, civil society entities, and private sector organizations in the fields of education, culture, and local economy. According to Ethos Institute (2008), such an approach is the result of a recent shift in management attitude regarding decision-making, impact management, and the role of the community in the company's business.

However, CVRD could have been more proactive in engaging *garimpeiro*s with programs such as those implemented by the Canadian company Placer Dome[3] in Las Cristinas in Venezuela (Davidson, 1995). In 1995, Placer Dome donated part of its property to artisanal miners and provided technical and social assistance to *garimpeiro*s (Veiga, 1997). This approach could bring long-term benefits to the company that go beyond the simple co-existence with artisanal miners. Without tenure, *garimpeiro*s invariably become nomadic, moving to new areas once the easily extractable gold is depleted. It has also been observed that *garimpeiro*s are no longer working as individual panners, and that garimpo 'entrepreneurs' are creating 'digging squads' and increasingly operating on a larger scale. As a result, *garimpeiro*s continue to silt the rivers, inappropriately use mercury and cyanide and leave thousands of pits unreclaimed year after year.

Development practices

In developing the mine and comprehensive logistics systems to support it, the state-owned CVRD established the communities that support their operations in a context of extremely remote locations, with a lack of infrastructure and tremendous land disputes that made the state of Pará notorious for violent land and access conflicts (Marín, 2003; Monteiro, 2003; Coelho et al., 2002).

In order to accommodate workers in this context, CVRD built a company town in close proximity to the project site. Access to this community has since been controlled and is mostly limited to residents and mine employees. Residence in the community is controlled by the company and seen as a privilege and a benefit offered to higher-ranking employees. According to Costa (2008), CVRD's development approach was paternalistic within the company town boundaries, providing all the facilities and services and imposing a code of conduct. On the other hand, CVRD took a 'hands off' approach to the spontaneous development that occurred in Parauapebas, a small village built in the early 1980s to accommodate contractors, municipal services, and the workforce for the development of the Pará–Maranhão railway project. Parauapebas was built adjacent to the mining concession and included basic infrastructure and services, attracting thousands of migrants to the area. By the mid-1980s, the population had reached 20,000. In 2007, it was estimated that more than 100,000 people lived in this community (Instituto Brasileiro de Geografia e Estatística, 2007).

Rapid and uncontrolled growth in Parauapebas was a consequence of high expectations of employment and business development in the region (Prefeitura Municipal de Parauapebas 2005). Currently, migration to Parauapebas continues, primarily from the north and northeastern states, where high unemployment rates and environmental challenges have created large poverty pockets. The local population is highly transient, and economic, social and environmental standards are quite low, with several neighborhoods established through squatter settlements (Costa, 2008).

CVRD's approach to development has resulted in social segregation created by the dichotomy between company town and squatter settlement. The company town is located in a privileged, protected natural area, with clearly superior basic infrastructure and services. This situation has been a predicament for CVRD, since Parauapebas has developed so chaotically that some employees and prospects have expressed unwillingness to move and raise their families outside of the company town (Costa and Scoble, 2006).

Lessons learned

The analysis of conflicts around the Carajás Iron Ore Complex provides lessons for policy-makers and resource extraction companies operating in Latin America, suggesting new approaches which have the potential for overcoming challenges. This case highlights the risks that are often present when there is transfer of ownership in a situation of social mismanagement by a government that either too vigorously promoted mining development or exercised insufficient control over conflict situations. These lessons highlight the importance of proper identification of a wide range of stakeholders to be involved in the engagement process, adequate timing and duration of the engagement processes, adoption of culturally appropriate mechanisms to conduct public consultation, proactive attention to indigenous rights, use of effective communication about technical and environmental aspects of the project, and identification of local players – including governments – to adopt an integrated approach to sustainable development that recognizes local challenges and meets the specific needs of local communities.

These aspects of engagement practices seem to be particularly effective when they incorporate the intent to build lasting relationships with all sectors of community rather than individual relationships, and when they are accompanied by transparency and accountability. In some cases, such as the CVRD's community development practice, the company has demonstrated that it is possible to overcome unsuccessful engagement experiences and legacy issues through improved community programs.

An objective analysis of this case reveals the company's challenges in focusing on engagement with all necessary issues and groups due to pre-existing conditions on the ground. This case required very careful management of engagement strategies and risk communication as poor governance and legacy issues increased tension with residents of indirectly affected areas and environmental groups. Adequate engagement is very difficult under situations of lack of governance and political change. It is clear that under such situations, the environmental agenda can be a catalyst for social and political frustration.

Contemporary efforts made by CVRD to engage with local communities under the influence of the Carajás Iron Ore Complex lacked a holistic view of sustainable development. The company might have avoided escalation of conflict by engaging as early as possible with the directly impacted communities and groups of interest and by taking a more proactive approach in supporting the aboriginal groups that were resettled because of its operations. The delicate situation exacerbated by previous state ownership may have made it impossible to avoid criticisms. Nevertheless, incorporating a wider range of stakeholders, building trust as well as early consultation, and proactively addressing local people's indigenous rights under the International Labour

Organization might have avoided the escalation of the social conflict. It is important to note, however, that the entire process of environmental and social responsibility of corporate mining in the Brazilian Amazon is still in its infancy. Companies, governments, and communities alike are still going through a learning phase.

Corporate investments in local infrastructure seem to have largely been made to meet operation's needs and growth pressures. The Carajás Iron Ore Complex was largely driven by national export interest (as opposed to local development needs), resulting in strong links with the federal government and the neighboring state of Maranhão, where the Ponta da Madeira port is located. From its inception, the operation failed to build an effective relationship with the government of the state of Pará. The operation's remote location, sparse and diverse local population, and shortage of mining skills of the Pará population in the late 1960s and 1970s underscores Pará's inability to effectively benefit from the development in Carajás, despite the magnitude of mining royalties being directed to all three levels of government.

Brazilian mining royalties are shared by the municipal (65 per cent), state (23 per cent) and federal (12 per cent) governments (da Silva, 1998). As an example of the mining royalties Parauapebas earns every year, in 2000, the municipal government received over US$12 million from iron ore mining alone (Barreto, 2001). Given that mineral production in the Carajás Iron Ore Complex has steadily increased since 2002, it is certain that mining royalty revenues have also increased significantly. In 2006, Parauapebas was the sixth richest municipality and had the seventh highest income per capita in Pará (Instituto Brasileiro de Geografia e Estatística, 2007). From 1991 to 1995, mining activity contributed 23 to 28 per cent of Parauapebas' municipal revenue (da Silva, 1998; Barreto, 2001). Royalty payments by CVRD equipped the city of Parauapebas with infrastructure, education facilities, and other community development projects. Notwithstanding the economic vibrancy of the city, the percentage of people living under the poverty line in Parauapebas is quite high: 43.55 per cent of the total population lives below the poverty line,[4] and the poverty rate of the rural population is almost 50 per cent.

CVRD-funded or -supported community development projects in Parauapebas have been perceived as a company strategy to combat the stigma effects of negative anecdotes about the state-owned mine. Although one cannot blame Parauabepas' inability to adequately benefit from mining royalties, CVRD as well as the municipal and state governments should be responsible for establishing effective public–private partnerships aimed at providing direct benefit to the local communities (Palheta da Silva, 2004).

As a result of these shortcomings, CVRD has more recently taken another approach to their relationship with local communities through the work being carried out by the Vale Foundation. In contrast to the situation in the 1960s, when no environmental or social impact assessments were completed for the project, a full assessment process took place for the development of the Sossego Copper Mine (operating since 2004) in the vicinity of the municipality of Canaã dos Carajás. This is reflected not only in the way the workforce accommodations were planned (integrated within the existing community, instead of a closed company town) but also in the way the company has partnered with the municipality to fund new infrastructure and local sustainable economic development projects with a view to diversify the local economy.

CVRD's new approach is an important trend, particularly because Canaã dos Carajás is expected to serve as the major centre for the development of CVRD's upcoming mines. It is intended to serve as a locus for the housing, education and health care for the mining workforce for decades to come (Costa, 2008). CVRD has also recently partnered with the Federal University of Pará and is partially funding the construction of a new campus in Marabá (120 km from Canaã dos Carajás) to develop mining-related courses that would ideally provide locals with the education and skills necessary to benefit from future employment opportunities in the region.

Partnerships with municipalities and communities impacted by CVRD's operations and the state of Pará would therefore be imperative for sustained and inclusive development and better management of conflicts in the region, such as those involving the artisanal miners and aboriginal groups. Partnerships that can bring about long-term and inclusive development will happen only if CVRD and municipal and state governments are able to work together to improve mutual trust, clarify expectations and responsibilities, define accountability frameworks, and carry out inclusive and legitimate consultations. Given the local context in which most companies operate in Latin America, development strategies also need to focus on assisting communities to develop governance and civic engagement capacity, tripartite initiatives (industry, government, community), and participatory development planning, in addition to the more common foci on economic and infrastructure development.

The above analysis of contextual and company practice drivers indicated that, in many cases, a crucial challenge for companies is to proactively identify contextual issues (governance, presence of indigenous people, environmental and social legacy) and successfully navigate through these issues by implementing a solid and encompassing CSR framework. In recent years, there has been a growing recognition of the need to adopt responsible corporate social practices among most mining companies. In many situations, companies recognize that failure to adopt CSR practices that fully respect the human rights of the communities neighboring their operations presents a significant risk to the capital invested. In addition, many international standards and principles with respect to CSR have emerged, including the International Finance Corporation (IFC) Performance Standards, the Voluntary Principles on Security and Human Rights, the Global Reporting Initiative, and the Equator Principles.

CVRD understands this and has set up the Vale Foundation as an umbrella organization for CSR programs. It has also joined GRI and puts some effort in monitoring and reporting of CSR initiatives. A final noteworthy observation is that CVRD, because it started as a state-owned company, has always been expected to contribute to socioeconomic development of the areas where it operated. This case therefore exemplifies the evolution of CSR standards for mining companies in Brazil. Looking at the historical trajectory of CSR, as described by Utting and Marques (2010), the notion of what constituted socially responsible business was quite different in the early days of CVRD when 'paying taxes, generating employment, employing workers for the long term, upholding labor rights and giving back to the community through philanthropy were the routine expectations' (Utting and Marques, 2010).

Notes

1. According to *MST* (2008), less than 3 per cent of the population owns two-thirds of Brazil's fertile land and 60 per cent of farmland remains unused, while millions of peasants survive by temporary agricultural work.
2. FUNAI is the Brazilian government agency in charge of protecting aboriginal interests and culture.
3. Placer Dome was purchased by Barrick Gold in 2006.
4. The term 'poverty line' is used here as it is defined by IBGE (*Instituto Brasileiro de Geografia e Estatística* [Brazilian Institute for Geography and Statistics]): the proportion of people who live in a family with a monthly income of less than one half of the minimum wage per capita.

10
Community and Government Effects on Mining CSR in Bolivia: The Case of Apex and Empresa Huanuni

Robert Cameron

Introduction

In a context of poverty, where communities are characterized by low capacity and desperation for sources of income and where government social services and regulatory capacity is largely absent or of low quality, what is the possibility that mining operations and accompanying Corporate Social Responsibility (CSR) programs will make meaningful contributions to sustainable development? While CSR is often put forward to help alleviate poverty and close governance gaps, it is also challenged by these very same characteristics and dynamics that are so often present where mineral extraction takes place.

As some observers note, while CSR is largely instituted by the firm to advance its own interests, the motivation for CSR and its impact depend to a great extent upon the enabling and disabling dynamics and characteristics of actors outside of the firm, such as home and host governments, social movements, industry groups, lending institutions, and local communities (Sagebien et al., 2008). Following this line of thought, this chapter explores how the Bolivian government and local communities motivate, assist, or impede the contributions of mining firm CSR to sustainable development.

Two mining operations were selected for this study. At the time of this research the first operation belonged to the US majority-owned transnational Apex Silver. The second operation is Bolivia's only 100 per cent state-owned mine, Empresa Huanuni. Although not the primary focus of this research, Empresa Huanuni's identity as a state mine offers an opportunity to explore influence of state ownership on CSR, and some observations in this regard are made.

Bolivia provides a relevant context in which to study the influence of community and host government on mining CSR and its relation to sustainable development. Bolivia is a country struggling to develop, and mining has long been an important sector of the Bolivian economy. The rural communities of Bolivia's Altiplano – a high, arid plain extending along Bolivia's western half – have a long and conflicted relationship with mining. On one hand, they have feared mining as a danger to their agricultural livelihoods, and on the other hand they have seen it as a source of employment and other benefits that are often lacking in rural Bolivia.

Moreover, recent political events in Bolivia have the potential to change the relationship between mining and communities. The 2005 election of Evo Morales – Bolivia's first indigenous president, who ran on a platform of community and indigenous rights and the nationalization of its resources – raises the possibility that mining companies operating in Bolivia will face a more exacting government and communities with increased leverage. This demanding environment may lead to greater CSR contributions to sustainable development.

Evidence from the cases examined in this chapter suggests that the dynamics and characteristics of both government and local communities have potential either to enable or to disable CSR's contribution to sustainable development, and sometimes do both simultaneously. Government adoption of international conventions – both past and present – on indigenous rights and voiced government support for communities in Bolivia have, at least in the case of Apex, provided communities with leverage and motivated CSR. Unfortunately, community leverage has been limited by poverty, lack of capacity and/or awareness and, in the case of Empresa Huanuni, the identity of the mine as a state mine. Furthermore, community focus on immediate tangible benefits from mining also appears to have disabled CSR's connection with sustainable development. The community push for employment stood out in this respect and, in the case of Empresa Huanuni, it appeared to be an important factor in the government's failure to enforce its environmental regulations. Finally, limited community and government capacity also disabled CSR's connection with sustainable development, either by limiting incentives for CSR in the first place or by limiting the contribution of actual CSR efforts. These shortfalls in government and communities suggest the need for contributions from other actors to improve CSR's connection to sustainable development.

Background

To put host government and community influences on CSR into context, this section provides background on Bolivia's mining history and current policy, and also describes some of the influence and effects of the Morales administration. Following this, the rural communities of Bolivia's Altiplano (where the majority of Bolivian mining takes place) are described. This section ends with a description of the two mining operations investigated in this study and a more focused description of the populations that surround them.

Since colonial times, Bolivia has been a mining country. Although mining has lost some of its dominance to hydrocarbon extraction, agriculture, and manufacturing, it remains one of the most important sectors of the Bolivian economy and continues to play an important political role. In the mid-1980s Bolivia's economy collapsed and much of the mining nationalization that had occurred as part of its 1952 revolution was reversed. As many as 30,000 state miners were left unemployed and joined the ranks of cooperative miners. To this day, cooperative miners make up the vast majority of miners in Bolivia and are an important political force.

During the 1990s, the Bolivian government began to increase efforts to protect the environment and communities from the negative effects of mining. With respect to community and indigenous rights, the most prominent government effort was the adoption of ILO's Convention 169 in the early 1990s.[1] While the adoption of this convention was a positive step for Bolivia's large indigenous population, its effectiveness was limited as the rules for its operationalization were not implemented. In this regard, Jordan (2008) notes that, in cases where ILO 169 conflicts with established law, without adequate mechanisms for implementation, ILO 169 may be disregarded (p. 24).

Similarly, while the increasing attention to environmental issues was an improvement over the past, environmental problems continued during this period. Illegal mining contamination is commonly observed in the numerous operations of small-scale/cooperative miners (Enriquez, 2002; Jordan, 2008). Moreover, illegal contamination is an obvious and serious issue in the case of Bolivia's prominent state mine, Empresa Huanuni.

In 2005, Bolivia's first indigenous president, Evo Morales, was elected. Morales ran on a platform that included nationalization of natural resources and equality for rural and indigenous communities. However, despite this platform the election of Morales has not led to the radical changes that many expected and hoped for. While there are joint ownership requirements for new foreign mines, existing private mines have not been nationalized. And while the Morales government has adopted another international convention on indigenous rights – the UN's 2007 Declaration of Indigenous Rights[2] – as with ILO 169, the rules for its operationalization are not yet in place. Morales did change the distribution of some tax revenue so that 15 per cent of mining royalties collected from a mining operation now go to the municipality where the mining operation is located. However, the overall contributions of mining revenue for local distribution and other policy initiatives will, at least in part, depend on how effectively they are managed. Unfortunately, according to the World Bank and Transparency International, the quality of governance is far from ideal in Bolivia (Transparency International, 2008; World Bank, 2008).

In spite of Bolivia's long history of mineral wealth and extraction, Bolivia is one of Latin America's least developed countries. This is even truer for the departments of Potosi and Oruru, where the majority of mining operations are located. These departments are among the poorest, least developed regions in Bolivia. In 2001, out of Bolivia's nine departments, Potosi had the lowest HDI (human development index) (0.514) and highest incidences of extreme poverty (66.7 per cent) while Oruro had the third worst HDI (0.618) and the third highest levels of extreme poverty (46.3 per cent) (UNDP, 2007).

If Bolivia's long history of mining has not made communities in mining areas rich, it nevertheless has transformed them. After centuries of contamination, these rural and largely indigenous communities simultaneously fear mining for its potential environmental effects and desire it for the employment and various services often missing in Bolivia's rural areas. In some cases, communities are neither purely miners nor purely farmers, but rather people who have both farmed and mined the land for generations as a means of survival.

Case study sites: Apex Silver and Empresa Huanuni

Apex's San Cristóbal mine is located in the remote southwest corner of Bolivia's Altiplano in the department of Potosi and in the municipality of Colcha K. Agriculture is the principal activity of the area; prior to Apex's entry to the area there was only limited experience with mining. Apex's project began its first explorations in 1995 and was operational by 2007. It is an open-pit, capital-intensive, high-tech operation extracting lead, zinc, and silver. It is Bolivia's largest mine, processing approximately 40,000 tons of ore per day and directly employing approximately 1000 workers. In 2008, at the time of this research, San Cristóbal was 65 per cent owned by the US firm Apex Silver and 35 per cent by the Japanese corporation Sumitomo.

The Huanuni mine resides on the same high, arid plain as Apex. It is located in the municipality of Huanuni, on the edge of Huanuni town in the department of Oruro. Huanuni is an underground tin mine and at the time of this work produced approximately 900 tons of tin ore per day and employed 5000 workers, most of whom originated from the surrounding population. There is a long history of mining in the Huanuni area dating back to before the Spanish conquest, but its scale became important in the first quarter of the twentieth century with the increase in international demand for tin.

After Bolivia's revolution in 1952, the Huanuni mine was transferred (as were many other mines) from private hands to the state mining company COMIBOL. With the exception of a brief period of joint ownership, it has remained under the ownership of COMIBOL since this time. In the mid-1980s, with the tin price collapse and Bolivia's economic crisis, COMIBOL contracted as many of its mines were closed or privatized and it shed many of its workers (Enrique, 2002). In terms of numbers, cooperative miners became the dominant mining presence as they gained access to areas no longer occupied by COMIBOL, and their ranks grew with the unemployed COMIBOL workers.

The long history of mining in Huanuni has left behind a legacy of environmental damage. The Huanuni River, which passes just below the Huanuni mine, has long been a dumping ground for mining waste. The river's contamination has had consequences for agriculture in the area and downstream, including the declining quality of pasture and cropland close to the river, and the contamination of some sources of irrigation water (Madrid et al., 2002).

The communities surrounding the mines of Apex and Empresa Huanuni match to a large extent the general description of communities given above. The major difference between the respective populations surrounding the two mines is the relatively greater urban nature and tighter links with mining of the Huanuni population. Notably, many residents of the Huanuni municipality refer to themselves as mining farmers, as many have long earned income from both agriculture and mining (Madrid et al., 2002). A brief comparison of the respective populations is given below in Table 10.1.

Table 10.1 Comparison of communities surrounding the Apex and Empresa Huanuni mines

Characteristic	Apex Silver	Empresa Huanuni
Population	10,000 inhabitants – rural	20,000 inhabitants – 15,000 urban (in Huanuni town)
Familiarity with mining and its economic importance	Little history of mining; principal activity is agriculture	Long history of mining; mining is principal economic activity followed by agriculture
Poverty level	90 per cent at or below the moderate poverty line	57 per cent of urbanites and 86 per cent of rural inhabitants at or below moderate poverty line
Indigenous character of communities	90 per cent Quechua	60 per cent Quechua, 13 per cent Amayra

Source: Adapted from Instituto Nacional de Estadística & UNDP (2005).

Government, community, and mining CSR in Bolivia: Key research findings

The discussion in this section is based on interviews carried out in Bolivia in 2008 by the author. Most interviewees came from one of four backgrounds – community members, Non-governmental Organizations (NGOs), various levels of government, or mining firms. Interview questions were focused on the following key issues: (1) the role of the Bolivian government in providing local communities with leverage;[3] (2) the role of the government in either assisting or impeding CSR efforts; (3) community characteristics that increased or decreased community leverage; (4) elements of sustainable development prioritized by communities; and (5) community characteristics that limit CSR contributions to sustainable development. The elements of sustainable development referred to are quality of life improvements that are available to all community members, the ability to extend benefits (such as education, health, infrastructure, employment, and livelihoods) beyond the life of the mine, and environmental protection. Although the prime focus of this study was community and government influence on CSR programs, the author also sought information, literature, and participant perspectives on the CSR programs of the mining operations of this study. Clearly there are differences in the quality and extent of the respective

CSR efforts, and some of these differences in part may be plausibly attributed to differences in government and community influences.

A comparison of Apex and Empresa Huanuni's CSR programs

Of the two mines in this study, Apex attracted the most extensive and sophisticated CSR efforts. Much of Apex's CSR appears to have had the input of affected communities and some of it has been designed and implemented in cooperation with multilateral agencies and, in some cases, the local government. With the possible exception of local hiring, the CSR programs of Empresa Huanuni appeared to be very limited or absent. Asked about Empresa Huanuni's social and environmental performance, many research participants were more likely to comment on the mine's serious and illegal pollution of the Huanuni River and the company's seeming deafness to the complaints of the affected communities than on any positive contributions the mine had made to the community or local environment. Some state mining representatives noted that although the Huanuni mine itself was old, Empresa Huanuni in its current form as a state mine was relatively new and just beginning to pay more attention to social and environmental issues. A brief comparison of the mines' CSR programs and policies is provided below in Table 10.2.

Government and community influence on CSR – Leverage

Communities may have a range of options open to them to exert pressure in pursuit of their demands. These can range from direct confrontation such as legal action, protest, and disruption, to less direct and less confrontational strategies like negotiation and bad publicity. The ability of the community to effectively use these strategies is dependent upon factors both internal and external to local communities. As already noted, this work focuses on just two of these actors – the government and the community.

Both case studies suggest that adoption of international conventions on indigenous rights and the populist position of Evo Morales have given communities greater leverage. However, the cases showed some important differences. Specifically, different communities or segments of communities seemed to have received more leverage than others from the election of Morales. Further, Empresa Huanuni's identity as a state mine may have reduced the leverage of some communities, due at least in part to the real or perceived position of the state as the most powerful authority.

Table 10.2 CSR comparison of Apex Silver and Empresa Huanuni mines

Characteristic	Apex Silver	Empresa Huanuni
Employment	Employment of locals with mine	Employment of locals with mine
Development of human capacity and alternatives to mining	Community Foundation: Foundation is community-controlled, company-funded; started with US$2 million in seed money from Apex. Foundation is to help develop alternatives to mining Training and initiatives for mining employment and alternatives to mining such as agriculture and tourism	As an enticement to ceding community land for a tailing pond, Empresa Haununi offered one community opportunity to develop a small business to employ community members; community rejected this offer
Infrastructure assistance	Infrastructure development: new town, roads, infrastructure	Some limited assistance in the form of technical help or loans of equipment. Possible future assistance for Huanuni town with one of either drinking water, sewage, garbage disposal
Participation	Policy of community consultation and joint planning where Apex provides expertise and finance	Admission by some company representatives that community complaints often ignored at least until the point where conflict may erupt; some indication that company is now working on improving its consultation practices
Environmental aspects	Obey and sometimes surpass Bolivian environmental law. However, not all of Apex's environmental practices would be permitted in mining districts of developed countries; this includes the use of a lagoon as a tailings pond and the open-air storage of silver/zinc/lead ore	Violates Bolivian environmental law through dumping of tailings directly into the Huanuni River

In the case of Apex it appeared that the government had made community 'consent' necessary. The requirement of this consent was not so much a legal requirement codified in regulations but is instead informal and appears to come from government reluctance to use force against community protest. Several Apex representatives maintained that if the communities wanted, they could shut Apex's mining operation down with mobilizations. One community member suggested that communities had been able to intimidate Apex executives with the use and threats of protest. For example, in 1997, 2000, and 2009 the community held marches to protest labor issues such as wages, poor treatment, and formal qualification requirements. On one of these occasions (1997) they managed to shut down operations (Contreras and Madrid, 2006; Minería de Bolivia, 2009). These protests succeeded, in at least some of the cases, in convincing the mine to negotiate community demands (Contreras and Madrid, 2006).

Given that much of Apex's relationship with its surrounding communities predated 2005, including successful community mobilizations in demand of increased benefits from the mine, not all government empowerment of communities can be attributed to the Morales administration. However, there were signs that Morales' arrival in power had contributed to the leverage of those communities near the Apex mine. Several community members suggested that the Morales government would protect communities from bullying by mining companies. One of these community members even went on to say that, with the recent arrival of Evo Morales to power, the community could ask for the mine to be nationalized. Finally, Bolivia's Director of Mining and Metallurgy noted that with Evo Morales in power, rural communities felt listened to and, unlike the past, there would be no use of force against communities on the part of the government. These changes in community leverage (perceived or real) were not the result of changes in formal laws of Bolivia, but instead seemed to result from Evo Morales' stance against exploitation.

On the more formal or legislative side, the government provided leverage to communities through its adoption of ILO 169 and the UN's 2007 Convention on Indigenous Rights. While the rules for the implementation of these rights had not yet been operationalized at the time of this study, there were indications that the rights nevertheless were called upon by communities to assert themselves. Speaking about the Bolivian mining scene in general, an ex-vice minister of mining noted that, despite the lack of exact rules for their use, the ILO Convention 169 and the new Declaration of Indigenous Rights were being invoked by community members (even if incorrectly) occupying mining operations on their land (Centre for the Study of Labour and Agriculture Development [CEDLA], 2008, p. 5).

There is also evidence that the government had provided leverage to the population surrounding Empresa Huanuni. However, there appeared to be large variations in this empowerment across the Huanuni population. On the one hand, that thousands of cooperative members (many of whom come from Huanuni town or other communities in the Huanuni municipality) had been able to access employment with Empresa Huanuni with violent protests and without the government use of military forces is a standout example.[4] On the other hand, Empresa Huanuni was illegally and severely contaminating the Huanuni River and did so despite the damage to downstream communities and their resulting complaints. Two members of one such community observed that while Evo Morales' arrival in government meant rural peasants might be better represented than before, this had not changed their relationship with Empresa Huanuni since the president favored Empresa Huanuni and the mining sector.

While the failure of the government to shut its own mine despite its illegal contamination may be attributable to government financial self-interest, it is also reasonable to assume that the social and political power of mine workers was a major factor. Miners are numerous, are known for their vocal and militant style, and they supported Morales' 2005 campaign. In fact, the most cited reason for not shutting down Empresa Huanuni was fear of the likely social unrest that 5000 unemployed Empresa Huanuni miners would create.

Perhaps surprisingly, given the populist stance of the Morales administration, the mere identity of Empresa Huanuni as a state mine may reduce the leverage of communities. In reference to the Huanuni community's silence on Empresa Huanuni's contamination, one community interviewee commented thus: 'It's like fighting the government. It's possible to fight a private firm, but if the population wants to mobilize, it must mobilize against the government, it must mobilize against the state.' By way of explanation, Hinajosa (n.d.), in his observations of Bolivian mining, notes that communities have found it difficult to complain against state mining because it has been a large employer, a big polluter, and a regulatory agent. Furthermore, Empresa Huanuni's identity as a state mine means that communities cannot use the threat of nationalization as a lever.

As discussed above, while communities were often able to exert their demands given government support in the form of its political position and adoption of rights for indigenous groups, the cases of Apex and Empresa Huanuni suggest that the leverage by communities was limited by internal characteristics of the community – poverty, lack of negotiating experience, and limited awareness of rights.

One community representative commented that his community had little choice but to say yes to the Apex mine because the community lacked the benefits that the mine might bring, such as employment, infrastructure, and health services. Another community member noted that while communities had been able to make demands of Apex, they had 'charged' Apex a relatively low price in return for access to their lands and water resources. This reflected the fact that the government had left an inexperienced community to negotiate with Apex on its own. Similarly, when asked why communities had not been more demanding of Empresa Huanuni and the departmental and municipal government for the use of its mining revenue, one government representative blamed the community ignorance of their rights.

Government and community influence on CSR – Government and community capacity and community demands

While there were some limited examples of joint planning and tax breaks for CSR, outside of the legal and tacit support for communities it appears that low government capacity impedes CSR rather than encouraging or assisting it.

Government regulatory efforts and expression of confidence in Apex's environmental compliance and performance did not seem to prevent environmental complaint or quell environmental fears. While the government expressed its confidence in the environmental performance of Apex, some community members and activists nevertheless expressed environmental fears and doubts about the government's impartiality or ability to regulate and report on Apex's environmental performance. In a context where there is distrust or doubt regarding the regulatory capacity of government, the benefit and perhaps motivation for compliance with environmental norms is reduced. Where the credibility of government regulation is questioned, the legitimacy derived from complying with such regulation is decreased. Perhaps for this reason, Apex representatives expressed the desirability and the difficulty of finding a third party capable of reporting on the environmental performance of their mine and gaining the trust of community members.

Similarly, the questionable capacity of local government to manage and distribute mining revenue negatively impacts the mining firm reputation and the local development that should result from companies paying their taxes. Both Apex and Empresa Huanuni representatives noted the considerable contributions of their operations to municipal coffers under the Bolivian royalty regime. However, research participants from both government and the mines expressed concerns about the local government's ability to manage this money. Apex was hoping to remedy this by building the capacity of the Colcha K municipal government to manage its revenue.

The connection between CSR and sustainable development was also hindered by the low capacity of communities to manage benefits provided by mining. The best example of this was the Apex mine, which arguably provided the most sophisticated and extensive CSR efforts. One Apex representative observed that the community was 'crazy for mining and were not attending to those activities that were sustainable beyond the life of the mine such as tourism and agriculture.' Similarly, another Apex member noted that some of the new infrastructure brought to the region by Apex as part of its CSR suffered from neglect as the community was 'too occupied with mining to care for it.' According to another Apex representative, the community's low capacity to participate in the planning and management of alternatives to mining, to manage other benefits such as infrastructure, or to take advantage of some of the upstream or downstream opportunities presented by mining made it difficult for communities to appropriate their own development and raised the possibility of dependency.

The issue of community capacity to care for CSR benefits was mentioned to a much lower extent with respect to Empresa Huanuni. This is likely due to the very limited nature of Empresa Huanuni's CSR, which provided few benefits outside mining and provided few opportunities for community participation. No matter what the level of community leverage and community and government capacity, however, the connection between CSR and sustainable development may still be limited if communities are not interested in elements of sustainable development.

In the case of Apex there were notable pushes by communities for CSR measures linked to sustainable development. Community members supported or asked for CSR investment that might last beyond the life of the mine, such as education and alternative forms of employment. For example, a representative from one of the communities near the Apex operation noted how the community had influenced Apex with respect to sustainable development: 'Once it [the mine] stops we have to be well re-established, we will have to sustain ourselves... There have to be sources of employment and thus we have to create more jobs in this region. I believe this has been the vision of the communities and therefore the firm [Apex Silver].'

However, despite the efforts described above, communities were very focused on immediate benefits that were unlikely to contribute to long-term sustainable development. The focus on mining employment is a prime example. In both cases, mining employment was seen as the most important and sought-after benefit of mining. This push for mining employment was especially strong in the case of Empresa Huanuni. In fact, the failure of the government to regulate Empresa Huanuni's illegal pollution may be in part attributable to the push for employment. Asked why Empresa Huanuni had remained open despite its environmental violations, a number of interviewees maintained that the government could not shut down Empresa Huanuni because the resulting 5000 unemployed miners would generate dangerous social unrest. While Empresa Huanuni is a large state mine, what is happening in Huanuni confirms an observation Jordan (2008) makes about low-tech traditional mining in Bolivia: 'Its [traditional mining's] nature as an employer of a large number of laborers in an economy where there is open unemployment and high levels of underemployment, does not facilitate any action by the state to enforce environmental regulation' (p. 118).

With respect to the environment, the communities at both mines had a somewhat ambiguous push. On the one hand, community members indicated the environment was important to them and expressed concerns about the environmental effects of mining. On the other hand, there were indications that the environment was used as a bargaining chip for economic benefits. This phenomenon was present at both mines, but the most glaring example occurred with Empresa Huanuni. In this case, members of a nearby community opposed the construction of a tailings pond on their land on the basis of environmental concerns, but they were willing to cede if they could get employment in the mine. While it might seem unethical or unwise for communities to use the environment in such a manner, the underlying explanation for this seemingly contradictory position of communities is not that their environmental complaints or concerns are disingenuous, but rather that immediate economic needs take priority over other concerns.

With respect to their demands, the communities of Apex and Empresa Huanuni had much in common. In a sense the differences between the two communities are to a large extent difference in degree rather than type. What stands out is that much of the population surrounding Empresa Huanuni seemed less interested in elements of sustainable development than the population surrounding Apex. The underlying causes of this relative disinterest are uncertain, but the relatively greater importance of agriculture in the area of Apex and the longer history of mining and relatively large presence of cooperative miners in the area of Empresa Huanuni may be factors. Lack of connection to place is also one possible explanation. One local NGO member noted that many of Huanuni's miners had little concern for or notion of sustainable development and little incentive to invest in the Huanuni area because they see themselves leaving after they have earned enough money to migrate to another area of Bolivia.

Conclusion

The results of this study suggest that local communities and the host (Bolivian) government respectively possess a number of characteristics and dynamics that can either enable or disable the contribution of mining CSR to sustainable development. This last section briefly summarizes these characteristics and dynamics and then provides a number of lessons and recommendations.

Factors that enabled the alignment of CSR with sustainable development included the increases in community leverage that resulted from government adoption of conventions on indigenous rights, refusal to use military force, the threat of nationalization, and the expressed support of the government for the indigenous and marginalized. There also appeared to be, at least in the case of Apex, a push for CSR efforts connected with sustainable development from the surrounding population. While this push was much smaller than the push for other more immediate benefits like employment, it nevertheless appeared to be significant.[5]

Factors disabling CSR's connection to sustainable development included high levels of community poverty and lack of awareness regarding their rights. These characteristics lowered community leverage and contributed to their focus on benefits whose connection to sustainable development was tenuous. This included the use of the environment as a bargaining chip, but the most prevalent example was the community focus on mining employment. In the case of Empresa Huanuni, this focus was strong enough to impede even government enforcement of its environmental regulation.

Low government and community capacity also appeared to disable CSR's connection to sustainable development. Concerns about the government's ability to act as a credible judge and regulator of the environmental performance of mining operations and to manage mining revenue may reduce mining firm incentives to increase contributions in these areas. For their part, communities were limited in their ability to participate in and manage the benefits that were available through CSR initiatives. This raised the possibility of dependence and the danger that when the mineral deposit is depleted, development will disappear. While one cause of Empresa Huanuni's relatively poor CSR performance was the potential unrest of many unemployed miners, its identity as a state mine did not seem to push the mine toward sustainable development. In fact, it may have lowered community leverage.

How might government and communities better enable the contributions of mining CSR to sustainable development? One answer is capacity building. Communities have the potential to tighten the connection of CSR to sustainable development if they can build leverage through negotiating skills and awareness of their rights. With increased awareness of sustainable development, communities might modify their demands away from immediate tangible benefits of mining to measures that can better convert the natural capital of mineral deposits to other forms of capital that can outlive a mining operation.

Changing the focus of communities away from tangible benefits may be more successful if combined with alternatives for earning income. In this study it was apparent that community desperation for sources of employment caused all communities in this study to focus on obtaining mining employment. In the case of Huanuni, this focus was severe enough to impede government enforcement of environmental regulations. Capacity building might also be extended to the government to manage and distribute mining revenue more effectively. As was observed here, there were concerns about government's ability to do this.

In a related observation, Bebbington et al. (2009) note that the institutions for governing inequality and inequity in Bolivia remain weak. Unfortunately, similar to what was observed in this study they note that Bolivian social movements have focused on nationalization and getting a bigger share of the mining rent as opposed to less immediate and less tangible benefits such as transformation of institutions. In this study, communities showed relatively less interest in improving governance of mining rents than in satisfying their immediate needs. Even if improvement in management of government revenue is what is needed, it does not appear that communities are very interested.

If both the host government and communities act in ways that disable CSR's connection to sustainable development, who might fill the gap? One possibility is that mining companies alone might increase their contribution to sustainable development by augmenting the immediate tangible benefits they provide with capacity building. This capacity building might not only be extended to communities but also to the host state as well, in what Kapelus refers to as 'leading from behind' (as cited in Hamann, 2003) – a process in which a mining company builds the capacity and/provides resources to the state while at the same time allowing the state to take to take credit for any resulting CSR benefits (Hamann, 2003).

However, there are areas where mining companies need to tread carefully or risk undermining the legitimacy of the government or processes where CSR is involved. In the case of Apex, mining representatives indicated their intent to build the local government's capacity to monitor the environment. Asked about these efforts, one community representative noted this would likely undermine the credibility of government environmental assessments.

This last consideration and the dynamics and characteristics of the Bolivian government and local communities that disable CSR's connection to sustainable development suggest the need for assistance from other contextual actors. As part of their loan requirements, financial institutions might place CSR requirements on mining firms to increase community leverage, national and international civil society organizations might use their skills and connections to increase community leverage and build community capacity for managing benefits, and home governments might also both place requirements or provide assistance, so that their companies are motivated to better contribute to sustainable development. Which type of organization will work best will depend on what is needed and the given details of the context.[6] Whatever the details of which contextual actors should participate and what they should do to increase the efficacy of CSR efforts, this study suggests that the connection of CSR to sustainable development is disabled by some of the same gaps it has been put forward to fill.

Notes

1. ILO 169 deals specifically with the rights of indigenous and tribal peoples. It confers, for instance, rights on indigenous populations to decide their own development priorities, and requires government to consult with and provide compensation to indigenous communities for damages resulting from development activities.
2. The UN's Declaration on the Rights of Indigenous Peoples is a declaration and, unlike conventions, is not normally legally binding. However, Bolivia adopted it as part of its national law.
3. Leverage refers to the ability of an individual or group to determine or influence events or decisions toward a desired outcome.

4. In this instance long, simmering tensions between cooperative miners and state-employed miners reached a flashpoint with high mineral prices in 2006 and cooperative miners attempting to take over the mine. In the resulting battle with state miners 16 people died. Without military intervention, the Morales government resolved the conflict by offering cooperative miners jobs with Empresa Huanuni.
5. The ultimate reasons behind this push are unclear, although there seemed to be awareness in the community that mines had a limited lifespan and that mining had environmental costs. This awareness is perhaps derived from mining's long and prominent role in Bolivian history and society.
6. For instance, while financial institutions might be able to influence private mining firms needing access to finance, it might be more difficult to influence state firms that have greater flexibility in their financial sources or where loan requirements might appear to be a violation of sovereignty.

11
Corporate Social Responsibility in the Extractive Industries: The Role of Finance

Allen Goss

Introduction

The idea that firms owe a duty to the wider communities in which they operate is now widely accepted among practitioners and academics. Nowhere is that duty more pronounced than in the mining sector. While few doubt the societal benefits of the extractive industry, much has been written about the negative impacts of mining operations to the communities in which they operate. This chapter examines the role of finance in the extractive industries. Drawing on theoretical and empirical research, I explore the potential for both public and private debt and equity to influence corporate behavior, asking how sensitive investors are to the activities of the firms in which they invest; whether investor boycotts influence corporate behavior; and whether banks and bondholders care about the environmental, social, and governance records of the firms to which they lend. In exploring these issues, this chapter sheds light on the potential for finance to shape the conduct of firms in the mining and extractive industries.

The mining process

Regardless of the resource being considered, the mining process can be roughly divided into the exploration, mine planning, and production phases – each having different capital requirements. The first stage is the hunt for mineral deposits and an evaluation of the economic potential of any ore bodies that are identified. Contingent on a positive result in the first stage, a mine is planned and constructed. The final stage is the actual exploitation of the resource followed by mine closure and site remediation. Each stage of the process is capital intensive, but the choice of financing depends on the activity being undertaken.

In the exploration phase, geologists review maps and surveys, attempting to pinpoint areas of likely mineralization. These initial efforts, called 'desk study,' are followed by exploratory drilling or trenching on the most promising targets. If the results of exploratory drilling are positive, more extensive drilling programs are undertaken. The speculative nature of these 'grassroots' activities cannot be overstated. While exploration can last two to four years, only one in 10,000 exploration projects are ever commercially exploited.[1] The consequence of positive results is the preparation of more in-depth studies of the geological and economic potential of the deposit. This necessitates more drilling and can take up to five years.

Contingent on a successful pre-feasibility report, a full feasibility report is prepared in the second phase. This study will typically include detailed economic evaluations for the proposed mine, as well as plans for the mine design. Mine planning and construction can take up to an additional five years. The final stage is the operation of the mine. The length of this stage depends on the amount of commercially exploitable material in the mine and the type of mining employed. After the resource has been fully exploited, the project concludes with mine closure and remediation. This may involve reclamation of the land after open-pit mining, as well as management of tailings ponds and groundwater quality for both surface and underground mines.

Just as the process of mining can be divided by activity, so can the companies that make up the industry. Junior firms are the smallest, followed by the mid-tiers (or intermediates) and the majors. Generally speaking, junior mining firms are engaged in exploration and mine planning. In rare cases, a junior may be able to fund its exploration activities from the production revenues of mines that have moved to production, but more often exploration is financed through successive rounds of equity financing. Cash flows are uniformly negative as the junior undertakes its exploration program and debt financing is not possible given the risks involved.

Intermediates are larger firms, with revenue of at least US$50 million and not more than US$500 million. These firms will be involved in all stages of the mining process, from exploration through to mine operation, although the focus is on planning and production. Cash flows will be positive but can be very volatile, as mid-tier producers have lower levels of geographic and resource diversification than more senior miners. The largest firms are the majors, with revenues in excess of

US$500 million annually. These firms may be active in the mining of several types of mineral at multiple sites and will experience the least cash flow volatility. Debt usage is common among the majors, although overall leverage is less than found in other industries.

The choice of financing

At the most fundamental level, all firms must choose to raise capital through some mix of debt and equity financing. Debt financing proffers significant tax advantages to the firm because interest payments on debt are tax deductible. Finance theory suggests that in the absence of bankruptcy risk, the optimal capital structure is to finance the firm exclusively with debt (Modigliani and Miller, 1958). As a practical matter, however, the tax benefits of leverage come at the cost of increased bankruptcy risk for the firm. It follows that the optimal capital structure for the firm will use debt until the marginal tax benefit of additional debt is offset by the increased risk of bankruptcy. Miners that use too much debt risk being unable to repay obligations when they come due, so the optimal amount of debt decreases in situations where cash flow volatility and sovereign risk are higher and increases with diversification, either geographically or operationally. Generally, the extractives sector has low amounts of leverage relative to other industries, reflecting site-specific sovereign risk and uncertainty over future commodity prices. It is no surprise that the diversified majors are the firms most able to access debt markets since they have the capacity to benefit most from geographic and resource diversification.

In addition, the extractives sector makes widespread use of project financing, where a mine can be viewed as an economically separable project.[2] Debt and equity used to finance the project is paid off exclusively with the cash flows from the project. Creditors do not have recourse to any assets of the sponsoring firm except those of the project being financed. Neither the assets nor the liabilities of the project are recorded on the sponsor company's financial statements, and the project is said to be financed off the balance sheet. I discuss the advantages of this financing alternative later in the chapter.

Financing at the exploration phase

Junior exploration firms undertake the majority of grassroots exploration. Exploration by the mid-tier and major firms is predominantly done around existing mine sites, either to confirm or extend the size of previously identified ore bodies. The risk inherent in mineral exploration effectively precludes the use of debt financing, and juniors are financed almost exclusively through successive rounds of equity financing. Without access to debt markets, continued exploration depends critically on producing positive exploration results in order to secure periodic injections of new equity.

The junior may undertake a public issue, listing on an exchange in order to raise capital on the public markets. Before a public issue can proceed, a prospectus is required. The prospectus is a comprehensive disclosure document intended to provide 'full, true and plain disclosure of all material facts relating to the securities issued'[3] including information about the issuer, the securities on offer, and audited financial statements. An alternative to costly listing requirements is to issue equity directly to sophisticated investors. Known as private placements, these capital-raising activities circumvent the prospectus requirements that accompany public issues. Because the purchasers of the equity are sophisticated investors, it is assumed that they can assess the value of the equity on offer without the aid of a prospectus. The purchasers of the equity are often hedge funds that create portfolios of speculative exploration companies. The private placements often include both shares and warrants, giving the purchaser the option to purchase more shares if the exploration is successful.

In Canada, another financing alternative is the issuance of flow-through shares. First introduced in 1984, flow-through shares allow the tax savings from exploration expenditures to be renounced by the exploration firm and passed onto the investor.[4] Juniors typically have no earnings and pay no tax, so the tax deduction is passed directly to the investor to offset personal taxes payable. The funds raised by the juniors through flow-through shares must be spent in Canada. Coupled with favorable tax rates, flow-through shares have helped make Canada one of the world's leading centres for mining finance.

Financing at the planning phase

As the exploration process moves toward feasibility and mine planning, capital requirements increase and cash-strapped juniors will typically be bought out by large-scale producers or enter into joint ventures with host governments and/or mid-tier or major producers. Major producers provide the capital to continue exploration in exchange for an equity stake in the project. As the likelihood of success increases, the support of communities and host governments takes on greater importance.

The role of host governments deserves comment here. Developing countries see the development of natural resources as an important driver of economic growth and foreign currency. By exploiting mineral wealth to meet foreign demand, they can transfer wealth to other sectors of the economy. In countries with export-driven mineral sectors, the goal of government is to maintain the stream of revenues through rents – income which is needed to subsidize other areas of the economy. Host governments can either take majority or minority stakes in joint ventures in order to bring projects to fruition. However, when local citizens oppose a development, the financial imperatives under which host country governments operate can put them in direct conflict with the citizens they represent. Furthermore, the need to maintain cash flows may prompt host governments to continue production of marginal mines (Bomsell et al., 1990). The capital intensity of the extractives sector generates high barriers to exit, so governments may favor price-cutting instead of abandoning an uneconomic project, exacerbating the price swings that are common in the commodities sector.

Another problem is the potential for corruption. Firger (2010) describes the case of Equatorial Guinea, where the majority of the citizens lived in extreme poverty while the rulers of the country profited from oil and gas revenues. This 'resource curse' led a group of NGOs (Non-governmental Organizations) to form the Publish What You Pay initiative in 2002 to push for greater transparency surrounding payments from extractive firms to host countries. The initiative aims to track payments made by miners to ensure that they are used for the welfare of citizens in those countries. Their advocacy has met with some success, notably in the US, where the Energy Security Through Transparency Act (ESTTA) was passed in July 2010. The act mandates the disclosure of all payments made by US listed extractives to host countries on a country-by-country basis. Since many of the largest firms in the extractives industry are listed on US exchanges, the law will help stakeholders in resource-rich countries battle corruption.

Financing the producer

The financing needs for producers are much larger than for exploration, but the increasing likelihood of success is often sufficient to permit debt financing. Options at this stage are syndicated loans, project finance, government agencies, or bond issues. Financing can also be arranged through forward sales of the commodity to be mined.[5]

Lenders typically include commercial banks, development banks, and governmental agencies. They recognize the risk to projects if stakeholder issues are not addressed, since delays and disruption to production jeopardize their ability to be repaid. In addition to credit risk, lenders face liability risks if they are forced to take equity positions following default. Finally, lenders face reputational risks if stakeholder issues are not addressed by mine operators. In projects financed by the International Finance Corporation (IFC), the private sector arm of the World Bank, funding is contingent on compliance with a set of Performance Standards designed to ensure environmental and social sustainability. The IFC is the largest multilateral source of debt and equity financing for projects in developing countries. Given the importance of the IFC in emerging markets, the Performance Standards have been instrumental in moving toward more sustainable mine development (Morgera, 2007).

The use of project financing has emerged as an alternative to traditional debt and equity investments by multinationals. Project finance treats each mine as an independent company, financed with equity from the sponsors and non-recourse debt serviced solely by the cash flows from the mine, without recourse to the assets of the sponsors in the event of non-payment. This method of financing is attractive because it increases the leverage of the investment, amplifying returns for a given equity position. It also limits sponsor losses to the equity invested in the project. The largest providers of project finance, including ABN AMRO, Barclay's Citigroup, and West LB, have developed a set of environmental and social guidelines similar to the IFC Performance Standards. Dubbed the Equator Principles, the voluntary code insures that stakeholder issues are addressed before the project proceeds. As of January 2011, 68 financial institutions representing over 71 per cent of emerging market project financing have adopted the Equator Principles.

The risk of host country expropriation heightens the risk to debt and equity investments. Geology does not recognize national boundaries and firms often work in areas of the world where legal regimes are weak and communities are underdeveloped. Political risk insurance is a critical component of mine financing and the presence of a government agency as a lender in a project may inoculate the firm against arbitrary seizure of assets or other actions by foreign governments. The critical importance of political risk insurance has provided further impetus for the integration of environmental, governance, and social factors in private sector projects, since adherence to the IFC Performance Standards is a condition for the provision of political risk insurance.

In Canada, Export Development Canada (EDC) is active in mining project finance, and also insures firms against political risk. The importance of such governmental financial support was illustrated recently when Canadian lawmakers and opposition politicians suggested EDC funding should be tied to firm-level CSR. Fierce industry lobbying led to the defeat of the proposed Bill C-300 in 2010, with miners arguing that the risk of losing EDC funding would be sufficient to consider relocating to another jurisdiction.

Can finance influence firm behavior?

One view of sustainability is that it represents a cost to the miner. Management must balance between competing objectives of environmental/social and economic performance (Walley and Whitehead, 1994). Sustainability is a cost to be controlled. A related argument sees environmental and social investments as agency costs where managers gain personal benefits from environmental or social initiatives and the costs are borne by shareholders (Barnea and Rubin, 2005). Here again, investments in sustainability are a cost except that it falls to the directors to rein in these costs.

The alternative view suggests that investments in sustainability can be positive net present value endeavors (Humphreys, 2001). The costs of improving environmental or community performance are offset by increased competitive advantage (Porter and Kramer, 2006). A key difference between the competing views is that management will focus on minimum compliance if CSR is a cost, whereas the focus will be on proactive environmental and social initiatives if CSR is a source of competitive advantage. The tension between civil society groups and mine operators has led many to wonder whether the providers of capital can influence the behavior of firms. Understanding that the risk faced by mining firms influences the choice of financing, I now examine each type of funding in turn.

Equity holders

There is some evidence that socially conscious investors will shun certain classes of stocks (collectively known as 'sin' stocks) because they want their financial holdings to mirror their beliefs. Hong and Kasperczyk (2009) find that alcohol, tobacco, and gaming stocks generate consistently higher risk adjusted returns than the market because share prices are depressed. Fabozzi et al. (2008) confirm these results using a wider definition of sin stocks. Critically, however, there is no evidence that share boycotts will induce changes in the behavior of firms. While it is true that 'sin' stocks generate abnormally high risk-adjusted returns because of depressed share prices, those industries continue to operate. There is no evidence that social investors cause more exits in the alcohol, tobacco, or gaming industries. Instead, 'clienteles' form with relatively homogenous investor groups owning stocks that reflect their beliefs.

In the context of the extractive industry, if mining firms engage in socially unacceptable practices, it is reasonable to expect that socially responsible investors will choose not to hold shares of those firms in their portfolios. In a high profile divestiture in 2009, the Norwegian government ordered its pension fund to divest its US$185 million position in Barrick Gold, citing environmental concerns. However, there was no significant impact on the share price, consistent with the notion that boycotts have little impact on the cost of equity capital. Barrick simply noted that, 'the [Norwegian government] have a right to decide which investments to make and which ones to hold, or sell – that is their business.'

Angel and Rivoli (1997) offer theoretical support for the ineffectiveness of boycotts. Using Merton's (1987) model of capital market equilibrium in the presence of incomplete information, they find that fully 65 per cent of all investors have to boycott a company's shares before it begins to impact materially on the cost of capital. In a related paper, Heinkel et al. (2001) model the costs of investor boycotts on firms that pollute. In their model, 20 per cent of investors need to participate in a boycott before it induces any reform among polluting firms. Both the theoretical and empirical evidence suggests that boycotts are unlikely to be effective in modifying firm behavior. More likely, shareholder clienteles will form, with those opposed to mining choosing not to hold the shares and those favorably disposed to mining holding the shares.[6] While the clienteles will align investors' financial holdings with their beliefs, they will do little to modify corporate behavior.

A more promising avenue through which equity investors might influence behavior is advocacy and shareholder engagement (Irene Sosa discusses two such cases of shareholder advocacy in Chapter 12 in this volume). The presence of large shareholders with the motivation and expertise to monitor management can influence corporate behavior. In a recent study of hedge fund activism, Brav et al. (2008) report that meeting company executives was the most popular technique to influence corporate behavior (48 per cent). However, such 'soft' advocacy is by its nature neither observable nor testable. Anecdotal evidence suggests that holders of substantial equity blocks might be successful in influencing miners to embrace sustainability, but these interactions cannot be observed, making empirical testing difficult.

More visible forms of advocacy, including shareholder resolutions, lawsuits, and takeovers are easier to test. Perhaps the most high-profile example of 'hard' advocacy is the annual focus list published by the California public employees retirement system (CalPERS) between 1992 and 2009. The fund identified 6–12 firms with poor governance and poor financial performance in the hope that public pressure would motivate these firms to adopt more shareholder-friendly policies (for example, repeal of poison pill provisions or staggered boards). Barber (2006) reports that inclusion on the focus list was associated with positive abnormal announcement day returns of 0.23 per cent between 1992 and 2005. Since investors reasonably expected CalPERS to work with management to improve governance, the abnormal return represents the value of those advocacy efforts. Barber estimates the economic impact of CalPERS advocacy at US$3.1 billion. However, updating the sample to 2009 shows that the effect has been muted in the past 3 years, with cumulative abnormal returns over the entire 17-year period falling to 0.09 per cent. Replicating such a strategy depends critically on two factors. First, the equity holder needs to control a large enough block of shares and second, the threat of divestment or takeover must be credible. If either condition is not met, hard advocacy is unlikely to be successful.

Debt holders

There is broad acceptance in the finance literature for the notion that banks act as delegated monitors of the firm (Diamond, 1984; Fama, 1985). In that capacity, banks have access to inside information about the firm and they use it to judge the creditworthiness of the firm. Once a loan is made, they use their 'quasi-insider' status to ensure repayment. Among the options available to banks to mitigate risk are demands for security, shortened maturity, or increasing the initial spread charged on the loan. In a study of US bank loans, Goss and Roberts (2011) report that banks punish firms with poor CSR records. However, banks provide little incentive for firms to make proactive CSR investments and actually punish CSR investments made by low-quality borrowers because of concern over agency costs. Overall, the evidence suggests that unless the actions taken by the miner impinge on timely repayment of debt, banks will not significantly influence the firm. Turning to public debt, Menz (2010) reports a higher risk premium for socially responsible bond funds, although the results were only marginally significant. Overall, firm-level CSR activities are not captured in bond prices. These findings suggest that mitigating credit risk remains the primary reason for monitoring the firm, with lender reputation and liability risk being second-order concerns. Only to the extent that the actions of borrowers negatively impact the ability to meet debt obligations will creditors monitor or influence firm behavior.

However, the creation of the IFC Performance Standards and their project finance counterpart, the Equator Principles, suggests that lenders are sensitive to the unique risks faced by borrowers in emerging markets. The importance of IFC financing has led to the integration of environmental and social criteria into the financing of projects in developing counties. Projects are independently assessed and classified based on their potential environmental impact. Those projects having the greatest potential for harm are subject to the greatest scrutiny. With both project funding and political risk insurance contingent on adherence to the standards, this regulatory approach holds promise that future projects will be more sustainable.

In mines using project financing, the Equator Principles are a voluntary set of guidelines designed to ensure that the projects meet minimum levels of environmental and social performance. If the Principles are effective and if 70 per cent of all emerging market non-recourse lenders are signatories, it follows that a substantial majority of the stakeholder issues in the extractive sector should have been resolved. Unlike the IFC Performance Standards, critics point to the voluntary nature of the codes and question whether lenders are simply paying lip service to the Principles. In an open letter to the Equator Principles financial institutions in January 2010, a group of NGOs criticized the signatories for a lack of transparency or accountability in the implementation of the Principles (BankTrack, 2010).

The role of exchanges

A final intermediary is the exchange on which the extractives list. Is it possible for the exchange to influence firm behavior though listing requirements or other regulation? The evidence here suggests that firms are sensitive to the regulatory environment. The reporting requirements of the ESTTA will not take effect until 2012, so it is too early to judge the responses of US listed extractives firms to the new disclosure rules. Firger (2010) argues that the ESTTA is unenforceable, since there is no incentive to accurately report payments and no way to verify the accuracy of the figures that are reported. Other commentators suggest that the reporting requirements will discourage resource rich countries from dealing with firms listed on US exchanges. While such speculation is premature, research conducted after the introduction of the Sarbanes Oxley act suggests that firms reacted to the more stringent requirements by migrating to other exchanges or privatizing (Engel et al., 2006). Neither outcome is consistent with the desire of regulators to restore faith in US capital markets following the accounting abuses of the early 1990s.

A similar example in the Canadian mining sector explains how an attempt to increase fairness in the capital markets following an insider trading scandal in Ontario led to the rise of the Vancouver Stock exchange as a world leader in mining finance.[7] In 1964, an insider trading scandal erupted in Ontario, involving Windfall Oils and Mines, a junior miner operating in northern Ontario. A major silver discovery in the area surrounding Windfall's property heightened speculation that the junior would also announce the discovery of significant mineralization. Stock prices rose tenfold in the weeks prior to the announcement of drill results. Unfortunately, the drill results were negative and the share price collapsed upon their publication. Subsequent investigations revealed that the principals in the company had been aware of the drill results before publication and had traded on the material non-public information. The scandal shook the faith of investors in the mining sector and led to a significant tightening of securities regulation in Ontario in the early 1970s that limited the returns available to mine promoters.

While the intent of the regulation was to restore public trust in capital markets and prevent future abuses, the result was the mass migration of junior mining issues from the Toronto exchange to the Vancouver exchange. By 1978 the level of financing for junior miners in Ontario had fallen to 15 per cent of the level enjoyed in the 1950s. At the same time, Vancouver saw equity financing increase from US$19 million in 1971 to US$322 million in 1985, with 84 per cent of listed firms being resource companies. Free of the restrictive securities laws in Ontario, Vancouver became a major centre for exploration funding not only for British Columbia, but for all of Canada and the western US. The Ontario Advisory Committee on Junior Resource Financing (1986) offered this blunt assessment of the changes in the regulatory regime following the Windfall scandal:

> The net effect of the legislation was that not only did it have an immediate detrimental effect on the fringe operators in the business but it provided a major impediment to bona fide resource professionals who needed a flexible financing system to accommodate the substantial risks of the exploration business. There is no doubt that the baby got thrown out with the bathwater. (p. 45)

The lessons regarding the role of the exchange are clear. Unilateral attempts to increase regulation are likely to be met with migration to less stringent regimes. Dr. Kernaghan Webb, Special Advisor to the United Nations Global Compact on the ISO 26000 Social Responsibility Standard and an expert on regulatory frameworks within the mining sector, has suggested that if regulations promoting sustainability are desirable, a solution may be to arrange for all major exchanges to act in concert, implementing consistent new rules and thereby removing the temptation for exchanges to lower listing requirements in a 'race to the bottom.' It is worth noting that Canadian exchanges have so far resisted the call for disclosure of payments by Canadian listed mining companies.

Conclusion

Like all firms, companies in the extractive sector need access to capital, but the unique challenges facing firms involved in different aspects of mining mean that their financing choices can be quite different. In the exploration stage, cash burn is a significant issue and neither the firm nor institutional equity investors are likely to welcome any expenditure that is not directly related to identifying mineral deposits. Individual equity investors with flow-though shares will be primarily motivated by tax considerations. As the project moves from exploration to exploitation of the resource, fresh injections of equity are required. Joint ventures or farm-in agreements are common, as are asset sales and takeovers. Debt financing may be present at this stage, but covenants are likely to be overly restrictive given the uncertainty regarding the ultimate viability of the mine. As projects move from exploration to exploitation, host governments are more likely to become involved, either as majority or minority equity holders. Their desire to use mineral wealth to support other areas of the economy may place them at odds with affected communities. Corruption is also a common problem, with the gains from resources being funneled into the hands of corrupt officials.

Project finance is possible and the presence of government or quasi-governmental agencies is beneficial when political risk is high. While mining firms are generally less levered than comparable firms in other industries, mines using project financing are held off the balance sheet of the sponsor company. The independent project companies can have considerable leverage, with providers of non-recourse debt typically providing from 60 to 70 per cent of the funds.

Looking at the ability of finance to influence managerial behavior, there is evidence that both public and private debt holders put a price on corporate social responsibility through higher initial spreads and lower credit ratings for poor performers. Banks monitor miners and can influence behavior through the application of restrictive covenants, but their interest extends only as far as loan repayment.

The impact of equity holders depends on the size of the holding and the type of influence. Boycotts of widely held miners are likely to result in shareholder clienteles, but will not generally induce changes in corporate behavior unless implausibly large numbers of investors are involved. On the other hand, large block-holders may be able to induce changes in managerial behavior through soft advocacy, but the empirical evidence on hard advocacy remains mixed.

Finance plays an important role in the mining industry, but using it as a tool to affect change in corporate behavior may be less successful than other available mechanisms. Prudent regulation has the potential to bring about positive changes, as evidenced by the IFC Performance Standards, but care must be taken to ensure that changes in regulation do not lead to the unintended migration of firms from high-regulation to low-regulation regimes.

The discussion suggests several avenues for future research. The extractives sector faces much higher risk than firms in the economy generally, making it an ideal setting to understand the link between CSR and finance. A key to uncovering that link is likely to be a better understanding of the motivations of both managers and the suppliers of capital. Regarding equity holders, what is the impact of hedge funds amongst junior explorers? Turning to the majors, what is the impact of 'socially responsible' investing on the shareholder base of these firms? Are there clienteles in the extractive sector, and how might they influence management? The evidence to date suggests that boycotts are unlikely to change corporate behavior. However, there is little research on the effectiveness of soft advocacy. If advocacy can affect meaningful change, it might provide a way to change corporate behaviors.

More research on the motivations of debt holders is also needed. The IFC Performance Standards and the Equator Principles are the standard for due diligence in stakeholder and environmental issues. But less is known about how well they work. Do projects negotiated under the Principles have materially lower levels of stakeholder conflict? Are yields on non-Equator projects meaningfully higher? If not, how can the Principles be improved to meet the goals of both non-recourse lenders and stakeholder groups? Finally, a deeper understanding of the sensitivity of the sector to the regulatory environment is needed. The Ontario (TSX) experience after the Windfall insider trading scandal suggests that careful consideration must be given to any changes in regulatory regime. The introduction of the ESTTA in the US will be an interesting case study on the sensitivity of extractive firms to changes in regulatory regime.

The link between CSR and finance is an important one and the extractives sector provides an interesting forum to explore that linkage. Understanding the motivations of each participant in the financing of extractive sector projects would deepen our understanding of how finance contributes to or mitigates the negative externalities identified by civil society groups.

Notes

1. For a high-level overview of the process, see U.S. Department of Agriculture (1983) 'Anatomy of a Mine: From Prospect to Production.' General Technical Report INT-35.
2. See John and John (1991) for a thorough description of project financing.
3. This definition is taken from the Ontario Securities Act R.S.O. 1990, S5. 56(1). The majority of the world's junior mining listings are on the Toronto Venture exchange (TSX-V). Other jurisdictions have similar prospectus requirements.
4. Although flow-through shares are predominantly issued in Canada, there have been similar programs in Australia. South Africa and Chile have also considered using flow-through shares.
5. The discussion on types of mine financing draws on information in C. R. Tinsley's 'Mine Financing' in *SME Mining Engineering Handbook*, Vol. 1.
6. Bauer and Smeets (2010) provide evidence of socially responsible clienteles in a study of European bank customers.
7. The discussion in this section draws largely on the 'Final Report and Recommendations of the Advisory Committee on Junior Resource Financing and the Competitive Position of Ontario, Vol. 1: Capital Markets,' July 1986.

12
Responsible Investment Case Studies: Newmont and Goldcorp

Irene Sosa

Introduction

Responsible investing (RI) is practiced by a growing number of institutional investors who are motivated by the goal of aligning investment practices with their values, managing their portfolio's exposure to environmental, social, and governance (ESG) risks, or both. This growth has been driven by several factors, including legislative changes, the recognition of materiality of ESG issues, a shift in understanding of fiduciary duty, and the entry into the RI sphere of high-profile institutions, such as the United Nations Environment Program.

Many responsible investors practice active ownership, which can include filing shareholder resolutions, voting proxies, or engaging in dialog with management. This chapter is particularly interested in the impact of active ownership on Corporate Social Responsibility (CSR). It examines case studies of responsible investors engaging with two large mining companies active in Latin America: Goldcorp Inc. and Newmont Mining Corporation. Both companies commissioned independent studies in response to shareholder resolutions, with different results. While the on-the-ground impact of both engagements still remains to be seen, this chapter concludes that, as a case of active ownership, Newmont's response to shareholder activism can be considered more successful than Goldcorp's. The chapter also seeks to assess the extent to which some of the elements that determine shareholder salience, including legitimacy, the business case for the shareholder proposal, and corporate culture, played a role in the two case studies. It concludes with an assessment of the potential and limitations of RI to influence CSR in the mining industry.

The growth of RI

Responsible investing (RI), or the integration of environmental, social, and governance (ESG) issues into investment analysis and decision-making processes, is an evolving and growing phenomenon. RI is practiced by pension plans, mutual funds, mission-based organizations, and other institutional investors who, generally speaking, are motivated by one (or both) of two main goals. Some seek to align their investment practices with their values; this approach is often referred to as Socially Responsible Investment (SRI), Ethical Investment, or Core SRI. Other investors take into consideration the financial impact of ESG issues on investment and seek to reduce their portfolio's ESG risk. This approach, often called 'Responsible Investment' (Eurosif, 2008) or 'Broad SRI' (SIO, 2009), has gained popularity in recent years, particularly among mainstream investors.[1]

In Canada by the end of 2008, RI assets had grown to over CAN$609 billion (or 19.9 per cent of the assets under management in the country), up from CAN$503.6 billion in 2006 (SIO, 2009). In the US in 2007, RI assets totalled US$2.7 trillion, or 11 per cent of the country's assets under management, an increase from US$2.29 trillion in 2005 (Social Investment Forum, 2008).[2] The growth of RI is also illustrated by the number of signatories to the United Nations Principles for Responsible Investment (PRI). Launched in 2006 by 20 asset owners, as of September 2010, the principles had 829 signatories managing over US$22 trillion in assets (UNPRI, 2010).

There have been several drivers of this growth. First, there is a growing recognition among investors that ESG issues can be material to a company's financial performance, especially in the long term. The United Nations Environment Programme's Finance Initiative (UNEP FI) has played a key role in fostering research that supports this view (UNEP FI & AMWG, 2006; UNEP FI et al., 2007). Furthermore, there is evidence that reducing the size of the portfolio by applying RI screens does not negatively affect financial performance. In Canada, the Jantzi Social Index® has slightly outperformed its non-screened counterparts, the TSX Composite and the TSX 60.[3]

Secondly, there has been a shift in understanding of fiduciary duty. In light of evidence of the potential materiality of ESG factors, some experts have argued that the consideration of these factors is no longer contrary to the fiduciary duty of asset managers, as critics of RI have long argued, but is indeed part of this duty. In a 2005 landmark report, a leading international law firm concluded that 'the integration of ESG issues into investment analysis, so as to more reliably predict financial performance, is clearly permissible and is arguably required in all jurisdictions globally' (UNEP FI & AMWG, 2005, p 13).

Thirdly, since 2000 several OECD countries, such as the UK, France, Belgium, and Australia, have introduced requirements for pension trustees to report the extent to which (if at all) ESG issues are taken into consideration in the investment decision-making process (SHARE, 2007). While the new legislation did not require trustees to consider ESG factors, the mere public pressure of having to report on this issue moved many of them into the RI arena.

Finally, the launch of the PRI initiative has given a significant boost to the RI sector. The principles were developed by some of the world's largest asset owners in partnership with the UNEP FI and the UN Global Compact. Signatories commit to incorporate ESG issues into investment analysis and decision-making processes; promote disclosure of ESG issues by the investees; practice active ownership that incorporates ESG issues; work together as signatories; report regularly on the principles' implementation; and promote implementation of the principles within the investment community (UNPRI, 2010). This last commitment has turned signatories into champions of the RI field.

The PRI initiative is overseen by a board composed of 11 asset owners and two UN representatives (UNPRI, 2010). The principles are aspirational and voluntary, although in August 2009 five institutions were delisted as signatories for lack of reporting.[4] There are several strengths to the PRI. First, they were developed by very high-profile institutions. Secondly, signatories include large asset owners and the climate of collaboration fostered by the PRI can give them significant power vis-à-vis companies and securities regulators. Finally, signatories have joined calls for addressing a limitation faced by responsible investors: access to information about ESG risk and what companies are doing to mitigate it.[5]

How does responsible investing work?

There are several ways in which investors can practice RI. They can screen out companies that are laggards in their industry or that are involved in activities such as weapons and tobacco production. They can invest in community economic development or clean-tech companies, which is often called 'impact investing' (GIIN, 2010). Finally, responsible investors can also practice active ownership by establishing dialog with management, voting proxies according to their ESG policies, or filing shareholder resolutions.

In 2009, 88 per cent of the respondents in UNPRI's 2010 survey[6] reported voting at least a portion of their proxies according to ESG principles, and between 30 per cent and 40 per cent of them reported practicing extensive shareholder engagement (UNEP FI, 2010).[7] In 2009, responsible investors filed 101 resolutions with Canadian companies, 12 of which were approved (SHARE, 2010). Those 12 resolutions addressed corporate governance issues, which seems consistent with studies that have found that these issues gather more shareholder support than social issues (Rojas et al., 2009).

The above modest rate of approval of RI resolutions is likely due to several limitations, including the relatively small size of the RI community and the strong tendency among institutional investors to vote with management, as illustrated by Taub Isenberg (2009) and SHARE (2009). Nevertheless, resolutions can impact corporations in other ways. First, some investors file resolutions as a way of starting dialog with management on a particular issue, which leads to the resolution's withdrawal. According to Gifford (2010), there is evidence that management will seek to accommodate investors' demands to prevent having shareholders vote on an issue. Indeed, 29 of those 101 resolutions were withdrawn, most of them after management made certain commitments (SHARE, 2010). Secondly, resolutions that are not approved but gather significant votes can be the basis of further dialog or, as Rojas et al. (2009) point out, make 'management uncomfortable enough to satisfy shareholder demands' (p. 245).

This chapter analyzes case studies of responsible investors practicing active ownership with two of the world's largest gold companies: Newmont Mining Corporation and Goldcorp Inc. The cases were chosen because of the companies' size, their operations in Latin America, their history of poor community relations, and the fact that both companies initially agreed to support shareholder demands. Despite these similarities, there are differences between the two cases that made the comparison interesting, as well as lessons for the stakeholders involved.

Case study 1: Newmont's community relations review

Newmont Mining Corporation has a long history of strained community relations at various sites, often caused or aggravated by poor environmental performance or fears about water scarcity. Community opposition has translated into violent road blockades, lawsuits, some of Newmont's employees being taken hostage, and one of its exploration camps being burned. Furthermore, in response to opposition from local residents, in 2004 the company was forced to withdraw its exploration permit for Cerro Quilish, Peru, reclassifying 2 million ounces of gold from 'proven and probable reserves' to 'mineralized material not in reserves' (Vecchio, 2006; Newmont, 2009a).

Since 2003, Newmont has taken significant steps to improve its community relations. The company requires its operations to implement over ten community-related standards, as well as to conduct audits, reporting, employee training, and stakeholder engagement (Newmont, 2009a, 2010c). Newmont is also a signatory to several international CSR initiatives (see Table 12.1). Nevertheless, controversies continued to haunt the company. As a result, in 2007 11 faith-based investors filed a shareholder resolution requesting that a committee of independent directors conduct a review of the company's community relations policies and practices. In an unprecedented move in the mining industry, Newmont recommended that shareholders vote in favor of the proposal, which received almost 92 per cent of the votes (Newmont, 2009a).

The independent, 18-month community relations review included interviews with more than 250 local residents and external stakeholders and over 100 company employees; an examination of company policies, standards, and procedures; and a review of five case studies (Newmont, 2009b). The review was conducted by independent consultants and guided by an eight-member independent advisory panel, which included one community representative and two members respected in the Non-governmental Organization (NGO) community.[8]

The final report of the review, released in March 2009, acknowledged that the company had strong policies and standards in place, but that there was a gap between the standards and site-specific practices (Newmont, 2009a). The company's president and CEO expressed a 'personal commitment' to following through on the review (Newmont, 2009b). In May 2009, Newmont's board of directors instructed management to develop an action plan to revise its standards, engage further with key stakeholders, and develop a set of metrics to monitor the company's performance on the issues identified in the report. By April 2010, Newmont had made significant progress on the action plan (Newmont, 2010b).

Whether the review succeeds in improving the company's community relations in the long run remains to be seen. Nevertheless, considering Newmont's support to the resolution, the thoroughness and credibility of the review, the CEO's personal commitment to implementing its recommendations, and the strong board oversight of this implementation, this case can be considered an example of successful active ownership.

Table 12.1 Newmont mining corporation

The Company	
Incorporated:	1921
Revenue (2009)	US$7.7 billion
Ranking:	World's second largest gold producer
Workforce (2009)	31,000 employees and contractors
Countries of operation:	US, Australia, Peru, Indonesia, Ghana, Canada, New Zealand, Mexico
Significant M&As in the last 10 years:	Merger with Normandy Mining and Franco Nevada in 2002
External CSR initiatives	Signatory to UN Global Compact since 2004
	Founding Member of International Council of Mining and Metals (ICMM)
	Signatory to the Voluntary Principles on Security and Human Rights
Executive CSR position	As of May 2010: VP Health, Safety and Loss Prevention; VP and Chief Sustainability Officer
	First Vice-President of Environmental Affairs named in 1991
Significant shareholders (>10% shares)	None
The Resolution	
Year of filing	2007
Filer	Christian Brothers Investment Services Inc.
Co-filers	Catholic Health East, Sisters of St. Joseph, Mercy Investment Program, Christus Health, General Board of Pension and Health Benefits, United Methodist Church, Presbyterian Church (USA), Missionary Oblates, Unitarian Universalist Service Committee, Evangelical Lutheran Church, Catholic Healthcare West
Text	'That shareholders request that a committee of independent board members be formed to conduct a global review and evaluation of the company's policies and practices relating to existing and potential opposition from local communities and to our company's operations and the steps taken to reduce such opposition; and that the results of that review be included in a report (omitting confidential information and prepared at reasonable cost) that is made available to shareholders prior to the 2008 annual meeting.'
Result:	Approved

Source: Newmont (2009b, 2010).

Case study 2: Goldcorp's human rights impact assessment

Goldcorp's Marlin mine in Guatemala has long faced strong opposition by local indigenous people who claim that they were not properly consulted before the mine was developed and were intimidated into selling their land. Local residents are also concerned about water pollution in the area and blame the mine for a high incidence of health problems. Residents of one of the two affected municipalities voted overwhelmingly against the mine in a referendum held in 2005. There have been protests, one of which resulted in a fatality, as well as vandalism of mining infrastructure (NISGUA, 2009; Law, 2009a). Local residents and NGOs have also brought complaints against the mine before the World Bank's Compliance Advisory Ombudsman, the International Labour Organization (ILO), the Inter-American Commission on Human Rights (IACHR), and the Organization for Economic Cooperation and Development (OECD) (Imai et al., 2007; CIEL, 2009; IACHR, 2010). In early 2010, the Committee of Experts on the Application of Conventions and Recommendations of the ILO and the IACHR recommended that the government of Guatemala suspend mining activities at the site (IACHR, 2010; ILO, 2010).

The Marlin mine was originally developed by Glamis Gold, a company with a long history of poor community relations at various sites. Following Goldcorp's acquisition of Glamis Gold in 2006, the CEO of Glamis Gold became the CEO of Goldcorp. Since 2008, Goldcorp has named a new CEO, hired a new Vice-President of CSR, become a signatory to the UN Global Compact, and improved its CSR reporting. Yet, the company has not adopted community relations management systems comparable to Newmont's and initiatives implemented at the Marlin mine, such as community consultation meetings and a community environmental monitoring committee, and have not improved Goldcorp's relationship with local residents. Table 12.2 lists Goldcorp's key statistics.

In February 2008, two shareholder resolutions were filed with regards to the Marlin mine. The first, filed by an individual shareholder, called on the company to halt any plans to expand the mine 'without the free, prior and informed consent (FPIC) of the affected communities.' Goldcorp refused to circulate this resolution, arguing that it did not relate to the business of the corporation (Breaking the Silence, 2008). The same individual investor filed another resolution in 2010, asking the company to adopt a policy on FPIC and apply this policy retroactively. Goldcorp recommended voting against this proposal, which was not approved (Goldcorp, 2010b; SHARE, 2010).

Table 12.2 Goldcorp Inc

The Company (Source: Goldcorp, 2010a)	
Incorporated	1994
Revenue (2009)	US$2.7 billion
Ranking	World's third largest gold producer
Workforce (2009):	11,500 employees and contractors
Countries of operation	Canada, Mexico, Guatemala, US
Significant M&As in the last 10 years:	January 2003, acquisition of Peak Gold Mines Pty Ltd April 2005, merger with Wheaton River Minerals Ltd December 2006, acquisition of Glamis Gold Ltd
Participation in external CSR initiatives	Signatory to the UN Global Compact since 2009 Member of ICMM since October 2009
Executive CSR positions	Vice-President, Corporate Social Responsibility (since March 2009) Vice-President, Sustainable Development of the Corporation (since 2007) Vice-President, Safety and Health (since 2008)
Significant shareholders (>10% of shares)	None
The Resolution (Source: Share, 2010)	
Year of filing	2008
Filer:	Public Service Alliance of Canada Staff Pension Fund
Co-filers	The Ethical Funds Company, The First Swedish National Pension Fund, the Fourth Swedish National Pension Fund
Text	'That the board of directors commission an independent human rights impact assessment for Goldcorp's operations in Guatemala. A transparent stakeholder consultation process should be established by which an independent party will be selected to carry out the human rights impact assessment.'
Result:	Withdrawn

The second resolution filed in 2008 was submitted by a group of institutional investors that had recently visited the mine and asked Goldcorp to conduct an independent human rights impact assessment (HRIA) of the mine. The resolution was withdrawn after the company agreed to commission the HRIA, which would be overseen by a steering committee composed of one representative from each of the company, the shareholders group, and civil society. The memorandum of understanding between Goldcorp and the shareholders does not commit the company to implement the HRIA's recommendations but, if the company chooses to ignore them, it has to publicly explain the rationale for doing so (HRIA, 2008).

The HRIA received strong critiques from local residents, who called the process racist, exclusive, and discriminatory. They alleged that the communities were not properly consulted at the planning stage and that an 'independent' assessment should not be led by a committee on which the company is represented. They also expressed concerns about the primary purpose of the HRIA, which was 'to improve the opportunity for Goldcorp to continue to operate profitably in Guatemala' (ADISMI, 2008). The criticism was echoed by NGOs (Rights Action, 2008; Coumans, 2009). Furthermore, the Catholic Church in Guatemala has initiated an alternative HRIA (Law, 2009a), which as of October 2010, had not been released.

In March 2009, one of the shareholders participating in the HRIA withdrew from the process, citing concerns about the lack of consent from the communities and that 'the interests of Goldcorp [were] being put before the interests of the local people' (Law, 2009b). That same month, the HRIA Steering Committee revised the scope of the assessment, concluding that 'the conditions necessary to engage local communities and organizations in open dialog do not exist in the current circumstances.' The revised HRIA would primarily review 'the potential impact of the presence and operation of the Marlin mine on human rights, relying on a review of company policies, practices, and procedures, secondary data analysis and expert sources' (HRIA, 2009, p. 1).

The final report of the HRIA was released in May 2010, 19 months after its initial target date. The report raises significant concerns about escalation of conflict at the site, as well as gaps in the mine's community consultation, grievance, and land acquisition mechanisms. It contains strong recommendations for Goldcorp, including adopting a moratorium on land acquisition. Nevertheless, the report warns that, due to the shift in focus during the assessment, the recommendations 'reflect the judgments of the assessment team, rather than the affected communities, and therefore may not be viewed as appropriate or adequate responses. This creates a requirement for the company to engage and consult...before moving forward with an action plan' (On Common Ground 2010, p. 8). In its response to the HRIA report, Goldcorp committed to, among others, expand its consultation through multi-stakeholder fora, train its staff on human rights issues, improve its security incident follow-up processes, revise its grievance mechanisms, and adopt a human rights policy. While Goldcorp agreed to revise its land acquisition procedures, it did not commit to a moratorium on land acquisition or to a formal, independent review of past acquisitions, as recommended by the HRIA (Goldcorp, 2010c). Goldcorp reports regularly to shareholders on these issues, but it is not clear how the company is engaging local stakeholders on the implementation of the report's recommendations. As of October 2010, neither the NGOs nor the local communities have issued a formal position statement regarding the HRIA report or Goldcorp's response. Local residents continue to demand that the mine be closed (La Nacion, 2010).

As a case of active ownership, Goldcorp's can be seen as having limited success. On the one hand, the company has accepted many of the recommendations and has been commended for posting the report on its website. On the other hand, Goldcorp continues to face local opposition and has not committed to a moratorium on land acquisition, which may remain a significant source of tension between the company and local residents. Finally, lack of effective consultation of affected communities by shareholders undermined community participation in the HRIA process, created further tensions within the same communities that it sought to help (Coumans, 2009), and may result in reluctance from local residents to participate in the implementation of the report's recommendations. It also pitted NGOs and investors – who have often been traditional allies in shareholder engagement – against each other.

Lessons learned

The above case studies illustrate the potential and limitations of active ownership and some of the elements that can play a role in determining its success. Gifford (2010) points out that individual and organizational legitimacy has an impact on shareholder salience. This seems to be supported by Rojas et al. (2009), who found that large institutional investors, such as mutual funds and public pension funds, are the most successful filers of social resolutions. In Goldcorp's case, legitimacy may explain why the company engaged with a group of institutional investors that had drawn significant attention through a trip to Guatemala while refusing to circulate a proposal filed by a lower-profile, individual investor. Furthermore, both Goldcorp and Newmont agreed to conduct independent studies requested by groups of institutional investors, which suggests that coalition building is important for increasing legitimacy (Gifford, 2010).

Another element that has an impact on shareholder salience is corporate culture and management's openness to engage (Gifford, 2010). Corporate culture may explain why Newmont recommended that shareholders vote for the 2007 proposal and commissioned a truly independent review, while Goldcorp engaged in a HRIA process that had limited success and whose independence was questioned. Goldcorp has undergone major mergers in recent years and only recently has begun to address CSR issues in a more systematic way. While Newmont is by no means perfect, at the time of the shareholder engagement it seemed to have a better understanding of CSR issues than did Goldcorp.

According to Gifford (2010), another key element in shareholder salience is the business case supporting the need for action requested by shareholders, which he describes as 'pragmatic legitimacy.' It should come as no surprise that Goldcorp recommended shareholders to vote against the 2010 resolution that called on the company to adopt a FPIC policy and apply it retroactively, which could have meant stopping operations at the Marlin mine. The resolution may have stood a better chance of success had the requirement for retroactivity not been included. On the other hand, the shareholder resolution filed with Newmont made reference to the financial impact that the controversies were having on the company, although it is not clear whether this was a significant factor in determining support from management and shareholders. The evidence available is not enough to draw conclusions on the role of pragmatic legitimacy on Goldcorp's HRIA resolution.

Even when investors draw management's attention to an issue, their ability to consult properly with affected stakeholders and consider the impact that their demands can have on these stakeholders will determine whether the engagement succeeds in improving CSR performance. As the Goldcorp case shows, failure to do so can reduce the effectiveness of active ownership, weaken social transnational networks, and damage the reputation of the investors involved.

Conclusions

Responsible investors will increasingly play a role in CSR. First, RI is a fast-growing field. While not every responsible investor practices active ownership, PRI signatories are doing so. Secondly, because their interests are aligned with the company's financial performance, shareholders can establish a dialog with management that other stakeholders such as NGOs, often labeled 'anti-mining,' cannot achieve. Shareholder engagement can give marginalized groups a new route of access to companies.

However, the impact of active ownership by responsible investors on CSR faces some limitations. First, the RI community is still relatively small and the ability of responsible investors to influence companies, particularly on social and environmental issues, is still modest. In this context, coalition building is important. Secondly, shareholder salience is determined by factors such as the investors' legitimacy, corporate culture, and the business case for their proposal.

Finally, the very alignment of interests between shareholders and the company's financial performance, as well as the need to frame active ownership in terms of a business case so as to have 'pragmatic legitimacy' vis-à-vis the company, can create an inherent tension between shareholders and other stakeholders. NGOs need to recognize this tension if they want to be able to use this route of access to companies. Shareholders, on the other hand, need to make sure that they consult extensively with NGOs and affected stakeholders to reduce this tension as much as possible.

Notes

1. In line with the United Nations Principles for Responsible Investment, this chapter defines Responsible Investment (RI) as the integration of ESG issues into investment analysis and decision-making processes, a broad term that seeks to encompass all of the above terms.
2. Both the Social Investment Organization(Canada) and the Social Investment Forum (US) use the term 'socially responsible investment,' but their definitions match this chapter's definition of RI.
3. The Jantzi Social Index® (JSI) is a market capitalization-weighted common stock index consisting of 60 ESG-screened large Canadian companies. The index also excludes companies with significant involvement in nuclear power, tobacco production, or military contracting. Since its inception on 1 January 2000 through April 30, 2010, the JSI achieved an annualized return of 5.92 per cent, while the S&P/TSX Composite and the S&P/TSX 60 had annualized returns of 5.91 and 5.79 per cent, respectively, over the same period (Sustainalytics, 2010).
4. Delisted institutions: DESBAN, Christopher Reynolds Foundation, Foresters Community Finance, Oasis Group Holdings, and Trinity Holdings (UNPRI, 2010).
5. Responsible investors worldwide are pushing for stronger ESG reporting requirements for publicly traded companies. Some OECD countries have expanded these requirements, but Canada is not among them (SHARE, 2007).

6. Almost 80 per cent of signatories eligible to complete the survey did so (UNEP FI, 2010).
7. Some 30 per cent for asset owners and 40 per cent for investment managers (UNEP FI, 2010).
8. Steve D'Esposito was the president and CEO of the NGO Earthworks for ten years. Julie Tanner was the Manager for National Wildlife Federation's Finance and Environment program for five years (Newmont, 2009b).

13
Anti-corruption: A Realistic Strategy in Latin American Mining?

Carol Odell

Introduction

Latin America is perceived as one of the most corrupt regions of the world (Zinnbauer and Dobson, 2008), and mining is perceived as one of the industries most likely to engage in corrupt practices worldwide (Riano and Hodess, 2008). Under endemic corruption, unethical companies gain significant short-term advantages from some of their corrupt transactions (Hellman et al., 2002). This is especially true for mining, which depends on the state for a series of concession approvals and operating permits, requires sizable upfront investment, and faces time pressure from financing, leaving it vulnerable to extortion (Marshall, 2001). However, there is mounting evidence that corruption is very costly to firms in the medium to long term and that aggressive prosecutions under international laws, including a recent fine of US$1 billion, are a potent disincentive (Labelle, 2008). Thus, overall the business risks of corruption for mining firms in Latin America appear considerable.

Anti-corruption efforts tend to focus on ethical issues; however, there is substantial evidence that where 'doing what is right' is perceived as contradicting 'furthering business interests,' then many companies adopt a strategic approach to ethics (Hellman et al., 2002). Recent surveys indicate both positive and less positive trends in corporate anti-corruption in Latin America (Miller Chevalier, 2008) and in the mining industry (Simmons and Simmons, 2006). While there is growing awareness of corruption as an issue, a sizable minority of companies ignore or pay lip service to the issue, while direct and indirect participation in corruption is relatively widespread (Miller Chevalier, 2008). Corporate social responsibility (CSR) leaders are, however, implementing promising approaches to anti-corruption in collaboration with international institutions and Non-governmental Organizations (NGOs), demonstrating that collusion with corrupt governments is not inevitable and that under emerging incentives, avoiding corruption makes business sense (Bray, 2007b). Indeed, this chapter argues that mining CSR is an important enabler of anti-corruption in the region. Future progress will require a tailored strategic approach, which leverages anti-corruption enablers, systematically evaluates outcomes, and promotes collaboration among interested actors in the social and environmental value governance ecosystem to counter emerging threats.

Organization of the chapter

This chapter provides insight into the relationships between corruption and mining in Latin America through the lens of emerging research, grounding this in high-profile, largely Peruvian cases. After introducing concepts, it summarizes evidence of corruption as overwhelmingly detrimental to national development and describes corruption patterns in the region in the light of their causes and implications for effective anti-corruption approaches. It then establishes that anti-corruption incentives are increasing for many mining companies in the region and investigates the impacts of these changed incentives on corporate anti-corruption implementation. It ends by proposing a typology of mining corporate corruption approaches and discusses what can be done to encourage anti-corruption behavior on the part of companies lagging in strategic investment in this area.

Framework and definitions

Research into corruption is concentrated in four key disciplinary areas: political science/law, economics, sociology/anthropology (Brown, 2006), and business administration. This chapter leans most heavily on economic studies, introducing other insights where appropriate. Economists regard corruption as rational behavior rooted in the assessment of incentives. 'New Institutional Economics' explores the incentives and barriers for rational corrupt behavior (Lambsdorff, 2007), insights used here to evaluate the costs, benefits, and risks of a variety of corruption responses for mining companies. A recent proliferation of empirical econometric analyses, which test hypotheses regarding corruption, are employed to explore the patterns and trends of corruption in Latin America, their causes and consequences.

Corruption is traditionally defined in terms of a dichotomy between societal and private interest as 'an illegal payment to a public agent to obtain a benefit that may or may not be deserved in the absence of payments' (Rose-Ackerman, 1999). Indeed (Hellman et al., 2002) argue that there is no such thing as private sector corruption separate from the state, and that private sector behaviors similar to public sector corruption are properly termed fraud or theft rather than corruption. International anti-corruption legislation also requires direct or indirect involvement of state representatives in corruption. However, mining company reports of corruption in Global Reporting Initiative (GRI) compliant reports and industry-focused research (for example, see Marshall, 2001) reveal that mining companies conceptualize both company–public and company–private incidents as corruption. This coincides with sociological research, which moves corruption beyond narrow legalistic definitions to incorporate 'the misuse of entrusted power for private gain' (Alatas, 1986; also adopted by Zinnbauer and Dobson, 2008, p. 1). Because this chapter focuses on corruption in the mining sector, broad definitions are used.

The social consequences of corruption

Empirical econometric studies over the past decade have effectively overturned previous notions of corruption as a benign mechanism for correcting governance failures. They have replaced this with an emerging consensus that corruption is overwhelmingly detrimental to society[1] by demonstrating that:

> corruption clearly goes along with a low GDP (Gross Domestic Product), inequality of income, inflation, increased crime, policy distortions and lack of competition...This suggests that countries can be trapped in a vicious circle where corruption lowers income, increases inequality, inflation, crime, policy distortions and helps monopolies at the expense of competition. These developments in turn escalate corruption. (Lambsdorff, 2005, p. 27)

In quantitative terms, the World Bank (2004) estimates the magnitude of economic losses from corruption at over US$1 trillion for 2001 (or 3.33 per cent of world GDP) in bribes alone, ignoring the associated costs of theft, embezzlement, and the transaction costs of corruption. The World Bank calculates that effective anti-corruption policies could increase per-capita incomes in corrupt nations by 400 per cent. The magnitude of these losses and increased understanding of the vicious cycle of undesirable consequences/causes of corruption have together increased the ethical imperatives for anti-corruption initiatives and increased demands for corporate oversight.

Patterns of corruption in Latin America

The Corruption Perceptions Index (CPI), collated annually since 1995, reports business perceptions of corruption worldwide, on a scale from 0 to 10, where 10 is the least corrupt and 0 the most corrupt (Zinnbauer and Dobson, 2008). Through the CPI lens, Latin America is one of the more corrupt regions of the world, ahead of Africa and the ex-Russian republics and level with Asia and the Middle East. Within the region, Chile and Uruguay have the lowest corruption (CPI = 6.9), while Puerto Rico, Cuba, and Costa Rica have moderate CPI (5.5–4.5). Corruption is high in all other countries (CPI<4) and highest in the Bolivarian countries, although very high corruption proceeded regime change.

Figure 13.1 also demonstrates that corruption changes slowly and shows a slight increase over time in the region. Over half of countries show no significant change over the ten-year period 2000–09, while corruption increased in approximately a third. The only significant reductions are in Uruguay, Colombia, and Paraguay.

Broadly, Latin American political parties, parliaments, judiciary, and police are perceived as the most corrupt institutions in cross-institutional research, while business, the media, NGOs, and universities are relatively less corrupt (Miller Chevalier, 2008). Corruption in the legislative and judicial branches is particularly problematic in creating 'State Capture,' defined as 'shaping the basic rules of the game (i.e., laws, rules, decrees, and regulations) through illicit and non-transparent private payments to public officials' (Hellman et al., 2002, p. 2) – an activity to which the mining sector is perceived as being especially prone (Riano and Hodess, 2008).

Explaining corruption patterns in Latin America

On a political/economic level, factors that correlate with increased corruption include the following: low GDP, high inflation, policy distortions (Lambsdorff, 2005), complex regulation of market entry and tariffs that appears to limit competition (Ades and Di Tella, 1999), and weak fledgling democracy (strong, long-term democracy decreases it) or strong, short-term presidential power (Shugart, 1999). Geographically,

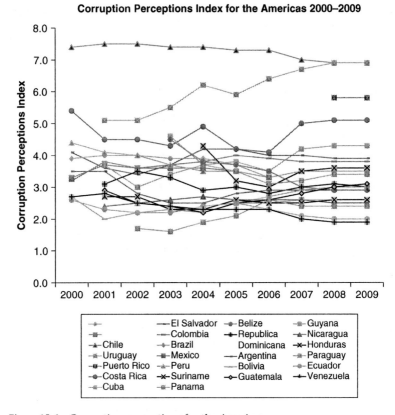

Figure 13.1 Corruption perceptions for the Americas
Source: Adapted from Transparency International, http://eitransparency.org, date accessed 16 February 2011.

remoteness from the major trading centers and natural resource dependence are also correlated with increased corruption (Wei, 2000).

Most Latin American countries have weak democratic systems instituted in the early twentieth century punctuated by dictatorial interludes. Powerful oligarchies have interests in limiting competition, most vividly witnessed during postwar import substitution policies. Most countries in this region have cumbersome, complex, and overlapping legislation. Economic dependence on natural resources varies widely: Ecuador, Bolivia, and Venezuela obtain approximately 30 per cent of GDP from mining; Peru and Chile are highly dependent (10–20 per cent GDP); while Colombia, Mexico, Brazil, Argentina, Guatemala, and Honduras obtain 1–7 per cent of GDP from mining (ECLAC, 2008).

All mining-dependent countries in the region, except Chile, have high levels of corruption (CPI<4), with Ecuador, Bolivia, and Venezuela three of the most corrupt. Low corruption in Chile is credited to a long period of political stability and relatively independent and capable regulatory agencies (ICMM, 2006b).

On a social level, poverty and inequality are strongly correlated with corruption along with limited press freedom and high crime rates (Lambsdorff, 2005). Latin America is noted as the region with the most extreme income inequality in the world, reflected also in inequalities in land tenure and education (De Ferranti, 2003). Inequality creates poverty and promotes criminal activity related to survival.

On a cultural level, data from the *World Values Study* (*WVS*), which tracks a comprehensive set of social variables across nations, have been correlated to corruption indicators (Lambsdorff, 2005). Inglehart and Weltzel (2005) simplify the WVS data, demonstrating that 70 per cent of the variables can be mapped onto two dimensions: the traditional/secular and the survival/self-expression dimensions. As can be seen in Figure 13.2, the most corrupt countries share survival values as a result of underdevelopment, lending support to the notion that endemic corruption is rooted in daily struggles against poverty, disease, and exploitation. Latin American values cluster in the region of strong traditional values with above-average self-expression values. High corruption here correlates with deference to authority and family and low social trust. Trust is a crucial variable in corruption research, as it is the only variable driving reduced corruption in longitudinal studies (Björnskov and Paldam, 2004). Chile and Uruguay have the most secular values and the least corruption, while extremely corrupt Venezuela shows average values for the region.

The causes of corruption are thus complex and entwined, and frequently operate in vicious cycles. Figure 13.3 diagrammatically shows the connections between manifestations of corruption and the key causal factors identified for Latin America. The figure hints at the challenges in tackling such an integrated phenomenon: changes instigated by any actor will generate reactions from other actors in the ecosystem. Perhaps for this reason, generic and direct attacks on corruption have limited effect, while tailoring to local circumstances, political will, and civil sector involvement are key (Fjelstad and Isaksen, 2008). In light of the region's inequity problem and the role of survival values in perpetuating corruption, pro-poor development, institutional strengthening, and initiatives increasing social trust are all promising indirect anti-corruption strategies.

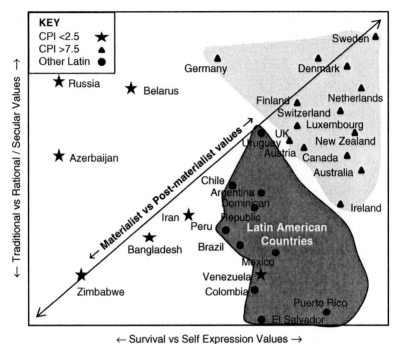

Figure 13.2 Superposition of corruption perceptions on world values
Source: Adapted from Inglehart and Weltzel (2005).

Mining trends in Latin America

Latin America is one of the most important mining regions in the world, a leading producer of aluminum, copper, gold, iron, and zinc and the focus of a quarter of all exploration spending in 2006. Currently, Peru, Chile, Brazil, and Mexico account for over 75 per cent of mining investment in the region (USGS, 2006). Mining development in these countries from the early 1990s centered on the sale of previously nationalized assets, while more recent development is increasingly exploration based.

Although mineral potential is critical in exploration decisions, many potential investments in the region at present are deterred by institutional concerns (Price, 2009). Argentina and Colombia are the most attractive future mining prospects; however, many other jurisdictions include promising deposits. The attractiveness of Venezuela, Ecuador, and Bolivia is limited by regulatory uncertainty and expropriation

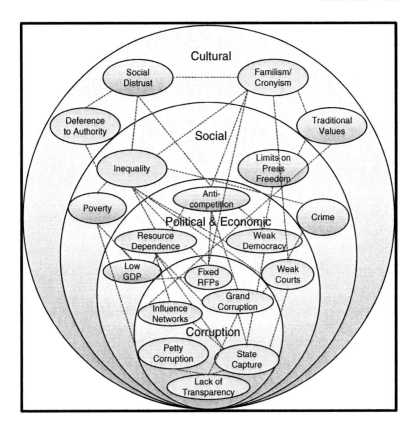

Figure 13.3 Corruption and the factors found to increase it in empirical econometric studies

concerns, demonstrated in the statement of one executive that 'In Venezuela, if you build it, Hugo Chavez will steal it. Ecuador is a close second for similar reasons along with tribal claims' (MacMahon and Cervantes, 2009, p. 36). However, Russian investors who are less risk averse have invested in Venezuela, and China is opening a small niche in base metal concessions that are too risky for Western companies. For example, Chinese companies own both the social conflict-plagued Majaz concession in northern Peru and the Toromocho concession in central Peru, a project which involves massive resettlement. Apart from these examples, in 2008 Canada continued to dominate mineral exploration in Latin America, accounting for 49 per cent, with the local share at 19 per cent and increasing (NRCAN, 2008).

The corporate costs, benefits, and risks of corruption in Latin America

While the effects of corruption on society are overwhelmingly negative, the effects on business are mixed. In the high-corruption context that prevails in Latin America, companies are able to gain competitive advantage through leveraging social, political, and economic influence and, through grand corruption, to secure concessions and contracts. One study of corporations in high-corruption settings outside the Americas shows that well-connected and bribe-paying companies grow at up to four times the rate of companies that are neither influential nor corrupt (Hellman et al., 2002). Mining projects are attractive to corrupt actors: they generate revenue flows large enough to insinuate sizable kickbacks for concessions and contracts, are regulated through complex legal frameworks incorporating sizable fines that provide opportunities for collusion between corrupt companies and bureaucrats, and involve sizable sunk costs leaving companies vulnerable to extortion over permits (Marshall, 2001).

According to the Global Fraud Survey, bribery has a 47 per cent prevalence in the mining sector (Ernst and Young, 2008). Worldwide, mining is considered the fifth most likely sector to pay bribes and the third most likely to engage in 'state capture' (Riano and Hodess, 2008). Typical bribe rates are 5–6 per cent of contract value (Schubert and Miller, 2008). While this implies multimillion dollar bribes at prevailing concession prices, expected profits make bribery economically attractive.

Despite this, empirical and experimental evidence suggests that companies underestimate the medium- to long-term, direct and indirect costs of participating in corruption. Lambsdorff (2002) demonstrates that the transaction costs of corrupt contracting are high and extend long after the corrupt contract has been finalized, due to blackmail risks. They include the costs of locating corrupt partners, negotiating conditions, and enforcing contracts. Costs are likely especially high in Latin America, where opportunism in corruption is rife, so results are unpredictable (Campos et al., 1997) and bribe-paying companies are typically targeted for repeat requests by public officials (Pedigo, 2009). In addition, as Kahneman and Tversky (1979) demonstrate through 'prospect theory' the uncertain future losses from corruption, compared with the relatively high shorter-term payoff, tends to lead to underestimation of these costs.

Indirect and longer-term costs include reputational damage accompanied by damaged share value,[2] while internal corruption wastes company resources through employee theft, kickbacks in return for contracts, and diminished employee and contractor quality. In addition, political and social instability associated with corruption generates increased security costs and country risk insurance payments. Finally, prosecution under international anti-corruption legislation is becoming increasingly aggressive, escalating the risk of significant economic sanctions and disbarment from future contracting.

In the US, investigations under the Foreign Corrupt Practices Act increased dramatically from 3 per year to over 20 per year during the first decade of the new millennium (Weiss, 2009). The most dramatic prosecution was the 2005 prosecution of the engineering giant, Siemens, which admitted paying US$1.4 billion in bribes worldwide between 2001 and 2007 (ibid.). The estimated cost to the company is US$4.1 billion, including US$1.6 billion in sanctions in Germany and US$2.5 billion for dismissals, restructuring, changed procedures, and training (Schubert and Miller, 2008). The size of fines and the fear of prosecution has led several major companies to voluntarily disclose corruption uncovered through internal audits in the last five years (Weiss, 2009), although Transparency International calls for more consistent prosecutions by developed countries if strong anti-bribery programs and high-integrity corporate culture is to result (cited in Schubert and Miller, 2008).

From a CSR perspective, corruption also limits the effectiveness of company social investments and contributions to national, regional, and local taxation, thus damaging the social license to operate that is so crucial to Latin American mining companies (Boutilier and Thomson, 2009). For example, a 'Public Expenditure Tracking Survey' that investigated the mining–canon tax redistribution program in Peru showed that 30–70 per cent of revenues to local and regional governments could not be accounted for, suggesting significant graft (Alcazar, 2001).

Mining responses to corruption in Latin America

The potential short-term gains of corruption, along with the costs of non-participation continue to tempt companies, although hard evidence of such corruption is sparse, and the high-profile cases of corruption that do come to light and the results of surveys indicate that it is a significant dynamic in the extractives sector. In a Simmons' and Simmons' (2006) Latin American survey, 43 per cent of mining companies believed they had lost business in the previous five years to bribe-paying competitors. Prominent resource-related scandals include the

2008 Peruvian 'Petroaudios scandal' (BBC, 2008), where audio recordings document bribery to secure oil concessions, and video footage from around 2000 of a Newmont negotiator pressuring renowned corrupt Peruvian presidential advisor Vladimir Montesinos to intervene on behalf of the company in the Supreme Court (Bergman, 2005). These cases illuminate high-level corruption associated with resource concessions and are perceived as the tip of a large iceberg. In addition, circumstantial evidence suggestive of corruption includes Doe Run Peru avoiding environmental upgrading for over ten years (Gestion, 2009). In Panama, another example is 'Concession Law 9' (Panama Legislative Assembly, 1997), which imposes a 2 per cent royalty on a mining concession, exonerating all other taxes – a taxation regime so beneficial to the company and against the national interest as to intimate state capture.

Although a majority of companies believe that their executives have a good grasp of corruption risks, reported anti-corruption actions in Latin America are broadly inadequate (Miller Chevalier, 2008). Corporate anti-corruption guidance has been published by NGOs, consultants, business associations, and international financial institutions (see for example Bray, 2007b). Despite this, however, here has been little independent evaluation of implementation, and strategies will probably require significant adaptation to local contexts.

Anti-corruption efforts typically start by building an anti-corruption culture and setting corporate priorities through policy. Miller Chevalier (2008) found that 71 per cent[3] of companies in the region had ethics policies, although almost half of foreign companies and 84 per cent of national companies had taken no steps toward implementation. Anti-corruption leaders in Latin American mining include Antamina, who have published their Code of Conduct online and began companywide online ethics training initiatives in 2008,[4] and Rio Tinto, where business integrity guidance training has been compulsory since 2005 for managers and companywide since 2007.[5]

Prevention is another anti-corruption strategy, and due diligence of third parties is performed by 39 per cent of international and 16 per cent of regional companies (Miller Chevalier, 2008). Due diligence is required under international law to prevent indirect corruption through contracting a third party to arrange bribes, a form of corruption that is commonly disguised in excessive consulting fees. Another preventive strategy is project risk assessment, which identifies red-flag corruption issues and coordinates adequate corruption-management responses and resources. Due diligence also prepares companies for setbacks, such as walking away from projects due to extortion. Simmons and Simmons

(2006) note that worldwide in 2005, half of extractives companies had avoided otherwise attractive prospects due to corruption risks. Contracts are another locus for prevention in that they can include contract clauses that prohibit or stipulate disincentives for corruption.

Oversight is also essential to reinforcing anti-corruption culture and detecting non-compliance, through auditing and whistleblower mechanisms. Xstrata's independently operated 'Ethics and Fraud hotline' available in several languages including Spanish[6] is a whistleblower mechanism, while Minera Los Pelambres, in Chile, informs that it uncovered a copper theft ring in 2008 through whistleblowers.[7] In terms of auditing, surveys emphasize that auditors need training and independence. Their work is aided by clear rules for allowable expenses and gift giving and accounting requirements for contracts in order to avoid hidden kickbacks as exemplified in Antamina's Code of Conduct. Finally, participating in sector and national anti-corruption initiatives builds anti-corruption coalitions. For example, in Peru, mining companies promoted certification under the Extractive Industries Transparency Initiative (EITI),[8] increasing transparency of revenue flows between companies and governments to dissuade corrupt payment/use of funds.

A typology of company approaches to corruption

The typology presented in Table 13.1 draws on Pedigo and Marshall's (2009) typology of corruption responses drawn from interviews about managing corruption in Asia with Australian executives. It describes five approaches as a flexible continuum, borrowing three categories from Pedigo and Marshall and adding two more, for a total of five: (1) *passive*, which ignores corruption; two corruption-embracing approaches: (2) *relativist*, applying local norms; and (3) *unethical strategic*, using risk assessment to identify 'those forms of corruption that suit...comparative advantages, generating substantial gains...challenging the premise that they are coerced [and] bribing in a highly strategic fashion' (Hellman et al., 2002); and two anti-corruption approaches: (4) *imperialist/universalist*, applying home country or international standards; and (5) *relationship building*, working with influential ethical local partners to limit extortion risks.

Below I discuss the typology in the context of the regional mining industry using a CSR framework presented by International Alert (2005) that focuses on three business issues: (1) *core business activities*: specific production activities, human resources, and contracting as well as critical secondary activities such as security; (2) *social investment*:

Table 13.1 Typology of corruption responses for mining companies in Latin America

Corruption management Issue	Unethical strategic	Relativist	Passive	Relationship Building	Universalist/Imperialist
	<Increasing investment in strategic corruption / Increasing investment of resources in anti-corruption>				
Characteristics of Approach					
Company Policies	No policy or generic policy selectively applied. Reports as PR select indicators.	No policy or generic policy which is ignored. No reports. No indicators.	No policy or generic policy which is ignored. No indicators.	Anti-corruption policy-ethics & sustainability. Clear reports/indicators.	Anti-corruption policy-ethics & inter national standard. Clear reports/indicators
Core Business	High-level confidential risk assessment of corruption. Pay more to avoid detection.	Tacit acceptance of corruption. Follow locally accepted practice to limit exposure risk.	Ignore corruption: use standard contracts. No due diligence. Follows employee ethics.	Associate with ethical influential partners & lawyers. Implement training, advice line, audits & incentives	*Implement training, clear rules and penalties, due diligence & auditing*
Social Investment	Confidential risks assessment re. alliances with key stakeholders.	Corrupt engagement where useful in short term.	Naive to local corruption. Negligent corruption.	Creative corruption avoidance while maintaining relationships.	Strict due diligence procedures & penalties.
Policy Engagement	Masked unethical lobbying in company interest.	Lobbying for state capture using local approaches. Corrupt syndicates.	Avoid politics or go with the flow. Negligent corrupt association.	Ethical lobbying on short-medium-term risks. Pro-poor development coalitions.	Anti-corruption leadership eg. EITI lobby.
Cost/Benefit/Risk Assessment					
Summary	Balances short-term gain & long-term risk. Arrogance & failure to adapt are key exposure	Short-term benefit in concession access. Significant longer-term exposure	Low-cost strategy with significant risks from inadvertent/ negligent exposure	Short-term costs from avoiding extortion. Medium-and long term benefits from reduced bribe demands and diminished losses and exposures	

stakeholder engagement to manage the operating environment; and (3) *policy dialog*: broader engagement to address issues of strategic corporate interest. In addition, the typology incorporates company policies – addressing corporate anti-corruption culture and cost/benefit/risk – which weighs these aspects of incentives in different corruption contexts.

Company policies

Clear anti-corruption policies and associated implementation measures only appear in the *Imperialist/Universalist* and *Relationship* approaches to corruption management. These approaches frame anti-corruption as an ethical and sustainable development issue, respectively, and hence policies differ in their emphasis on incentives versus sanctions and the rationale for social investment.

Core business

Integrating ethics into core business requires analysis and investment in redesign of procedures, training, support, oversight, and auditing, as well as accepting short-term, potentially sizable losses in the interest of long-term cost and risk savings, practices that tend to be exclusive to the two anti-corruption typologies. Losses for these companies can be reduced through integrity pacts: intercorporate agreements to follow ethical principles in government interactions, thus creating a coalition to challenge corrupt deals. Several important Peruvian law companies currently operate such a pact, although it has a low profile nationally. The UN Global Compact, which many mining companies have adopted, operates as an international pact and requires compliance with the United Nations Convention Against Corruption.

Relativists regard petty corruption as inevitable and assess larger-scale corruption on a case-by-case basis, while *passivists* minimize upfront costs and resolve surprises as they arise. The *unethical strategic approach* uses sophisticated risk-assessment to optimize business opportunities and limit corruption risks. These companies may spend more on bribes to avoid detection and seek legislative loopholes and influence in rule making, while controlling employee theft and fraud through internal auditing. The biggest risks of unethical strategy are overconfidence and unanticipated change, as was demonstrated by the Siemens failure to foresee international legislative change.

Social investment

Social investment and stakeholder engagement together provide a myriad of smaller-scale corruption opportunities for companies. Local leaders may request direct or indirect personal or political benefits and may attempt extortion by threatening protests. For example, in Peru in 2007 I observed a mining company backtrack on merit-based local hiring to favor local leaders and avoid a prolonged protest, demonstrating a *relativist* approach on this issue, but risking further extortion.[9]

Transgressions that may be legal but are seen as corrupt by important stakeholders also occur in community relations. Gambetta (2002) explains these as contraventions of social, religious, moral–ethical, public opinion, or public interest norms. For example, in Peru Barrick Gold is viewed by many stakeholders as unethical for avoiding the payment of over US$100 million in taxes by acquiring their Pierina concession using a regulatory loophole, resulting in prolonged protests (Ramirez, 2008).

Passive and *relativist* companies are likely to lag in social investment and to follow local norms. *Unethical strategic approaches* to social investment risk being condemned as manipulative. For example, a local scandal arose in the community of Huallanca, Peru, during the 2006 elections, when community electronic espionage revealed that a mining company had surreptitiously paid election expenses to a preferred candidate. The community was outraged that the mining company would directly attempt to influence election results and the candidate lost.

Policy engagement

The EITI is an example of a direct anti-corruption policy engagement strategy. Miners work with government to pass laws requiring publication of all payments between government and extractives companies and stipulating multi-stakeholder oversight. The effectiveness of the EITI in reducing corruption is yet to be evaluated, but given that large bribes are typically hidden in payments to consultants, it seems unlikely the EITI will uncover corporate kickbacks. Its benefits are likely indirect, dissuading would-be bribe-seekers, promoting transparency, and raising the profile of the anti-corruption debate.

In policy engagement, ethical companies balance short- and long-term corporate interests. The prevalence of 'state capture' in the mining sector suggests that many companies use a variety of legal and illegal mechanisms to leverage economic and other benefits. *Unethical strategic* companies optimize this behavior, but lobbying is ubiquitous and ethical lobbying is poorly defined. As has been discussed above, government transfers need to be pro-poor to effectively combat inequality, while company lobbying is typically pro-rich (De Ferranti, 2003). The Peruvian mining industry's 'voluntary contribution,' which arose from negotiations between the Peruvian government and mining companies to avoid a promised windfall tax, is an interesting initiative in this respect. Companies pay up to 3.75 per cent of profits to funds that are co-managed with local committees to be used to further achievement of the Millennium Development Goals. The contribution has been criticized for reducing the potential tax burden on companies and for perpetuating company control over resources (Bebbington, 2007). However, the voluntary contribution directed almost US$1 billion to pro-poor development initiatives in regions where governments have proven incapable of spending allotted funds and producing gains which appear significant, although they have not been fully evaluated.

More positively, companies have potential to promote pro-poor development indirectly through local hiring, training, and small business development to produce goods and services for the mine. However, in doing so they may cross the interests of influential partners and affluent locals seeking personal benefits from the project. Overall, ethical companies need to adopt a multifaceted approach to policy engagement, partnering with influential ethical individuals and NGOs in order to avoid counterproductive inconsistencies. Finally, in all anti-corruption efforts Bray (2007b) emphasizes the importance of cultivating diplomatic skills and respecting local progress to avoid being viewed as arrogant.

Cost–benefit–risk

In extremely high-corruption settings, it may be impossible for ethical companies to compete, while unethical approaches are increasingly untenable due to rapidly evolving unpredictable risks for Western companies. However, Bray (2007a) argues that in Nigeria, where the CPI recently improved from 2.3 to 2.9, ethical partnership and avoidance of corruption are feasible with patience, creativity, and determination, suggesting that there is potential for a similar approach in most Latin American nations.

In low-corruption settings, a *passive* approach is probably adequate, while in high- and medium-corruption settings, an approach combining the strengths of *universalist* and *relationship* strategies should optimize benefits for ethical and Western companies, while at the same time counteracting the disabling effects of corruption on social and environmental value creation and limiting exposure to threats.

It appears, then, that corporate home place and country of operation are emerging as important determinants of corporate corruption management approaches in the region. The level of acceptability and prevalent forms of corruption are determined by the country operating environment, and the corporate response to this environment is determined by the corporate ethics management system. The extent of the resources that companies invest in anti-corruption depends on the importance given to ethics by corporate managers, which in turn rests on the norms in the corporate home place and the extent of anti-corruption incentives imposed by the home place jurisdiction.

Discussion and conclusions: Enablers and disablers of anti-corruption

In Latin America, the ethical imperative for anti-corruption among mining companies is clear; however, research into corruption management in the region indicates that ethical practice is far from being integrated or universally applied in the industry. In order to advance anti-corruption in mining from corruption-embracing to anti-corruption approaches, collaborative strategies are needed. These collaborative strategies should leverage and publicize anti-corruption enablers while counteracting context-specific dynamics that disable anti-corruption implementation.

Aggressive international prosecutions of corrupt companies have recently emerged as an important enabler of anti-corruption in Latin America, creating a credible threat to corrupt practice. The economic importance of mining in the region imbues a multiplier effect to this threat, as mining companies use rigorous due diligence to hire ethical contractors in their own self-interest. However, only US and European companies have, as yet, been prosecuted. The dissuasive effect would be enhanced by more consistent prosecution of more cases by more nations, including in particular Canada and Australia due to their dominance in the exploration sector.

Requirements for due diligence under international anti-corruption legislation have the potential to extend anti-corruption incentives to exploration companies focused on the short term by making the purchase of a concession with a corrupt history more risky. However, in order to prevent disadvantage for ethical companies, such processes should limit the costs to buyers if proactive disclosure is made as a result of due diligence.

Another important enabler of anti-corruption is the increasing awareness of the hidden costs of corruption, generated in large part by dedicated academics and non-governmental organizations. This awareness is complemented by a greater emphasis on longer-term risk management by financial institutions, advancing the incorporation of risk assessment into mining decision-making.

These enabling factors must be understood in the context of an industry that has a number of anti-corruption disabling factors, particularly in regions where corruption is socially and culturally embedded. The large revenue flows and vulnerability to extortion of mining operations attract corruption. Effective transparency initiatives, increased auditing, and integrity pacts directly counteract these incentives; however, they may be ineffective in high-corruption settings where indirect approaches are more tenable.

Anti-corruption in Latin America is disabled by both a low trust environment and by high social inequity, which has remained unchanged despite more than a decade of pro-poor development. In this environment, indirect approaches to anti-corruption aiming at changing the underpinning resource, capacity, and income inequity can be integrated into core business, social investment, and policy dialog strategies. Investment in local education, training, hiring, and contracting programs is increasingly recognized as a positive way of leveraging corporate needs in the service of local marginalized populations and may have anti-corruption spinoffs. The Millennium Development Goals focus of the Peruvian 'voluntary contribution' is another example. Community relations strategies of maintaining a commitment register, including informal land-owners in resettlement and communicating respectfully with stakeholders, also gradually build social trust. These actions should be combined with a reflexive analysis of corporate lobbying activities to prevent pro-rich corporate lobbying from neutralizing pro-poor initiatives.

Most of these initiatives require a longer-term focus and deeper, more integrated economic, political, and sociological analysis than is typical in corporations, along with a sophisticated stakeholder analysis to identify emerging opportunities. One way to address these complex needs is for companies to partner with governments and NGOs that have more expertise in these areas; however, increasing corporate social capacity and altering corporate structures to encourage interdepartmental innovation will also be necessary. Academics also have an important role to play by clarifying the interactions of corporate CSR strategies within the governance ecosystem to delimit anti-corruption enablers and disablers and, more broadly, guide social and environmental value creation. They will, however, need to find theory-light, pragmatic ways to communicate their findings if they are to be applied by busy executives.

Notes

1. See Lambsdorff (2005) for a review.
2. For instance, Lee and Ng (2006) find that a 1-point decrease in CPI correlates with a 10 per cent decrease in share price.
3. Figures presented here are derived from statistics presented in the report.
4. See http://www.antamina.com, date accessed 16 February 2011.
5. See http://www.riotinto.com, date accessed 16 February 2011.
6. See http://www.xstrata.com/, date accessed 16 February 2011.
7. See http://www.pelambres.cl, date accessed 16 February 2011.
8. See http://eitransparency.org, date accessed 16 February 2011.
9. Observed during doctoral fieldwork in the community.

14
Sustainable Juruti Model: Pluralist Governance, Mining, and Local Development in the Amazon Region

Fabio Abdala

Introduction

The Sustainable Juruti model is an initiative of Alcoa, local government, and community-based associations aiming to promote a sustainability agenda for the municipality of Juruti, west of Brazil's Pará State, due to the installation of a large-scale bauxite mine developed by Alcoa (Alcoa, 2009). Such enterprises in the Amazon are challenged to overcome social and environmental liabilities, as shown by the collapse of the Serra do Navio project in the neighboring state of Amapá which, after a period of 45 years of economic boom and welfare resulting from the mining of manganese, became a symbol of unsustainability for large-size mining projects (Tostes, 2007).

The governance model on which the Sustainable Juruti project is based can be understood as an unusual governance experiment in the region, promoting pluralistic decision-making processes with multiple local and non-local actors and based on new paradigms of sustainability. Such an approach to governance demands a capacity for systems thinking, moderation, and conflict resolution skills in the interest of designing positive agendas in the midst of intense divergence of interests and political projects.[1]

Governance and pluralism

Krahman (2003) distinguishes the concepts of government from governance as two ideal types based, respectively, on the concentration and fragmentation of political authority. In the absence of a unified authority, governance structures and processes allow government and non-governmental organizations to coordinate their needs and interdependent interests through the co-design and implementation of policies. Governance results from the establishment of arrangements for policy making beyond the government sectors, leading to increased inclusion and empowerment of civil and private actors in providing services to the public, and with a goal to enhance the effectiveness and efficiency in regulating social and economic issues.

This diffusion of political authority is associated with external and internal factors related to the state. On the one hand, the rise of civil society organizations (CSOs) and corporations in politics is a result of the strengthening of these actors in the institutional arena, developing a political culture that has promoted non-state public goods and collective action. In the context of globalization, companies are more easily able to cross borders in pursuit of maximizing their interests, and have been increasingly successful both locally and globally. On the other hand, this kind of governance is related to a dual crisis of the state: first, limitations of the state apparatus in providing social welfare to the whole population and according to the demands of hegemonic economic sectors, and secondly, limitations on the social responsibility of the state, or its ability to be accountable to the public and the consequent loss of social legitimacy. Both factors result in reduced performance and effectiveness of actions undertaken exclusively by the state.

The ideological foundations of the concept of governance emphasize a liberal view of empowering civic groups and the insertion of market principles and practices into government administration. Among its major vectors are demands for efficiency, quality, and transparency in the services provided by the state; streamlining of activities; adoption of projects management; focus on results and consumer services; reduction of the state apparatus; and outsourcing, privatization, and public–private partnerships to produce public goods and services in contrast to bureaucratic practices that focus more on processes and departmental management. In same sense, a governance approach also entails some degree of depoliticization of decision-making processes, thereby reducing the importance of state factors, as well as increased need for the state to share its authority with private actors and communities.

The establishment of governance can be guided both by an elitist approach and as pluralistic decision-making (Alford and Friedland, 1992; McCool, 1995; Clemons and McBeth, 2001). The governance of sustainability policies tends to follow more a pluralistic praxis of the political process, where the decision-making system is permeated by a broader set of actors and interest groups. In the long term, organizing and coordinating differences among political actors is necessary to involve more directly the forces that shape social conflict and cooperation, incorporating different interests, identities, institutions, and values. Pluralists assume that centralized decision-making, planning, and large-scale structures are too rigid, creating bureaucratic difficulties for operation in the new environment of diversity and complexity that comprises sustainability governance.

In this way governance implies a coordination process among different groups of actors with different needs, conflicts, and disagreements in search of mutual benefit, fairness, and effectiveness. Players involved in the game are able to generate more cooperation when they can adjust their strategies in response to the interests of other participants as a function of the learning generated in the relationship, which allows a system of governance to change its policies and conditions for improved participation in collective action.

Pluralistic governance would imply the implementation of consultative processes, negotiation and collaboration, conflict management, and collective learning. Such schemes may be temporary and incremental, and while they may not necessarily surpass conflict, they seek to minimize it. Further, decisions are interactive, allowing players to continually move at different levels of agreement and disagreement, and produce adjustments, new problems, and opportunities for collective action (Wollenberg et al., 2005).

Considering the social–private–public participation in the Sustainable Juruti initiative and in control over the council, funding, and system of indicators, this model can be understood as a pluralist governance experiment aiming to promote local development.

Mining and sustainability: Assumptions and intervention tripod

Under the impact of new development paradigms, mining is challenged by two new factors: first, that social license to operate is a key element of its sustainability, and secondly, in the long run, mining development cannot sustain itself as an 'island of prosperity' in an environment of poverty and institutional instability. Therefore, it is necessary for mining projects to generate multiple benefits (Enriquez, 2007).

In face of those challenges, the Sustainable Juruti model was designed based on three assumptions to guide the construction of a local development agenda (Monzoni, 2008):

1. The concept of social coordination spaces, thus allowing for broad and democratic participation of the community in building the agenda toward a common future.
2. A focus on local territory, which considers the host city as a generator of development in the region.
3. A dialog with reality, which shapes the agenda in view of the local demands, regional public policies, and contextualizes the agenda within global and business initiatives focused on sustainability.

These assumptions give the background to a tripod of interventions that promotes multi-institutional partnerships (public authorities, communities, civil organizations, enterprises), whose components and purposes are as shown in Figure 14.1. The components of the Sustainable Juruti model are as follows:

1. The Sustainable Juruti Council is a permanent forum for dialog and collective action between society, government, and businesses, in which the common future of public interest is discussed. It prioritizes actions and formulates a long-term agenda.
2. Sustainability indicators provide the tools for monitoring the development of Juruti and its surroundings and feed the council with qualified information for the decision-making process.
3. The Sustainable Juruti Fund complements local funding for development actions based on council-prioritized indicators, and mobilizes resources to generate financial wealth for present and future generations.

This model was conceived and designed from the core challenge that presents itself in Juruti, which has been defined as:

> the installation of a large enterprise in a region of high biodiversity, social organization and public authority unprepared to face a horizon of large and rapid changes, as well as the lack of financial resources to meet the demands of the population. According to this scenario, such insertion carries potentially significant, long-term implications on a wide range of players, with huge social, economic, political and historical differences, as well as on the environment. (Monzoni, 2008, p. 8)

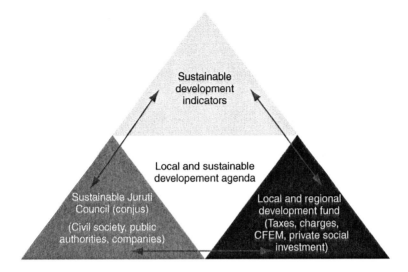

Figure 14.1 Sustainable Juruti model

In addition to addressing the economic and socio-environmental impacts of the mining development, the promotion of the Sustainable Juruti model would be a strategy to generate mutual benefits for all social, public, and private sector stakeholders.

Companies would benefit as tensions with communities, governments (local and regional), and other stakeholders are reduced, thus greatly contributing to the social license to operate. Likewise, there would also be benefits for the image and reputation of the company, from a local to a global level, by identifying it as the leader of an innovative model that seeks sustainability.

There would be social benefits for the current generation as it takes part in the strategic decisions about the allocation of the company's socio-environmental investments, connecting them with those promoted by the government, which in turn would benefit from investments in institutional development, planning tools, and social participation provided by the council and the indicators. There would be environmental benefits stemming from the investments in conservation and sustainable use of natural resources through the fund.

Further, there would be benefits for future generations in that part of the mineral income would be intended for an endowment fund, thus creating a financial legacy (savings) beyond what is invested in the short term.

Implementation, participation, and control

The governance of the model is based upon a partnership between civil society organizations, communities, businesses, and government, for which the Sustainable Juruti Council (CONJUS) is a privileged institutional forum. Created in 2008, this is a permanent forum for dialog and collective action among partners oriented for the sustainable development of Juruti.

The CONJUS Board of Directors is made up of 15 members who are representatives of the groups shown in Table 14.1.

The Council is characterized by pluralism and diversity of participants. It is organized into eight 'theme chambers' to discuss key issues in the municipality including security, land issues, production, and rights, amongst others. Currently, based on the diagnostic generated in the first results of the sustainability indicators, the CONJUS is in the process of preparing the *Agenda 21* of Juruti, a comprehensive blueprint of action to be taken locally by public and private organizations aiming at sustainable development.

The sustainability indicators are the result of the system for monitoring the development dynamics of Juruti and surroundings. This system is implemented jointly with Alcoa, the City Administration, and local civil society organizations, with technical support from the GV Foundation (FGV). It keeps records on Juruti's baseline indicators as well as the recent changes, creating a 'thermometer' tool for municipality management.

Table 14.1 Sustainable Juruti model

Social	1.	Fishermen's colony Z42	2.	Children's Pastoral Commission	3.	Guardian Council
	4.	Rural Workers' Union	5.	Women's Association	6.	Commercial and Business Association
	7.	Union of Rural Producers	8.	Juruti Association of Disabled Citizens	9.	Construction Workers' Union
Government	10.	City Administration	11.	City Council	12.	Emater
Companies	13.	Alcoa	14.	GRSA	15.	Hotel Garcia

The ultimate goal of the sustainability indicators is to help make the public aware of the changes they have been experiencing, to collectively define the directions they desire to follow and to evaluate the paths taken – thus strengthening the good actions in progress or, if necessary, correcting the course of its development. The sustainability indicators allow for CONJUS to guide the lines of action for the Sustainable Juruti Fund (FUNJUS), which also represents a partnership amongst the city administration, local civic organizations, Alcoa (first sponsor), and the Bank of Pará, with the technical support from the Brazilian Biodiversity Fund (FUNBIO).

FUNJUS is a long-term financing tool oriented for the sustainable development of the territory by encouraging the creation of value in the four sustainability pillars or types of capital: human, social, environmental, and economic. FUNJUS is a promoter and catalyst, a lever for local development. It adds to but does not replace the government as a financial backer of public policies. Given the recent transformations in the city, it considers the impacts, imbalances, and perspectives for the stability of Juruti (FUNBIO, 2010). In late 2009 while in the pilot phase of implementation, the first public notice of FUNJUS was issued. It is financing 21 community projects, mobilizing R$ 550 thousand from an overall budget of R$ 2 million.

First results

Up until 2006, Juruti had a population of 42,000, 60 per cent living in the rural area (IBGE, 2010) with low incomes and with no access to basic social services (health and education) – the tenth worst Human Development Index score in Pará State (UNDP, 2000). In addition, there is the vulnerability of the Amazon environment, both in terms of the fragility of forest ecosystems and the absence of the government. With the development of the Juruti Mine, these social and institutional weaknesses have become largely explicit. The company, in turn, was not adequately prepared to deal with the social and environmental conflicts that erupted under the leadership of communities and public agencies, even jeopardizing the issuance of the installation and operation license.

In 2006 and 2007, the MPE (Office of the State Public Prosecutor) filed a civil action against the project. At the same time, relations with the municipal administration deteriorated and the communities organized several protests against the project, including rallies promoted by the 'Movement 100 per cent Juruti' with the slogan 'Alcoa out of Juruti.'

The economy was growing, along with urban and environmental problems arising from the mine, and critics wondered what to do with the mineral income from installing the mine. How could that income be invested in strategic sectors of local development and how would it benefit the local communities? Will they dig out the bauxite, leave a hole, and go away? What legacy would be left for future generations? These questions had poor or incomplete answers and there lacked room for dialog among the stakeholders on local development, as well as 'critical mass' to give direction to what was coming up.

As the sustainability tripod was implemented and the stakeholders invested in dialog and building consensus, they began to organize ideas in the form of action plans targeted at the economic, social, and environmental sustainability and created means to implement the actions.

Through the Sustainable Juruti Council, it was possible to start a partnership involving the community, the government, and the private sector as a process for cooperation and for seeking joint solutions. In 2008, the directorate was elected among the city administration, Alcoa, and the 'Movement 100 per cent Juruti.' The level of conflict between the Juruti Mine, communities, and the local government was reduced, and also a place was created to settle controversies, give transparency to various interests, and generate agreements between parties.

The indicators (more than 80 items in the human, social, environmental, and economic fields) qualify the dialog among companies, communities, and government, seeking to balance the social and institutional interests with technical information. For such outcomes to be achieved, technical workshops, community meetings, Internet research, and public hearings were held, which mobilized more than 600 participants from 115 communities and 71 institutions, with 90 contributions through an online survey. In this process the most important factors that arose were: (1) health, education, and safety; (2) water, agriculture, financing, technical assistance, and land tenure; and (3) transportation, communication, energy, and waste.

In view of such outcomes, the parties started to develop a local sustainability plan based on *Agenda 21*, with responsibilities in common but differentiated – for example, the companies contribute, but do not replace the role of government and society.

The starting fund seeks to demonstrate the feasibility of new ways to produce (fish, vegetables, services), to preserve nature (fishing and lake management agreements) and to develop society (environmental education, comprehensive care for adolescents at risk). Today, there are 21 projects implemented in urban and rural areas, mobilizing about R$ 550 thousand, as shown in Table 14.2.

Table 14.2 Projects funded by FUNJUS, 2009/2010

	Type-1 Projects – Community Associations – Up to R$10,000		
	Applicant	Project	Objectives
1.	APROFASP	Cheiro Verde	Improve the structure of producers' vegetable gardens
2.	São José Producers Association	Assistance to Poultry Farmers	Renovation and expansion of the poultry breeding farm for chicken breeding
3	CTPJ	Structuring and Adequacy of flour mill	Construction of a flour mill with structure for production of improved-quality cassava derivatives
4.	Araçá Preto Producers Association	Support for Organic Vegetable Growers	Strengthen the productive activity through the implementation of community garden and legalization of the association
5.	APROs	Farming of Tambaqui fish in net tanks	Breeding Tambaqui fish in net tanks (200 kg of fish)
6.	São Braz Producers Association	Breeding of Native Bees	Implementation of bee keeping in the Amazon
7.	São Pedro Producers Association	Income generation for Rural Family Production through the breeding of laying hens	Generate income from the rural household production through the breeding of laying hens
8.	São Benedito Community Association	Amarrando Sonhos	Income generation through handicraft and extractivism for family sustainability
9.	ASPROFAGU	100% Fingerlings Breeding of Tambaqui fish in net tanks	Breeding Tambaqui fish in net tanks (200 kg of fish)

Table 14.2 (Continued)

10.	Nova Galiléia and Nova Esperança Producers Association	100% CURUPIRA – Breeding of Tambaqui fish in net tanks	Breeding tambaqui fish in net tanks (200 kg of fish)
11.	Santa Terezinha and Lago Preto Producers Association	Consolidation and strengthening of associations in the plateau region	Consolidation and strengthening of APROFASP, APROSEIS, ASPRUFARSHI
12.	Grupo Unidos Producers Association	Consolidation and strengthening of associations in the Lago Juruti Velho region	Consolidation and strengthening of associations CTPJ, ASPROFAGU, ASPEFANGE
13.	Santo Hilário Producers Association	Structuring of the family vegetable gardens of the communities in the Santo Hilário Region	Improve the structure of the producers' vegetable gardens

Type-2 Projects – Community and Civil Associations – Up to R$ 50,000

	Applicant	Project	Objectives
14.	MOPEBAM	Fishing Agreement at Lago Grande and Curai	Develop and implement Community Fishing Management Plans
15.	Pró-Tartarugas Association	Puxirum of Curumins – First Meeting of the Turtle Club	Hold the 1st meeting of the Turtle Club
16.	APRAPAEB	Juruti Pescados	Enable partners to implement the Project through courses that contribute to the development of skills and expertise of those involved, in order to ensure the success of the fish-breeding and marketing business
17.	APRAPAEVID	Pesca Milagrosa – fish breeding in net tanks	Breeding Tambaqui fish in net tanks
18.	ACEJ	5S Program – Juruti	Creating the culture of Total Quality by implementing the 5S

			Program (senses of the Use, Sorting, Cleaning, Hygiene and Health, and Self-discipline) in 30 companies in Juruti
19.	Bairro Santa Rita Community Association	Da Rua à Cultura	Rescue children and teenagers of the Santa Rita district and other districts, with the consent of the child protection agency, to integrate the program
20.	Bom Samaritano Association	Building and feeding with quality	Provide students of the Rosa de Saron School with quality service in the distribution of school meals by erecting a cafeteria and kitchen
21.	Juruti Artisan Association	Tucumán	Provide capacity-building for handicraft workers and infrastructure for making biojewels

Conclusion

The Sustainable Juruti model is demanding in terms of: (1) qualifying leaderships for dialog using appropriate social technologies, (2) dedicated staff and resources, and (3) mining operations that can be adapted to social and institutional demands.

Benefits may be either internal or external to the enterprise. Examples of internal benefits include anticipation, management, and (consequently) the reduction of risks arising from social and institutional conflicts generated by the project; strengthening of the license to operate; employee satisfaction through being part of a company recognized positively by society and its peers in the market; and the company's strengthened brand and reputation in the market. The external benefits include private social investment and corporate responsibility in line with the development goals set by local communities and the government, creating convergence and economies of scale; a space for direct, transparent dialog between the stakeholders and the company without patronage; and creation of a financial and institutional legacy generated by the mineral income that stretches beyond the closing of the mine.

The impacts and benefits of this type of corporate approach, jointly with communities and governments, can be observed in the short term, but will be more consistent and durable in the medium and long term. Short-term results include reduction of conflicts; increased trust amongst businesses, society, and government; a mechanism for dispute settlement; promotion of joint initiatives and mobilizing partnerships (public and private, local, regional, and national) to promote local sustainability; increase in corporate social responsibility and private social investment; increase in public transparency on private investments, budget, and use of public resources; and forwarding of social demands for public services to government agencies, thus allowing for directing private social investment in a complementary and strategic manner.

In the medium and long term, the highlights include participatory governance of the territory; increased living quality and levels of institutional and social development; increased presence of the state through increased control, supervision, and services; gradual withdrawal of investments from the mining project in providing public services that are typical of the state); social and economic inclusion of communities; development of local value chains (agribusiness, forestry, and services) with greater local autonomy in relation to the mineral chain; establishing conservation units of full protection and sustainable use; endowment fund for investments in local development; and mine closure without causing a collapse in local society.

Note

1. It should be noted that this chapter results mainly from participating observations of the author, an Alcoa employee. Although there is a risk of the author's close relationship to the subject matter, this chapter is based on the author's engagement in experimenting with innovation through sustainability in the mining business. Further, the opinions expressed in this chapter are the sole responsibility of the author and do not represent the opinions of Alcoa.

15
Energy and CSR in Trinidad and Tobago in the Second Decade of the Twenty-first Century

Timothy M. Shaw

Trinidad and Tobago (T&T) has been an energy producer – from oil to gas – for over 100 years. During that century, the relationship between the state and corporations has evolved, particularly since independence and now in a new century/decade/regime. At the end of the first decade of the new millennium, T&T presents a set of overlapping insights into (1) 'emerging economies' and 'developmental states'; (2) the prospects for small-island developing states (SIDS); (3) opportunities and constraints of 'globalization(s)'; and (4) prospects for Corporate Social Responsibility (CSR) in such a context of energy concentration. The contribution of energy to Gross Domestic Product (GDP) has grown exponentially, from a quarter in the late 1980s to almost half today, with liquefied natural gas (LNG) and petrochemicals being a growing proportion particularly since the early 1990s (Guyadeen, 2010, p. 91).

This chapter seeks to juxtapose the above discourses/genres informed by cycles in the national economy of T&T, especially its dramatic growth from the turn of the century until 2008 when an all-too-familiar downturn began (CMMB, 2009; EIU, 2009). This growth is likely to extend into the new decade, reflective of structural shifts and cycles in the global economy (Cooper and Subacchi, 2010) with T&T being the only one of the 'big six' in the region likely to enjoy any growth in 2009 (Business Monitor, 2009). But I also reflect on an unexpected change in regime in mid-2010 – a different generation and gender, as well as party in government,[1] which led to the cancellation of a major new sector under development: an aluminum smelter that had been intended to extend the established industrial usage of gas for beneficiation. It also led during the new regime's honeymoon period to a new openness about CSR, including deciding to apply to become a candidate country in the

Extractive Industries Transparency Initiative (EITI), establishing a high-powered representative national committee to this end chaired by the previous chair of Transparency International in T&T, Victor Hart.

T&T in comparative perspectives

In this chapter, first, we suggest the possibility of extending the notion of 'developmental state' from Asia (Chang et al., forthcoming) and Africa (Mkandawire, 2001) to the Caribbean, given the impressive growth levels achieved in T&T (along with one to two 'dependent' 'overseas' territories like Bermuda, British Virgin Islands, and Caymans) (Clegg and Pantojas-Garcia, 2009) at the start of the century. Moreover, T&T has limited its dependence on unprocessed energy revenues through diversification into production of related downstream, semi-manufactured raw materials like methanol, nitrogen fertilizers, and urea, and of heavy industrial goods like steel (Mottley, 2008) along with its ambitions to become the center for the region's financial and service sectors, as indicated below. Current development might be sustainable if such diversification is maintained and extended (Tewarie and Hosein, 2006), but direct and indirect dependence on natural gas may be more debilitating than recognized: a curse if it reduces the imperative and will to diversify. We turn below to the possibilities of and prospects for diversification.

Second, in terms of going beyond established small-island development state (SIDS) perspectives, T&T serves to both reinforce and transcend orthodox dichotomies (Cooper and Shaw, 2009). The pioneering work of Briguglio et al. (Briguglio and Kisanga, 2004) on the vulnerability of SIDS in the 1990s has come to be balanced by consideration of the resilience of at least some SIDS, like T&T (Baldacchino and Milne, 2006; Kisanga and Danchie, 2007). In turn, the orthodox SIDS approach needs to extend its purview to include several rather successful non-independent overseas territories (OTs) like Bermuda, symbolized by international banking in the Caymans (Vlcek, 2007, 2008) and significant Chinese and other offshore companies in the British Virgin Islands, taking into account the relative affluence of a not-insignificant proportion of such OTs (Clegg and Pantojas, 2009).

Furthermore, third, the burgeoning literatures on the alleged 'resource curse' (Auty, 1993) and on conflict over resources like energy (Klare, 2001, 2004), exacerbated by demand from Brazil, Russia, India, and China (the BRICs) (Cooper et al., 2006), rarely mention T&T. Nevertheless, it is presently the fifth largest producer of LNG, the price of which rose 50 per cent in the first half of 2008, only to fall to historically low levels in 2009. Further, T&T is the largest global producer of methanol, as well as being the primary source of LNG for the US, even if imported gas is but 3 per cent of its supply at present.

Fourth, such revisionist perspectives may become imperatives rather than options because of the current global conjuncture (Shaw, 2010). This is characterized by the decline if not demise of US hegemony and its converse: the rise of the BRICs and the 'Next-Eleven' (N-11) – Bangladesh, Egypt, Indonesia, Iran, Mexico, Nigeria, Pakistan, Philippines, South Korea, Turkey, and Vietnam (O'Neill and Stupnytska, 2009) – along with at least some developmental states like Singapore and Taiwan. Given the parallel erosion in the salience of the Washington Consensus, might any Beijing or Delhi 'consensus' open up unanticipated policy space for small, including island, states? What are the implications for SIDS like T&T, both shorter and longer term? Such dramatic shifts in global tectonics should inform emerging debates about appropriate architecture for sustainable global governance in the embryonic G8/G20 nexus, especially when it includes the cooperating trio of BRICs in addition to Russia as in recent summits (Cooper and Subacchi, 2010).

Fifth, we also seek to bring in notions like supply chains, logistics, networks, and standards that are compatible with the omnibus concept or nexus of globalization. The latter could not have emerged in its contemporary form without the former trio. Earlier globalizations may have relied on sail and then steam, along with the underwater telephone cable. Contemporary globalization requires sea and air containers, computer-tracked logistics, and networks that sustain extensive and ubiquitous supply chains (for example, coltan from the Congo for cell phone connectivity). T&T can aspire to being an airfreight hub if its claims in terms of airline or container hubs are less persuasive. Such concepts around supply networks should not be the monopoly of either corporate managers or business schools, as they are central to contemporary development as indicated by the burgeoning perspectives of Stefano Ponte et al. (Gibbon et al., 2010) *inter alia*. Meanwhile, the parallel analysis of 'varieties of capitalism' – the proliferation and expansion of Southern multinationals – has begun to get some, if not yet enough, attention (Goldstein, 2007).

Sixth, regional MNCs, especially Trinidadian companies like Ansa McAl, Caribbean Airlines, Guardian, Republic Bank, and Trinidad Cement (TCL), have begun to define a new non-state regionalism through such networks, virtual and tangible (*Business Trinidad and*

Tobago 2009–2010, 2010), as suggested by Trevor Farrell (2005). However, there is another side to such globalization and regionalism. T&T is in a unique space between Latin America and the US/EU, one that the global drug trade has come to appreciate. Serving as an entreport for the drug supply chain complicates economic and security relations with the two largest partners of T&T – the US and EU. A crucial question is how to maintain the legal flow of goods and services while containing the illegal, which also exacerbates inflationary pressures because of myriad forms of money-laundering (Vlcek, 2007, 2008).

Finally, in the run-up to the December 2009 Copenhagen conference both regional and global corporations in T&T were under pressure from consumers, shareholders, managers, and other stakeholders to advance CSR/corporate codes of conduct (CCC), particularly in terms of ecology. The South Trinidad Chamber of Commerce (STCIC) pioneered CSR policy in part because of its centrality in its sector of concentration – energy. But other chambers and sectors are increasingly doing likewise, even if T&T had eschewed being considered under EITI rules at least until the application from the new coalition government in late 2010 (refer to concluding section below). CSR in the South, especially around energy and mining (Canel et al., 2010), is the other side of the rapid proliferation and expansion of Southern MNCs (Goldstein, 2007; van Agtmael, 2007; Sirkin et al., 2008). But after a flurry of comparative analyses mid-decade (Haslam, 2004; Peinado-Vara, 2004), interest in CSR seems to have waned, perhaps in part because most economies and companies are small and thus being engaged in CSR is demanding. However, the seventh CSR Americas hemispheric conference did take place in Uruguay at the end of 2009, indicating a continued regional engagement with CSR issues.

Given the continuing – albeit uneven – global crisis, a debate has arisen about whether T&T is in 'recession.' Clearly its economy was growing less strongly at the end of the decade than in the middle, and the Central Bank has warned that return to growth will be slower and later in T&T than some economies. Lay-offs, declining imports – especially of cars – and fewer house sales led to another deficit in 2009/10. The EIU 2009 country report stated that: 'After slowing in 2009 to 3.5 per cent, GDP growth will weaken further in 2009 to 0.9 per cent as the global recession bites. A mild global economic recovery will help to lift GDP growth to 1.9 per cent in 2010' (EIU, 2009, p. 2). As LNG prices have been persistently low at the turn of the decade because of the unanticipated impact of shale gas in the US, a return to growth – let alone boom – is not anticipated.

I turn next to a trio of different rankings that place T&T in order of levels of growth and competitiveness.

The political economy of T&T at the start of the second decade of the twenty-first century

First, according to the World Bank's *World Development Indicators* for 2007, T&T has been in the top 15 growth economies for the decade 1995–2005 (no. 13 globally), while it has also had a stable population of some 1.3 million. Its gross national income (GNI) per capita has almost doubled in five years: from US$5230 in 2000 to US$10,300 in 2005. In 2007 its purchasing power parity (PPP) GDP per capita was US$23,507 (UNDP, 2009). T&T also scores quite well according to the UNDP's 'United Nations Development Programme' Human Development Index. In 2009 it was in the middle of the 'high human development' group, ranked 64 out of some 182 states: below Barbados and Venezuela but above Malaysia, Russia, and Mauritius (p. 71). But whether even T&T can sustain growth with or without development given the global crisis remains problematic (CMMB, 2009; IMF, 2009).

Second, in terms of the different criteria in the annual *Global Competitiveness Report 2007–2008* from the World Economic Forum, T&T is 84 out of 139, having been 84 out of 132 a year before. The report indicates that 'the most problematic factors for doing business' were crime and theft, poor work ethic, corruption, and inefficient government bureaucracy. The only other Caribbean state ranked was Barbados at 43, but Venezuela was ranked 122.

And third, according to a specialized 'small state competitiveness index' compiled by Monique Pollard (2007, p. 52), which takes into account merchandise exports as per cent GDP, per cent share of first product in export value, and recent manufacturing growth rate, T&T was, at least before the global crisis, 8 out of 40, behind Estonia, Bahrain, and St Vincent but ahead of the rest of the Caribbean as well as Mauritius, Qatar, Cyprus, and Botswana, several of the latter set being widely recognized to already be 'developmental states' (Mkandawire, 2001). Further, Joiner and Wignaraja (2007, p. 133) propose a 'small states manufactured export competitiveness index' (SSMECI) in which on a per capita basis at the start of the century, T&T comes 4th after Malta, Botswana, and Estonia. Moreover, unlike some Caribbean islands like Jamaica or Puerto Rico, migrations, diasporas, and remittances are relatively less crucial for T&T (UNDP, 2009; World Bank, 2006), though many of the professional upper middle classes hold two passports and have transnational extended family in the US and Canada, and less frequently in the UK.

Given its small population, T&T is one of the highest energy producers in the world on a per capita basis. It trails other underpopulated, big energy exporters like Qatar, Kuwait, Brunei, Norway, and Equatorial Guinea. Its per capita income has risen rapidly along with the price of oil and gas, especially since the middle of the decade. T&T's 12 per cent growth in 2006 was in part a function of mega-projects coming onstream, so construction and manufacturing sectors were booming along with offshore energy. But unlike some of these small states with big economies, it has not created a huge sovereign wealth fund (SWF) (Xu and Bahgat, 2010), just a modest stabilization and heritage fund (SHF). And because of the dramatic technological impact of gas extraction from extensive 'shale' formations in the US and Canada, while the price of oil has increased because of unrest on several parts of the Middle East region at the start of 2011, that for natural gas has remained stubbornly low

According to the IBRD, FDI per annum in T&T has risen from US$700m in 2000 to $1.1 billion in 2005, historically primarily from the US, but now increasingly from the UK too, although bilateral trade with the UK remains very limited. But this may change with the operationalization of the new, not uncontroversial, Economic Partnership Agreement (EPA) between CARIFORM (CARICOM and Dominican Republic) and the EU, including the UK; Canada is also in talks for a Free Trade Agreement (FTA) with CARICOM (as well as with the EU). Meanwhile, inward remittance flows have also doubled: from US$38 million in 2000 to US$87 million in 2005, but still just US$69 per capita in 2007. However, unlike Haiti or Jamaica or several states in Central America, T&T does not appear in the top twenty lists for remittances (World Bank, 2006).

It remains to be seen whether the impact of the unpopular Caribbean EPA with the EU is as negative as its detractors predict. Clearly its impact on T&T will be very different than for most of the region as the country is not dependent on agricultural exports or manufactured imports, remittance and tourism flows, and so on (IMF, 2009). Meanwhile, in terms of accountability, T&T's score and ranking on the Transparency International (TI) Corruption Perceptions Index (CPI) have remained stable; in 2010 it was ranked 73 out of 178, below Brazil and way below Barbados at 17, tied with Japan, but just ahead of China (Transparency International, 2010a).

Diversification of the economy of T&T?

The Energy and Related Industries sector, which is comprised of mainly foreign companies, plays a dominant role in the promotion of social and environmental programmes in T&T. That sector accounted for more than half of the total monies spent... It should also be noted that four companies alone accounted for over 30 per cent of the total monies spent. (STCIC and UNDP, 2007, p. 55)

Trinidad has diversified since independence both within and outside the dominant energy sector. Within the oil and gas industry, it has invested significantly in downstream use of natural gas – its erstwhile 'Point Lisas strategy' (Mottley, 2008) based on the name of the industrial park where the set of beneficiation plants is located, between the two major cities of Port of Spain (PoS) and San Fernando. Initially this strategy involved production of other gases for the ubiquitous petrochemical sector, but subsequently it has included the manufacture of steel as well as nitrogen fertilizer (Guyadeen, 2010). Outside the energy sector, it has diversified its tourism, including ubiquitous and profitable cultural industries (Ho and Nurse, 2005; Nurse, 2007, 2009) and begun to develop a service sector with an initial focus on finance (TTCSI, 2007–09).

Trinidad has produced oil for over a century, having earlier extracted oil products like pitch for boat keels from its unique lake of asphalt. The pre-WWI oil rush led to the first exports of oil in 1910, financed by the Union Bank of Halifax (NS) which was established in PoS in 1902, a predecessor to RBC (Royal Bank of Canada). Despite this history, T&T did not start exporting LNG until 1999. While it has major reserves of gas, depending on rates of extraction, tax regime, and new reserves, it could export for 25 to 50 years. It now has four trains for LNG production at Point Fortin, down the coast from Pont Lisas, generating 10 per cent of global production, at least until Qatar's Ras Laffan began to come on stream at the turn of the decade. Atlantic LNG is a global consortium of British (BG and BP) and Spanish (Repsol) corporations with Trinidad's own national gas company. Interestingly, perhaps symptomatic of their deficit in terms of image or reputation as large CO_2 emitters, all three partner multinational corporations (MNCs) from the North are signatories to the UN Global Compact (United Nations, 2009) and are 'supporting companies' for the EITI (EITI, 2009). And while most of its oil in the western Gulf of Paria is heavy, BHP Billiton has discovered lighter reserves off the east coast. There are already seven methanol plants in Point Lisas, including M5000, the world's largest, with another seven plants being constructed within the Ammonia Urea Melamine (AUM) Complex (see more below). With the cancellation of the aluminum smelter after the mid-2010 election, if Trinidad does not diversify further there will be lower domestic demand for gas and hence a disincentive to explore for new reserves.

As the STCIC and UNDP (2007) report indicates, the half-dozen major oil and gas companies are the leaders in CSR. Atlantic LNG emphasizes community, education, and health, particularly in the Point Fortin area in the southwest corner of the country; BP focuses on arts and culture, education, and enterprise development and capacity building; and BHP Billiton on the Turtle Village Trust and on reducing CO_2 emissions and flare and vent volumes. Several of these MNCs list payments to the state (25 per cent revenue from BP's affiliate in T&T, BPTT, alone), contribution to the local GNP, and so on (since start-up in 1999, by October 2010 Atlantic LNG had loaded 2000 cargoes and provided 75 per cent of LNG exports, almost 60 per cent of US imports of LNG). Atlantic LNG rebranded itself in late 2010 with a heightened CSR dimension.

Trinidad has diversified from sugar and tourism to oil then gas and now ammonia, methanol, and urea, as well as steel and nitrogen fertilizer, but what will its balance be over time? Simultaneously it has abandoned most agricultural commodities like cocoa and rice. This movement raises many questions: how to maximize sustainable development and financial reserves? How to avoid an Asian-type crisis: what lessons can be learned from the mid-1990s from the NICs? How to balance exchange with the US/EU with that with the BRICs/NICs, especially given the uneven impact of the current global crisis that has affected the transatlantic world most negatively? How to diversify sources of foreign investment from established economies like the US/UK/EU to the booming energy and mineral demand of the BRICs and onto the N-11 (see below in this and end of next section) (O'Neill and Stupnytska, 2009)? And what balance between regional exports of manufactures and global exports of gas and oil along with ammonia/methanol/urea and fertilizer/steel?

Trinidad has moved away from oil toward gas as the latter's price rise begins to parallel that of the former (Guyadeen, 2010). The price of gas stagnated at US$6–8 per MMBtu between March 2006 and March 2008. Since then it has both risen and fallen dramatically, especially between the start of 2008 and mid-2008 (from US$7 up to US$12.50 per MMBtu) but stagnated at below US$3 in mid-2009 – a 7-year low. This low was a function of declining demand and increasing supply as enhanced drilling techniques reveal large new non-traditional reserves, especially in shale rock formations in North America.

As oil reserves in T&T decline, it is moving to process offshore oil. Its second refinery, to be constructed by local capital, is to refine Brazil's newfound light crude for the US market. Because of US concerns about safety, in future some T&T LNG – in this case generated by Canadian Superior and Global LNG – will flow through an underwater pipeline from an offshore import buoy to New Jersey/New York, in addition to LNG via New Brunswick, Canada, mentioned below.

Price of and demand for Trinidad's LNG is in part a function of gas' clean image, but supplies are growing as new sources are found (for example, shale gas in Canada in British Columbia and Quebec, and in several parts of the US to match tar sands in Alberta), and Qatar has become a major exporter along with Australia and one or two West African countries. On the other hand, facilitated by its long-standing connection with markets on the US east coast, T&T companies, especially Repsol, have a new market in eastern Canada as supplies to a new regasification plant in New Brunswick come from Point Fortin. Repsol owns 75 per cent with Irving owning the remainder of the Canaport plant outside Saint John, NB, which ships gas south of the US border. T&T delivered two shiploads of 140,000 cubic metres per tanker in its first month of operation – late June to July 2009 – and four shiploads by the end of August, over its first ten weeks of operation. But the annual Ryder Scott estimates of gas reserves continue to decline – by another 10 per cent in mid-2009, indicating proven reserves for a decade and possible reserves for two.

CSR in T&T in the twenty-first century

As an aspect of becoming 'developed' under the previous People's National Movement (PNM) government, private sector companies and agencies in collaboration with UNDP and the UN Global Compact undertook a national CSR mapping exercise in 2007 (STCIC and UNDP, 2007). This was followed in late October 2009 by a Caribbean consultation in T&T on CSR (UNDP, n.d.). In early October 2010, the T&T Manufacturers' Association (TTMA) with UNDP launched the UN Global Compact in T&T to advance CSR (TTMA, n.d.b). Meanwhile, the national business school at the University of the West Indies (UWIO) created a Centre for Corporate Responsibility at the turn of the decade.

CSR does not exist in the informal or illegal economy, although drug lords often play the role of 'good' Robin Hoods for the communities that host or hide them. CSR also tends to be weaker in the SME/MME regional light industrial sectors. Conversely, it is strongest in the large-scale energy sector, particularly around MNCs who practice CSR on a global scale, in part in response to global attentiveness and scepticism. The CSR 'system' in the hemisphere is concentrated in the large economies, especially Brazil and Mexico (Haslam, 2004) – for example, the BRIC and N-11 countries respectively – and involves a set of heterogeneous actors, many of which are well-established institutions. According to Haslam:

> The system of CSR promotion and advocacy is well-established in Latin America and the Caribbean. It is supported financially by external agents such as the OECD, IDB, OAS, private foundations, international NGOs and the home offices of multinational enterprises; and it is run through a network of local civil society organizations, government offices, academic institutions and of course private companies. (Haslam, 2004, p. 14)

At present, aside from remittances and a few non-national regional investors like Sagicor from Barbados, T&T has a relatively limited range of sources of FDI and/or franchises reflective of the character of its gas-based economy: Australia (BHPBilliton), Canada (First Caribbean Bank, Flow, Methanex, Petro-Canada, Potash Corporation of Saskatchewan, RBC, Scotiabank), Germany (DHL), India (Mittal), Norway[2] (Norsk Hydro; now Yara), Spain (Repsol/YPF), Switzerland (Nestle), UK (BG, BP, Shell), US (Citibank, Coca-Cola, FedEx, Hilton, Holiday Inn, Hyatt, Marriott, UPS, and several freight and passenger airlines). Some of these interests are advanced though bi-national chambers of commerce such as the American–T&T Chamber and now an Indian–T&T Chamber, along with national or regional chambers like the T&T Chamber of Industry and Commerce and the Energy Chamber in Point Lisas (previously the South Chamber in San Fernando). But how much control does T&T have over such investments? The recent case of nationally owned and significant Royal Bank (RBTT) being taken over again by RBC of Toronto, its owner four decades ago, is suggestive of minimal leverage.

Meanwhile, a half-dozen downstream industries had been planned and were being constructed – some like the downtown office towers by Chinese building companies who import their own Chinese labor – to consume a further 500 cubic feet of gas daily. These included the Alutrint aluminum smelter; Essar steel complex (Indian); Ausa urea, ammonia, and nitrogen (UAN) plant; AUM; and gas-to-liquids (GTL) plant which turns gas into clean, sulfur-free gasoline/diesel. However, the first was cancelled after the change in government in mid-2010 and the last crashed for technical reasons. In early 2011, the government announced a contract to build a US$1.9 billion ammonia and downstream derivative project to be constructed over the next four years with Methanol Holdings Trinidad Ltd (MHTL).

The Alutrint aluminum smelter project had been the most controversial, leading to a series of legal as well as economic, environmental, and political reviews. Alutrint was a 60 per cent holding of National Energy of T&T, supported by a US$400 million loan from China's EXIM Bank, which tied the employment of Chinese labor to its conditions. Alcoa of the US and Sural of Venezuela were associated with the project at times, but both pulled out. Anti-smelter CSOs (civil society organizations) have appealed to as well as against the Environmental Management Authority (EMA) over its Certificate of Environmental Clearance, leading to a critical judicial verdict in mid-2009 that the EMA was appealing when the new coalition government of Kamla Persad-Bissessar cancelled the project (EMA, 2009). Both bauxite after independence and now smelters raise myriad developmental as well as environmental issues, having been the focus of the nationalist, dependency school of thought in earlier decades (Pantin, 2005, 2010).

2020 as deadline for the status of developmental state?

T&T can advocate and even exploit diversification and developmentalism, but it is very open to changes in the global economy, especially shifts in technology and changes in ownership, raising questions about the short- to long-term impacts of the BRICs and the new 'second world' (Khanna, 2009). If Canada's mineral sector can experience seismic changes mid-decade with national icons like Alcan, Falconbridge, and Inco being taken over by Australian (RioTinto), Brazilian (Vale), and Swiss (Xstrata) MNCs, what chance is there for the independence of national capital in T&T (Ramsaran, 2003, 2006; Bronfman, 2007)? And what implications for patterns of CSR, Canadian (FOCAL, 2005), European, and otherwise, including now the BRICs?

Further, what catalysts exist for development of industries other than oil and gas? The state under the PNM regime of PM Patrick Manning had designed and articulated Vision 2020, but what about the dreams and ambitions of civil society (including NGOs, diasporas, media, women, religious communities, think tanks like CaPRI) and private sectors (regional MNCs, SMEs, Chambers of Commerce), some of which are reflected in vigorous op-ed discussions in T&T involving UWI faculty? And with what implications for the shape of CSR policy? At the national level, the change of political regime in May 2010 to a coalition government headed by Ms. Persad-Bissessar, along with growing consciousness and concern around climate change – and at the global level, the uncertainty and instability around the financial crisis – have together changed the context for such discussions and decisions.

Diversification was advocated in the national Vision 2020 for which Singapore as a NIC was the explicit and implicit model. Like the NICs, T&T has produced a Vision 2020 scenario, raising the question of what contents/prospects, especially as the new coalition government has abandoned such dreams given the global recession? And with what implications for defining and evaluating CSR? Conversely, the STCIC and UNDP (2007, p. 56) report on CSR in T&T sees it advancing Millennium Development Goals (MDGs), although Vision 2020 laments that the latter lacks 'clarity of the specific objectives and key intervention areas of the country's development plan.' In such a context, companies have found difficulty aligning with the Vision. The national economic strategy emphasizes diversification away from energy toward 'seven sectors,' in addition to financial and other services, including cultural industries (Ho and Nurse, 2005; Nurse, 2007, 2009), some of which are already under development, others being more distant, even fanciful, and none of which seems to result from any particular rationale or methodology but rather appear to be randomly selected. These include the following:

(1) food and beverages, beyond rum, beer, and milk products. This sector is already valued at some TT$1 billion and as the economy has grown and prices risen, it has expanded at some 8 per cent per annum in mid-decade. Its major and minor players employ some 10,000 workers and include national, regional, and global companies and brands, such as Angostura (which now owns Athertons in Jamaica but was part of CLICO), Jaleel, Bermudez Biscuit, Coca Cola, National Canners, and Nestle;
(2) maritime sectors other than energy, especially around a trio of non-energy ports;
(3) yachting, which began to develop in the 1990s as part of the 'conversion' or decommissioning of Chaguaramas as a US base;
(4) film including TV, with a half-dozen local stations employing 500 workers; a dozen production companies already, a five-year-old T&T

Film Festival, and the advantage of the global *lingua franca*, English, being the national language (Nurse, 2007, 2009);
(5) entertainment beyond Carnival, which generates US$150 million per annum, employing 3500. The sector as a whole – musicians, technicians, promoters, etc. – already generates US$350 million per annum, with over 10,000 employees. And this sector has gone global with carnivals centred on diasporas in the UK, US, and Canada, and Soca music is now a category in the US Grammy Awards (Ho and Nurse 2005; Nurse 2007, 2009);
(6) seafood, with almost 20 plants employing 6000 workers; and
(7) printing and packaging, where T&T is already a regional leader and exporter with 150 print firms, 50 in packaging, and a pair of glass and can makers employing 4000 and generating TT$300 million already with links to agriculture, education, food and beverage, media, tourism, etc. (*Business Trinidad and Tobago 2008–2009*, 2009, pp. 60–4).

Such developments, even if not all are or even could be realized simultaneously, pose significant challenges for national and regional infrastructures: how to create robust supply chains and logistics to facilitate diversification? The investment in physical and virtual infrastructures will be considerable, including skilled manpower. Hence the importance of *de facto* PPPs like Caribbean Airlines/Piarco international airport, container piers in Point Lisas and PoS, the Flow cable communication network and so on. Increasingly, the private sector dimensions of parallel Organization of American States (OAS) and Summit of the Americas (SOA) processes advance such PPPs (Feinberg, 2011), which are broadly compatible with hemispheric (CSRAméricas, 2009) and now national systems of CSR (TTMA, n.d.a).

T&T: Which global sectors?

The driver of the national economy in the new century is clearly energy (EIU, 2007, 2009). Aside from dwindling oil production, its main resource is now LNG (Guyadeen, 2010). And its primary market for LNG is the US, now both directly and via Canada; it is a major player in the US in a new sector, since hitherto gas was pumped nationally or overland from Canada. T&T has attempted to exploit its limited reserves to facilitate downstream industrialization. Whilst locating and extracting new reserves is challenging, industrial leaders anticipate several more decades of activity (Guyadeen, 2010).

The global natural gas market is very fragmented, especially between the dominant Pacific Basin and the newer but booming Atlantic Basin, raising questions around distinctive forms of 'new regionalisms' around the price, supply, and technology of energy. The former is centred on Indonesia and Malaysia as exporters and Japan as primary importer, though demand from other NICs like Korea is growing. As the BRICs develop globally, especially China and India, so demand from and supplies to them will burgeon. Thus flexible 'netbacks' from T&T have been delivered to Asia as US demand/price have at times been weak. Meanwhile, at least according to EIA (Energy Information Administration), whilst its reserves are finite, T&T is still a leading LNG exporter: Equatorial Guinea became the 14th LNG exporter before the end of 2007 (Frynas, 2004; Frynas and Paulo, 2007; Shaxson, 2007; EIA, 2010).

Meanwhile, LNG meets but 3 per cent of US supplies of natural gas. Yet T&T is its major external source, approaching 70 per cent (EIA, 2003), with additional regasified supply now also being piped via Canada. And LNG demand is set to grow further in the US (EIA, 2003) with all four of the current US onshore regasification plants being on the Atlantic coast, though burgeoning supply from shale may slow such ambitions.

Given that LNG is essentially a twenty-first century sector, T&T has been agile in terms of meeting US market needs despite being a small player compared with LNG exports from Indonesia, Algeria, Malaysia, and now Australia and Qatar. But the character of the dependency relationship remains problematic – is the US dependent on T&T gas or T&T on access to the US? 'Interdependence,' then, might be a better term to describe the relationship.

Finally, the 'resource curse' (Auty, 1993; Klare, 2001, 2004) means exponential corruption (Transparency International, 2010b), especially in 'fragile' states or regimes that drown in windfall profits from dramatically rising prices of energy and mineral exports such as in Sudan or Venezuela (Frynas, 2004; Frynas and Paulo, 2007; Shaxson, 2007). Despite all its good intentions and high-level support from the G8, how can an EITI really expect to contain such tendencies when the attractions are so glaring and fleeting? EITI may have support from Northern states, civil societies, energy and mining companies, and international organizations (EITI, 2009), but the sums involved have become so huge that no careful accounting or evaluations can really eliminate personal enrichment, as was apparent even before energy prices soared in the case of Enron. However, the new EITI national committee in T&T includes representatives from civil society and the energy sector as well as government, with a well-respected, high-profile chair who previously headed TI in the country.

T&T has now applied to be considered as a 'candidate' to become a 'compliant country' under the EITI, reflecting the new UNC–COP People's Partnership government's more positive inclinations than the previous PNM administration. EITI candidacy involves meeting four sign-up indicators, such as seriousness of intention to implement, collaboration with civil society and companies about the EITI process, and nominating a senior individual to animate the application – in the case of T&T, Victor Hart, previously chair in the country of TI. Since the development of EITI in mid-decade, only Azerbaijan has yet been so classified as compliant out of the 30 registered for the two-year process, including Equatorial Guinea, Gabon, and Sao Tome (Shaxson, 2007). EITI is in part an extension of CSR: an attempt to contain corrosion or corruption via corporate financial 'donations' to national leaders in exchange for privileged access to natural resources.

In short: how to sustain the relatively advantageous niche of T&T as a 'rainbow' 'democratic, developmental island state' to 2020 and beyond (Ramsaran, 2003, 2006)? Can it improve its prospects of meeting all the MDGs, involving improved ranking on the Human Development Index (HDI) (UNDP, 2009) as well as GCI (Global Competitiveness Index) (Schwab, 2010)? And can CSR and EITI, either together or separately, advance development with growth? Such questions present profound challenges to both analysis and policy in a variety of overlapping disciplines, departments, corporations, and civil societies.

Notes

1. The 'People's Partnership' is a coalition centred on the established United National Congress (UNC) and fledgling Congress of the People (COP).
2. Norway is major animator and advocate as well as host of EITI.

16
Mining Companies and Governance in Africa

Ralph Hamann, Paul Kapelus, and Ed O'Keefe

Introduction

In this chapter we contribute to the overarching theme of this book with a focus on the African context, arguing that corporate responsibilities can be better understood and acted upon when seeing corporations as part of a broader governance system. Recognizing, of course, that the African continent provides for a wide array of circumstances (as is the case in Latin America), we use illustrative case studies from South Africa and the Democratic Republic of Congo (DRC) to highlight common themes in the interplay between mining corporations and governance. These two countries can be seen as representing, in the African context, polar extremes of socio-economic development (see, for instance, UNDP, 2010), as well as the extent to which the state is able to implement and enforce commonly binding rules, which we call 'statehood' following Boerzel and Risse (2010).[1]

As a point of departure, we revisit our previous analysis of corporate responsibility challenges in African mining, in which we discuss the role of mining companies in either damaging or enhancing the local governance system of which they are a part (Hamann et al., 2005). We argued:

> Rather than see the company at the centre of a range of stakeholders, it ought to be considered part of an intricate and dynamic web of interrelated role-players involved in (un)sustainable development at the local level. This is referred to as the local governance system, characterised by varying degrees of collaboration potential and complexity. It is argued that corporate contributions to the emergence of

more sustainable patterns of local governance require proactive and creative approaches to enhancing the collaboration potential and responding to complexity. (p. 63)

A model of this 'web of interrelated role-players' is illustrated in Figure 16.1. Key players in this system are the company (or companies), local government, traditional institutions, and other civil society groups. A frequent challenge encountered by mining companies in diverse African contexts (including South Africa and the DRC) relates to often conflict-prone relationships between the constitutional state – if indeed such a state is present in the area in question – and traditional leadership structures such as chieftains, tribal councils, and village elders. Such conflicts are a defining feature of many African countries' struggles to free themselves from the legacies of European colonialism (Mamdani, 1996), and mining companies seeking to prospect or develop a mine will often unwittingly bring such tensions to the surface, as the stakes in decision-making become so much higher with the onset of prospecting and mine development.

In addition, in the model illustrated in Figure 16.1, we describe as an important aspect of the local governance system the local development context, recognizing that the interaction between a company and its 'stakeholders' is bound to be influenced by a host of factors, which we characterized with reference to the sustainable livelihoods model of community development, and which has also been put to good effect in analyses of mining projects (see for example Bury, 2004).[2] Finally, the model in Figure 16.1 allows for the characterization of the degree to which the interrelationships between the actors in the governance system are conducive to interest-based cooperation in pursuit of sustainable development in the area, with reference to four preconditions for such collaboration as identified by Covey and Brown (2001).

The suggestion that companies will need to understand and, to the extent possible, contribute to relationships between stakeholders, rather than just relationships between themselves and stakeholders, is a far-reaching expectation levied upon company managers. However, while it is an outcome of analyses of mining companies' experiences described in Hamann et al. (2005), and it is also an important element in the case studies from South Africa and the DRC considered in this chapter, it is a significant departure from most approaches to stakeholder theory (Donaldson and Preston, 1995). Though Gulati et al. (2000) do talk about the ability to manage networks as an important resource for company strategy, they do so still from a firm-centric perspective.

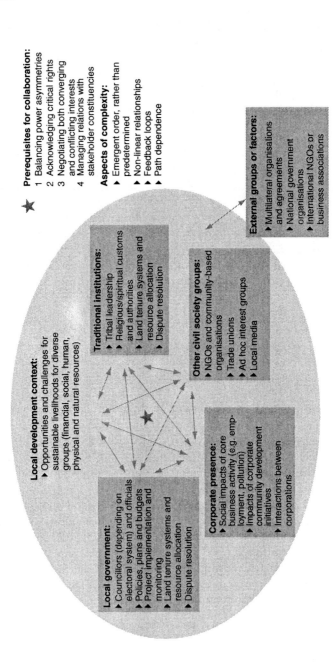

Figure 16.1 A schematic representation of the corporate presence as part of the local governance system in Africa
Source: Hamann et al. (2005), p. 65.

Our suggestion is likely to be especially challenging to those who feel that even the basic tenets of stakeholder theory are problematic (Jensen, 2010). We argue that this challenge and discrepancy is largely because predominant discussions of stakeholder theory and CSR focus on contexts in which firms' environment is stabilized and regulated by rules enforced by the state.

The reason why a proactive management approach to relationships beyond the firm boundary is so important in many African contexts is because the state is frequently unwilling or unable to fulfill functions that might be expected in more developed economies. Boerzel and Risse (2010) define such areas as having 'limited statehood' – 'where political institutions are too weak to hierarchically adopt and enforce collectively binding rules' (p. 113). In such circumstances, companies may recognize the state's inability to enforce collectively binding rules or to provide collective goods as a prominent risk to effective operations, through an increasing burden of disease among workers, for instance, or corporate reputation, through being associated with human rights abuses, for example. Therefore, we emphasize a two-way relationship between organizations and governance, whereas most of the organizational literature focuses on the impacts of governance on the organizational level only. Firms are not only influenced by the organizational field of which they are a part, and which in areas of consolidated statehood is commonly influenced fundamentally – even if only indirectly – by formal rules enforced by the state. Particularly in areas of limited statehood they also have a tangible role in directly or indirectly influencing the governance context in which they operate. This is schematically illustrated in Figure 16.2. Firms' influence on governance can be supportive of transparency and effectiveness – for instance, through transparent assistance to local government processes or structures – or it can be damaging and corrupting. Furthermore, this influence is often most direct at the local level, as discussed in each of the three cases below, but it also manifests at the national level, as considered briefly in the DRC example.

In this chapter, we seek to develop further our understanding of corporate responsibility in African mining by applying and building on the model in Figures 16.1 and 16.2 in four short case studies, each of which emphasizes particular themes. First, we briefly consider a case in which Anglo Platinum was embroiled in a human rights controversy surrounding the resettlement of communities near one of its mines in Limpopo, South Africa. It highlights the risks associated with giving insufficient attention to the cultural differences between the

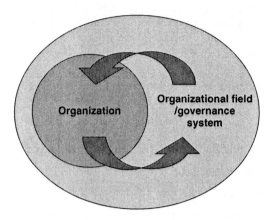

Figure 16.2 A schematic illustration of the two-way relationship between organizations and the governance system of which they are a part
Source: Authors.

company and affected communities, with particular reference to the challenge of establishing and maintaining legitimate representation for local stakeholders (the fourth prerequisite of critical cooperation identified by Covey and Brown (2001), mentioned in Figure 16.1). The second case study also focuses on Anglo Platinum, but considers more recent efforts at a different mine to improve company–community relations on the basis of more fundamental benefit-sharing agreements and the preparation of a long-term community development plan.

Third, we consider the experience of AngloGold Ashanti in the DRC, following allegations in 2005 that it was complicit in gross human rights abuses by supporting one of the rebel groups in the region. Many of the key challenges faced by companies in South Africa, particularly with regard to supporting legitimate representation structures (that is, the fourth precondition for critical cooperation identified by Covey and Brown, 2001), are exacerbated in the context of the eastern DRC. Dealing with some of these challenges requires that the company engages even at the national level, with the Extractive Industries Transparency Initiative (EITI) being an important mechanism. In the fourth and final section, we consider some of the success factors of the EITI on the basis of a survey conducted in Ghana, Nigeria, and Cameroon. Here, too, the notion of legitimacy is an important one, with the international brand of the EITI complementing its simplicity and focus in support of a relatively high adoption rate and erstwhile successes.

Human rights and the role of culture: Case study of the Mogalakwena resettlement

One of the most significant developments in recent years informing our understanding of CSR has been the work of the United Nations Special Representative of the Secretary General (SRSG) on business and human rights. For a start it has contributed to human rights becoming a crucial framework for clarifying corporate responsibilities, particularly with regard to the complicated relationships between mining companies and the often impoverished communities surrounding their mines. The final report of the SRSG's first three-year term, published in April 2008, emphasizes the role of governance: 'The root cause of the business and human rights predicament today lies in the governance gaps created by globalization – between the scope and impact of economic forces and actors, and the capacity of societies to manage their adverse consequences' (Ruggie, 2008, p. 1). With regard to corporate responsibilities, it highlights the need for companies to develop due diligence on human rights issues:

> Companies should consider three sets of factors. The first is the country contexts in which their business activities take place, to highlight any specific human rights challenges they may pose. The second is what human rights impacts their own activities may have within that context – for example, in their capacity as producers, service providers, employers, and neighbours. The third is whether they might contribute to abuse through the relationships connected to their activities, such as with business partners, suppliers, State agencies, and other non-State actors. How far or how deep this process must go will depend on circumstances. (Ruggie, 2008, p. 10)

Actually putting such systems into place, however, is a complex endeavor. A simplistic understanding of the 'community' surrounding a mining operation, and of what may be paradigmatically different perspectives on development and quality of life between affected communities and corporate decision makers, can also undermine even well-intentioned corporate efforts to support local sustainable development activities (Banerjee, 2001; Kapelus, 2002). The complexities of relationships among stakeholders at the local level clearly necessitate collaborative and highly context-specific approaches to stakeholder engagement. Yet setting up effective partnerships with government and civil society – while necessary for achieving sustainable development outcomes – is itself a huge challenge, particularly where public sector governance is weak (Hamann, 2004). Biggs and Smith (2003) suggest that major changes in institutional culture are needed when private entities are given the responsibility over development activities that are traditionally the role of government. Acknowledging the limitations and skepticism about participatory and collaborative approaches, they note that the degree to which these theoretical approaches succeed in practice often relates to the organizational cultural context in which they are undertaken and the leadership shown in support of a cultural transformation. The dominance of and momentum behind a particular corporate culture can be a barrier to genuine action on sustainable development and collaboration, as shifting the culture requires a shift in worldview as well as the development of new kinds of internal capacities for community engagement. The pains of this shift (or attempts to avoid it) may help explain tendencies toward 'greenwashing' and viewing sustainable development and CSR as a community relations issue, narrowly defined, and not fully integrated into business models or corporate strategic planning.

These tendencies, and their repercussions, may be illustrated by an analysis of a conflict between affected communities and Anglo Platinum at the Mogalakwena mine near Mokopane (formerly Potgietersrust) in Limpopo, South Africa. Anglo Platinum, the world's largest platinum mining company and recipient of many awards for its sustainability and reports and practices, found itself in 2008 faced with public controversy framed through a human rights lens in its handling of the resettlement of affected communities at Mogalakwena. The issues were initially written up by ActionAid (2008), an international human rights Non-governmental Organization (NGO), in a March 2008 report. The South African Human Rights Commission (SAHRC) then investigated the issue, releasing its report in November 2008 (SAHRC, 2008). It was SAHRC's first investigation of the question of human rights and mining, and while its report noted many ways that the resettlement process could be improved, it did not 'actually accuse Anglo Platinum of human rights violations' (interview with Anglo Platinum employee, 26 January 2009).

The company's initial reaction to the SAHRC findings was thus dismissive and defensive, challenging the critiques' value given their inability to conclude legal non-compliance. This was complemented by contemporaneous attempts to take legal action against some of the company's opponents. Though legitimate from a purely legal perspective, Anglo Platinum's aggressive response, focused narrowly on flexing legal and financial muscle, worked against the company by further entrenching rather than resolving the social issues at the heart of the conflict. Indeed, the SAHRC's report itself highlighted that many of the problems around the Mogalakwena mine may have been attributable to the company's emphasis on legal compliance, rather than a more holistic approach to community engagement focused on dialog, negotiation, inclusiveness, and human rights.

This complicated and drawn-out case, which is discussed in more detail in Farrell et al. (2009), illustrates the significant costs imposed on a company if it does not get its relations with surrounding communities 'right.' Not only did the company suffer significant reputational damage among a range of stakeholders, including national government and international investors, but it also experienced operational delays due to the difficulties with the resettlement. As in many other instances of this sort, an important aspect of this conflict was the contrast between the organizational culture of the mining company and that of the affected communities. Many of the company's troubles may be linked to a lack of willingness or ability to first recognize this difference, and second attempt to better understand the culture and governance logics operating in the communities.

For instance, the company needed to establish legally recognized representative bodies in order to enter legal agreements with the affected communities, so it set up 'Section 21 committees' with members voted in at a public meeting. The problems came when the legitimacy of the committees became undermined by community perceptions that committee members were not acting in the communities' best interests, a perception enhanced by the stipends paid to committee members by the company (which also created a perverse incentive for committee members to drag out proceedings). Tensions grew because no governance rules had been established for the re-election of committee members. The company thus lost its ability to communicate with legitimate representatives of the communities – the fourth, vital factor highlighted by Covey and Brown (2001) as a prerequisite for critical cooperation between companies and stakeholders.

The need to bring specialized skills to bear in such circumstances, and the risks of not doing so, were not sufficiently realized by corporate decision makers. This was illustrated by the fact that the resettlement project manager requested additional staff with social science training, but his request was turned down. Yet the Mogalakwena experience has contributed to potentially important changes within the company, including a larger community engagement department, with more staff 'on the ground,' as well as a company-wide initiative seeking to assess and address perceptions of the predominance of a hierarchical, patronizing culture within the organization.

Benefit sharing and long-term planning: Case study of Amandelbult mine and the Baphalane Ba Mantserre

Anglo Platinum's Amandelbult platinum mine is the second largest in the world, producing some 600,000 ounces of the precious metal per year and employing more than 13,000 people. It was opened in 1973, with the life of the mine estimated to be in excess of a further 75 years. Ten years prior to mining commencement, in 1963, the community living in the area was forcibly relocated by the government under the oppressive 'Group Areas' laws of the Apartheid government. They were taken to a farm referred to as Mantserre, 20km away. Following the transition to democracy in 1994, the Baphalane ba Mantserre community lodged a claim for restoration of their land rights in terms of the Restitution of Land Rights Act 22 of 1994, which they won in 2003. In the same year the community and the mine entered into a 50-year lease agreement, but community resentments remained amid claims that the agreement was forced upon it and that payments are too low.

These tensions came to the fore in 2008, with the conflict being highlighted in the same ActionAid (2008) report referred to above. Protest marches were organized and key company stakeholders, including international car makers, were petitioned by community groups. Partly in response to these pressures emanating from the community, but also as part of a broader shift in policy within the company motivated by a range of experiences, including the Mogalakwena resettlement mentioned above, the company negotiated a more fundamental array of benefit-sharing arrangements with the community. The rationale for this is to increase the level of support from and benefits accruing to host communities, which was motivated not only by a desire to improve company–community relations, but can also be seen as an implicit response by companies to provincial and national political debates focused on broadening the access to benefits accruing from mining. In particular, the ANC Youth League has for the past few years lobbied extensively for what they call the 'nationalization of mines,' with reference to the 1954 ANC Freedom Charter, as well as to other African countries' policies (Botswana is frequently mentioned, with specific reference to the government's 50–50 joint venture with DeBeers to mine that country's extensive diamond deposits).

The company's willingness to negotiate more fundamental benefit-sharing agreements contributed to an improved relationship between Anglo Platinum and the Baphalane Ba Mansterre (as of end 2010), coupled also with a revised and revitalized approach to community development. In contrast to the 'corporate social investment' (CSI) activities that have commonly characterized mining companies' attempts to support socio-economic development in mine-affected communities, which generally involved the company deciding upon short- to medium-term investments in education, health, or related projects (see Hamann, 2004, for a critique of this approach), the emphasis has moved to a longer-term, strategic approach with priorities being identified by the community trust.

Advised by a consultant (one of the authors of this chapter), the community trust embarked on an ambitious initiative to develop a 20-year 'community competitiveness plan.' Understanding itself as a unit that strives to attract investment and enhance its human and other resources has helped the community trust create a long-term strategy, seeing the company as an 'anchor tenant' with which to develop a mutually beneficial relationship. This reframing of the community's view of itself is meant to help it adopt a more proactive, strategic, and self-confident approach in interacting with the company, as well as other stakeholders, including different government organizations. Though it has required greater than usual investments from the company, including financial resources and other support given to the community trust, as well as significant time spent in negotiation, it should help the company by increasing the level of support among community members for the mine, predicated upon the longer-term vision for the area, and a greater sense of stability and cohesiveness within the community. The company also expects to benefit from improved supply of local labor and goods and services, which are aimed for in the community's strategic plan.

The community's longer-term development plan and its associated partnership with the mining company also have the potential advantage of facilitating the identification of appropriate performance indicators for monitoring and learning. This is in contrast to the previous situation, in which the three- to five-year CSI timeframes and reporting frameworks gave rise to superficial monitoring approaches, in which success was assessed with reference to funds spent, rather than actual development impact.

The Amandelbult mine therefore suggests a mechanism for the company supporting a dynamic within the community that engenders greater self-reliance, confidence, and self-determination, premised in part on more direct benefits from the mine flowing to the local community and a long-term development strategy. Amandelbult is not the only mine at which this is being trialed, and other companies have also been putting in place similar approaches. Their success and implementation lessons will deserve careful scrutiny in coming years.

Local and national governance in the DRC: Case study of AngloGold Ashanti in Ituri district

The DRC has for many years been the epitome of what the Organization for Economic Cooperation and Development (OECD) calls 'weak governance zones.' The DRC's human development index is among the lowest in the world, and it has fallen throughout most of the period in which this index has been measured, from 1975 to 2003, with a slight rise since then (UNDP, 2010). The conflict that engulfed much of the country from 1996 to 2001 (though conflict is still occurring in the north-eastern DRC) claimed almost 4 million lives and involved numerous other African countries. Parts of the country experienced the highest estimated conflict-related mortality rates in the world (Guha-Sapir and van Panhuis, 2003).

Primarily due to this conflict, international investment in the DRC has been very limited. The only sectors that experienced some investment have been extractive industries and related service providers. Given the DRC's large deposits of gold, copper, cobalt, and other metals and minerals, and given also the increasing global demand and the concomitant decrease in accessible deposits in more stable countries, resources companies have been the most likely to discount the political and economic risks associated with doing business in the DRC. In some cases this increased risk has been rewarded with significant returns. As noted by one mining company CEO: 'It's the holy grail of the copper industry – companies are saying: to hell with the political risk, we just have to be here!' (Clive Newall, CEO of First Quantum, quoted in Global Witness, 2006).

One of the most prominent cases illustrating the challenges and complexities faced by companies operating in 'weak governance zones,' and in the DRC in particular, was that of AngloGold Ashanti (AGA). The company has a license to explore a gold deposit in Ituri District in the north-east of the country. This has been one of the most volatile areas in the country, even after the large-scale deployment of the UN peacekeeping force Mission des Nations Unies en Republique Democratique du Congo (MONUC) in 1999 and the establishment of a transitional government in 2003.

In January 2005 the staff of the AGA exploration team in Ituri District made a US$8000 payment to the Front des Nationalistes et Intégrationnistes (FNI), which was controlling much of the Ituri District during the civil war, and which had been accused of committing extensive human rights abuses (HRW, 2005). In addition, the AGA exploration project provided FNI with accommodation and access to transport, and paid levies on cargo flown into the local airport. The payment and assistance provided established a relationship between AGA and the FNI and provided the FNI with legitimacy.

In June 2005 Human Rights Watch (HRW) released a report, *The Curse of Gold*, which provided details of the financial and material assistance provided by AGA to the FNI and stated that AGA had obtained permission from the militia group to enter the area, despite being warned against this at the start of operations by MONUC. For many observers this represented an example of corporate irresponsibility, and it was used to argue that companies such as AGA should not be operating in such areas in the first place (HRW, 2005). The company admitted that a mistake had been made, but it argued that the interests of local people and the DRC in general are better served by its remaining in the area (AGA, 2005).

At the local level, ensuring that indeed the local people are better off has been a challenge. The social complexities in the area are significant. Continued risk of violent conflict, ethnic tensions and political factions, the legacy of the war, high levels of poverty, tensions between the company and the workforce, and the significant challenge of artisanal miners (of whom there are about 100,000 working in the area) are closely interrelated and give rise to a very complex operating environment for the company. The company has not been willing or able to invest significant financial resources in the area in the absence of any revenues from mining, as prospecting continues. Furthermore, the local stakeholders forum, which the company helped establish to facilitate a more participatory and deliberative approach to local rule- and decision-making (on issues such as the expenditure of the company's and other donors' financial assistance), has struggled to develop the necessary legitimacy and effectiveness. In particular, the process of electing representatives onto this forum has contributed to ethnic tensions and has also raised the ire of local traditional leaders, who feel their power (which is partly underpinned also by national legislation) usurped by the forum.

At a broader level, the company's attempts to contribute to improved governance have involved its participation in the EITI. This is a global multi-stakeholder initiative focusing on the management of revenues and more transparent governance in resource-rich countries. Its key stipulation is that companies and host governments report on and account for all payments made by companies to host governments. The actual implementation of this objective may vary from country to country, but there generally needs to be an aggregating body that receives the information from the companies and the government. In other words, companies are not actually required to report such information publicly. (In contrast, the NGO coalition, Publish What You Pay, asks companies to make all its payments public.)

The DRC government committed itself to the EITI in 2005 and a national EITI committee was set up with financial support from the government. Nevertheless, EITI requirements were slow to be implemented. Instead, companies such as AGA have been constrained from publicly reporting on their payments to the government by a law that requires any such information to be vetted by parliament. Hence, in 2008 AGA provided figures for payments to all relevant host governments except that of the DRC on its company website. There was thus increasing pressure on the DRC government to implement its EITI commitment more rigorously (see, for instance, Publish What You Pay Coalition in the Democratic Republic of Congo, 2007). The EITI is considered in more detail in the next section.

The EITI: Lessons from Ghana, Nigeria, and Cameroon

The EITI is frequently cited as an example of successful mining company engagement with national-level development issues. Indeed it remains one of the few institutions that is successfully adopting a multi-stakeholder approach, at the national level, to addressing the linkages between extractive industries and development. There are now 27 'Candidate Countries' that have committed to adopting EITI, and 5 'Compliant Countries' that have been validated as fully meeting the criteria set by EITI (including Ghana and Liberia in Africa). In this section we briefly outline emerging lessons from a validation exercise for three African countries – Ghana, Nigeria, and Cameroon – conducted by one of the authors of this chapter.

One factor in the success of EITI to date is its simplicity and focus. EITI is primarily a transparent and trusted process for disclosing payments from extractive industry companies to government. It is therefore relatively easy for extractive companies to understand what is required, particularly by parts of the company not directly involved in development issues.³

EITI arguably has a strong international brand, which is assured through the certification process and which has broad support from government, civil society, as well as the private sector. This brand has been built up through a long period of engagement at an international level, as well as often at a national level. It is therefore attractive for extractive industry companies to associate their brand with that of EITI. This may not be the case for engagement on other national-level policy processes – such as those attempting to deal with artisanal mining and its frequent conflict with larger companies – which will often be more contested and not have an international profile.

Whilst EITI is a voluntary initiative for countries and companies to adopt, it is increasingly being incorporated in national legislation. The Nigerian NEITI Act was the first such national legislation of EITI, and several countries (e.g., Ghana) have followed suit or are in the process of developing legislation. This therefore leaves extractive industry companies with little choice but to participate in the process.

A further point working in favor of the EITI is that it has relatively low requirements for company engagement, particularly for those companies already committed to disclosing such information. The main engagement requirements are during the development of reporting templates and in provision of the required information. Company representatives are also required to be represented in the multi-stakeholder group overseeing the EITI process in each country. This will often be done through a small number of representatives, often through an industry body such as a Chamber of Mines. A key constraint to many companies' engagement in national level policy processes is often the lack of appropriate staff who are able to spend the time required. This is exacerbated when extractive industry companies have most of their staff based outside the main cities where these meetings often take place.

EITI also requires particularly little additional commitment from those companies that already meet relatively high standards for CSR and transparency. EITI requires companies to provide information that in many cases is already publicly disclosed, particularly where the host or home country of the company has higher financial disclosure standards, or among those companies that have already committed to transparent public reporting. For instance, in Ghana, the five main mining companies were publicly disclosing most of the information required for EITI reporting prior to EITI being implemented.

So while the simplicity, focus, and legitimacy of EITI represent important reasons for its adoption, its implementation often results in more far-reaching and complex discussions about the value chain associated with mining. In other words, while the EITI only requires disclosure of payments from companies to the government, it often has the effect (at least in the countries surveyed) of increasing interest in where and how the revenues were generated by the companies, and how the revenues are actually used by the government. For instance, in Nigeria, the EITI process has highlighted some discrepancies in payments from companies to government, but has also illuminated serious failings in the system upstream and downstream from this point, such as the lack of wellhead monitoring and the basis for accounting for oil production levels.

In Ghana, the EITI process has been a key factor leading to a review and subsequent increase in the royalty rates paid by mining companies. In Ghana, as well as other EITI countries, the process has also led to calls for disclosure of previously confidential stability agreements between companies and government, in order to determine the terms under which revenue payments are made, as well as associated concessions and conditions. However, it has also led to calls for greater accountability for the way in which extractive industry revenues are used and distributed by government.

Whilst EITI provides a useful, and in many cases a positive, example of extractive industry engagement in national-level development processes, it is striking that this is one of only very few cases existing. There continues to be a significant disjuncture – especially in many African countries – between national development policy processes and extractive company activities. There are surprisingly few forums for multi-stakeholder collaboration between key role-players at the national level, and similarly there is often a dearth of coordination of policy processes focused on mining or trade and industry, on the one hand, and community development, on the other.

Conclusion

The case studies discussed in this chapter give further impetus to the suggestion that companies do not just respond to their external governance context, but because of a range of operational and reputational drivers feel the need to contribute to more effective and legitimate governance arrangements beyond the firm boundary. In other words, they seek – and are trying to learn how – to contribute to relationships between stakeholders, rather than just relationships between themselves and stakeholders. We have situated this analysis theoretically by explaining the need for companies to do so as a result of 'limited statehood' in their operating environment in many parts of Africa – that is, because 'political institutions are too weak to hierarchically adopt and enforce collectively binding rules' (Boerzel and Risse, 2010).

In such circumstances, companies recognize the state's inability to enforce collectively binding rules and to provide collective goods as a prominent risk to effective operations or corporate reputation. For instance, in Anglo Platinum's experience in Mogalakwena, abiding by government rules with regard to resettlement just was not enough. Neither was the government able to effectively mediate or arbitrate between conflicting interests once the situation had become intractable. The company was – at that stage – unable to foster legitimate community representation structures to facilitate a negotiated outcome. On the other hand, the company's more recent efforts to diffuse tensions with the host community of its Amandelbult mine have shown more promise, including in particular the negotiation of benefit-sharing agreements and support for the preparation of a long-term community development plan, which gives legitimacy not only to the community trust that is negotiating with the company, but also to the company itself in the eyes of community members.

The third case considered the experience of AngloGold Ashanti in the eastern DRC, an area characterized by excessively long and severe violent conflict and extreme poverty, and correspondingly complex operating conditions for the company. Many of the key challenges faced by companies in South Africa, particularly with regard to supporting legitimate representation structures (that is, the fourth precondition for critical cooperation identified by Covey and Brown, 2001), are exacerbated in the context of the eastern DRC. Dealing with some of these challenges requires that the company engages even at the national level, with the EITI being an important mechanism. In considering some of the success factors of the EITI, we emphasized again the notion of legitimacy, with the international brand of the EITI complementing its simplicity and focus in support of a relatively high adoption rate and erstwhile successes.

In conclusion, therefore, many mining companies in Africa are going well beyond the predominant expectations of stakeholder theory in seeking to foster a more favorable operating environment. These efforts deserve more considered attention and, where appropriate, critique or support.

Notes

1. As a very partial proxy for statehood we may compare the two countries using indicators on corruption, for instance, with South Africa ranked 54 and the DRC 178 in the 2009 assessment by Transparency International (2009).
2. This way of approaching challenges and opportunities for local development in areas affected by mining is distinct from, but closely related to Asset Based Community Development, which has also attracted much interest in recent years (see Cameron and Gibson, 2005).
3. Other national-level processes have been far broader, such as the Guinea mining sector Community Development Framework developed in 2006–07 to establish a common framework to address 15 different issues at a national level (see http://commdev.org/section/projects/framework_sd_guinea) – arguably this broader scope has been a constraint to its widespread and committed adoption.

17
Conclusion

Nicole Marie Lindsay and Julia Sagebien

As with all emerging ideas, particularly those focused on phenomena that are complex and in flux as is the case here, the usefulness of the Social and Environmental Value Governance Ecosystem (SEVGE) model will depend on its further elaboration over time, no doubt bringing in new perspectives and ideas not addressed here. However, we'd like to conclude our initial overview and introduction to the ecosystem governance model with a brief discussion of the insights and observations of the authors collected here.

Recalling that the purpose of the SEVGE model is to realign analysis of CSR in the mining industry in developing parts of the world by shifting the firm away from the centre of our analytic model, we discuss below how various actors and groups of actors engaged in negotiation over mining development interact with each other in ways that can both enable and disable the creation/protection of social and environmental value. We refer to this field of interactions as a governance ecosystem to highlight the interdependent and dynamic modes of governance (to use an idea introduced in Chapter 5 by Campbell, Grégoire, and Laforce) involved in shaping development in mining-dependent economies.

SEVGE components

At its most simplistic level, the SEVGE model has two parts: the hub, which represents the total amount or quality of social and environmental value in a given geographical or development context, and the actors who seek to shape the overall direction of development in ways that either enhance or detract from the social and environmental value represented at the hub.

In the following sections we outline the primary actor/institution groups and the mechanisms, tools, and strategies used by each one in either enabling or disabling the creation of social and environmental value, drawing insight from the previous chapters.

The hub: Social and environmental value

Social and environmental value, like the related concept of sustainable development, resists easy definition or measurement, and defining Social and Environmental Value (SEV) in the mining industry especially presents a significant set of challenges. Indeed, critics such as Kosch (2010) have argued that the idea of sustainability in the mining industry can be seen as a 'corporate oxymoron,' used to 'conceal harm and neutralize critique' (Kosch, 2010, p. 92).

By nature, mining cannot be considered sustainable in the strictest sense of the term. Ore is a non-renewable resource and standard extraction processes are ecologically destructive and contaminating. Historically, net social benefits fall short of promises by mining proponents. However, industry and supporters of mining argue that by minimizing harmful environmental impacts, mining could be viewed as a process of capital conversion (World Bank, 2003; MMSD, 2002) whereby there is a trade-off between ecological capital in one geographic area for economic capital and, through effective redistribution, social capital in other areas. One problem with this view, as highlighted by Coumans (Chapter 7) and others, is that it fails to acknowledge the uneven distribution of harms and benefits created by mining activity.

Thus the concept of SEV, if it is to be useful and relevant to the significant challenges of guiding governance of mining in the twenty-first century, must maintain a critical component in valuating long-term ecological and social well-being. This will require a significant shift in current thinking.

Writing from a business perspective, Porter and Kramer (2011) point out that a dramatic change will be required to realign business perspectives of social responsibility as shared value rather than as damage control or public relations, suggesting that the corporation as an entity be redefined in such a way that its primary purpose is the creation of shared valued (versus solely economic value).

However, from a more business-oriented perspective, Porter and Kramer (2011) point out that a dramatic change will be required to realign business perspectives of social responsibility as shared value rather than as damage control or public relations. They define the principle of shared value as that 'which involves creating economic value in a way that also creates value for society by addressing its needs and challenges' (Porter and Kramer, 2011, p. 64), This, they argue, will require new forms of collaboration between business, government, and non-government actors, and may entail movement toward a 'new conception of capitalism' (ibid., p. 76). The analyses presented in some of the chapters collected in this work suggest that persistent problems such as poverty, weak governance, and environmental degradation owe much to the dynamics of global capitalism and, indeed, a new conception of political economy could be an important part of the solution.

Our notion of social and environmental value shares some aspects of Porter and Kramer's (2011) notion of shared value, though we shy away from the 'win–win' scenarios they imply. We believe that compromises and losses in some areas are necessary to allow for gains in other areas that will create optimum system-wide SEV. Deep social and environmental value may well lie beyond what any triple bottom line, or win–win, firm-centred stakeholder model can capture, if only due to the mere fact that SEV must be collectively defined.

Of course, even given less strict interpretations of the meaning of sustainability, the social and environmental value of mining is difficult to measure. On the environmental side, the Mining, Minerals and Sustainable Development (MMSD) final report, 'Breaking New Ground,' (2002) notes that some forms of capital (for example, the ozone layer) cannot be substituted or renewed and that there are cases where achieving sustainability will be costly. The World Bank's Extractive Industries Review (EIR) (World Bank, 2003) also notes the tensions involved in defining and achieving sustainability. From a social perspective, the EIR made a number of recommendations on the World Bank's lending policies to ensure that it contributed to poverty reduction through sustainable development. However, despite these efforts, precise definitions of sustainability and realistic methods necessary to achieve it remain elusive.[1]

Despite the persistent ambiguities, however, common elements of sustainable development definitions include goals around poverty reduction, equitable resource distribution, respect for ecosystem limits, and capacity-building for long-term social, environmental, and economic health of communities (Banerjee, 2003; MMSD, 2002). Devising more precise and measurable outcomes for sustainable development and SEV is an ongoing, albeit very urgent, process that requires considerable learning, knowledge-sharing, and genuine dialog among system actors.

Corruption as a systemic disabler

There is broad agreement in the development and business literature that corruption has negative consequences for economic and human development, and that it adds risks, uncertainty, and costs to business transactions. Similarly, the SEVGE model considers corruption to be a systemic disabler not just because of its ubiquity but because it destroys the trust between actors necessary for any complex collective endeavor (such as the definition, maintenance, and protection of SEV).

Although there is no easy remedy for the elimination of corruption surrounding the mining industry, as both Odell (Chapter 13) and Smillie (Chapter 6) highlight, many system actors are becoming involved in creating multilateral initiatives designed to curb corruption. However, as Odell points out, in the mining industry in Peru these initiatives are still emergent and require greater collaboration between key actors (corporate, government, and civil society), as well as integration throughout the industry. Framing her discussion in terms of anti-corruption enablers, Odell points to a growing awareness of the sometimes hidden transactional costs of corruption for corporations, as well as increasing international legal action against corrupt companies and international anti-corruption legislation requiring due diligence for exploration companies as contributing toward a multilateral anti-corruption movement.

These high-level efforts, however, can be disabled by contextual factors on the ground, including the historical social and political imbedding of corruption in certain regions and the high revenue flows of extractives operations, which make them particularly attractive and vulnerable to extortion. Further, the low trust and high social inequality of many developing regions further entrenches corruption, making anti-corruption efforts difficult. In order to counter these underlying systemic problems, Odell recommends that anti-corruption efforts integrate social, political, and economic capacity-building initiatives aimed at reducing the inequalities and social ills often underlying high-corruption environments.

Drawing lessons from the experience of the Kimberley Process (KP) in both Africa and Latin America, Smillie points to the need for robust and flexible multilateral initiatives to combat corruption in the diamond industry. Despite the initial successes of the KP in several African nations, it could not adequately address underlying social and political dysfunctionalities due to organizational constraints that prevented the initiative from adapting to changing demands and needs of its anti-corruption goals. As the initial successes gave way to new challenges, the KP was hampered by the lack of democratic decision-making capacities that would allow it to develop third-party monitoring mechanisms and penalties for non-compliance – two vital ingredients in effective anti-corruption measures. Thus, as Smillie points out, despite its wide recognition, legitimacy and participation among governments of diamond-producing nations, industry, and civil society organizations, the KP 'has been unable to enforce even the most basic chains of custody in the countries worst affected by conflict diamonds' (this volume, p. 111).

A potential solution, Smillie suggests, is the creation of a 'Kimberley Process-Plus' involving a more stringent set of standards and a revised decision-making apparatus. Although voluntary in nature, Smillie suggests that with broad support and participation, a revised process could set the standard for a new norms cascade (to draw a concept from Dashwood in Chapter 2) in which '"[v]oluntary" would gradually, or even quickly, become "compulsory" in the same way as participation in the KP itself is voluntary – and yet compulsory for any government wanting to participate in the world's legitimate diamond trade' (this volume, p. 112).

The actors, mechanisms, enablers, and disablers

Customers, governments, Non-governmental Organizations (NGOs), and affected communities, suppliers, and so on are not just 'stakeholders' of the firm. Rather, they are all legitimate actors in their own right embedded in historical social, economic, and political contexts. Thus, firm stakeholder interactions are also not just two-way interactions. Each actor has his/her own complex vision, agenda, processes, initiatives, and potential interactions with other actors, only one of which is the firm. These interactions in and of themselves create *enabling* or *disabling* dynamics and multidimensional interactions that contribute toward or limit collective outcomes in terms of social and environmental value.

Each of the system actors/institutions possesses a specific set of mechanisms, tools, and strategies available for achieving its goals and addressing its needs, some of which may not align with the economic interests of the firm. For example, governments make laws and regulations, while financial institutions responding to civil society pressure may monitor the CSR behavior of their clients through reporting and/or third-party verification requirements, and NGOs partner with communities to advocate through electronic media campaigns, civil disobedience, or public demonstrations. All of these activities have a similar goal – influencing and regulating the behavior of business firms in order to achieve social and environmental value, although their goals and articulated reasons for doing so may differ significantly.

It is important to recognize also that each actor is 'role bound' (limited), but at the same time can be efficacious in the use of specific mechanisms. For example, advocacy NGOs and unions can provide specialized support in community resistance efforts. Similarly, the fiduciary duty of financial institutions predisposes them toward lack of transparency; however, they have substantial power to determine the rules of the game for companies interested in securing capital from them, and they can thus require social and environmental impact evaluations and regular audits from their clients to an extent perhaps greater than that of the home/host state. This client relationship, in turn, requires credible public and confidential reporting mechanisms and verifiable adherence to industry standards. Moreover, the information made public through reporting requirements can contribute to the necessary evidence for well-informed discussion of the need for state-legislated hard law rather than voluntary corporate measures.

Although it is not within the scope of this conclusion to analyze the full range of actors, relationships, and contextual dynamics within the SEVGE model (and indeed, much more work needs to be done on many of the actors discussed here), we identify 11 actor groups that we consider most relevant in analyzing the mining industry: supranational governance, home/other government, host government, affected communities, industry associations, financial institutions and investors, reporting and transparency initiatives, advocacy and development NGOs, customers, business firms, and the media/public sphere. However, it should be noted that these 'actors' and their 'mechanisms' are often not mutually exclusive, especially when it comes to the multi-sectoral partnerships that increasingly characterize development initiatives and governance related to mining industries.

Supranational and multilateral institutions

Only a handful of the chapters collected in this volume refer to the roles played by supranational and multilateral organizations such as the United Nations, Organisation for Economic Co-operation and Development (OECD), International Labor Organization (ILO), and Inter-American Commission on Human Rights (IACHR) among others. Dashwood (Chapter 2) points out that early pressures on the mining industry for improved environmental performance came from the United Nations Environment Program (UNEP). This pressure came at least in part through agreements with host governments such as the World Heritage Convention, which restricted access to areas considered of high natural or cultural value. Smillie (Chapter 6) cites the attention of the United Nations to the issue of conflict diamonds in Africa as being an important part of discussions among governments, industry, and NGOs that led to the emergence of the KP Certification Scheme.

More research needs to be done in analyzing the role of supranational and multilateral institutions and their contributions to SEV, particularly in documenting the efficacy of various initiatives promoted at this level of governance. Generally speaking, we note that the mechanisms, tools, strategies, and initiatives used by multilateral and supranational institutions include: research and policy development; capacity building; participation in the creation of standards and guidelines; and development of codes around human rights upheld under international law. Specific initiatives include the UN's Global Compact, Special Advisor on Business and Human Rights, Multi-stakeholder Dialog on Partnerships for Sustainable Development, Millennium Development Goals, and UN Environment Program (UNEP); OECD's Guidelines for Multinational Enterprises; ILO's Convention #169 (specific to the rights of indigenous and tribal peoples); and IACHR decisions and recommendations.

SEV-enabling dynamics at this level include the legitimacy of institutions such as the UN and OECD on an international level, their capacity for multilateral consultation and partnerships to create effective and comprehensive guidelines and standards, as well as to make policy recommendations. Supranational and international organizations can also provide research and consultation to promote SEV, and dedicate material and human resources focused on SEV issues. Some agreements/decisions promoted at this level are binding (for example, member states are required to comply with ILO and IACHR decisions) and they can have important oversight, monitoring, and research functions that might contribute to knowledge and track issues over time.

SEV-disabling dynamics potentially include the difficulty of monitoring compliance and enforcing binding decisions, as well as the tendency for many standards and guidelines to be non-binding (UNEP and OECD). Further, all mechanisms at this level of governance are dependant on participation of member states or business signatories, and many states and businesses are not bound by membership.

Host governments

A great deal of existing research on the mining industry in Latin America and other developing regions focuses on issues related to the capacities of host governments to manage, monitor, and regulate the industry, particularly given the rapid expansion of mining investments in many parts of the world. Webb (Chapter 3) points out that although regulation may exist, weak enforcement capacity in many countries reduces the efficacy of legal mechanisms. Further, for companies operating in these jurisdictions, even compliance with the environmental regulations may fall well below the expectations of local communities, NGOs, and wider public opinion. In his analysis of home and host governments, Boon (Chapter 4) finds that greater enforcement and coherence of regulatory systems is a key element in reducing conflict related to mining development. However, efforts to improve the efficacy of host government regulation may be hampered by external restrictions placed on host governments through the investment protection clauses that are a part of many trade agreements. Further, the results of Boon's field research in Peru indicated that leadership and coherent policies at many levels of government were lacking, and that coordination problems between different levels of government and companies added to a lack of capacity and resources to disable effective implementation of regulations.

Many researchers have pointed out that lack of regulatory and enforcement capacity, under-resourcing, and mismanagement of revenues are persistent problems in many host governments. However, Campbell et al. (Chapter 5) are careful to point out that these problems are frequently the result of implementation of neoliberal economic policies (such as those carried out in Peru by the Fujimori government), and that an over-focus on 'capacity building' focused on the technical and administrative aspects of governance risks ignoring the key underlying dynamics of neoliberal modes of governance that undermine that capacity in the first place. Smillie's (Chapter 6) observation that the KP failed most dramatically in weak governance states such as Angola and the DRC reinforces Campbell et al.'s argument. Governments that cannot pass and/or enforce laws in accordance with their ruling mandate and commitments cannot effectively protect or enhance SEV, even given strong external supports such as those present with the KP.

Legacies of conflict can also disable ability for host governments to rule effectively and in the interests of SEV. da Silva et al. (Chapter 9) explain how Brazil's 1964 military coup set the historical trajectory of resource exploitation in the Amazon through the establishment of the Carajás Iron Ore Complex as a part of an immense agro-mineral project. The mine, later privatized, carried with it a legacy of political and social conflict resulting from its operation as a part of the military regime that lasted into the 1980s.

Alternatively, Cameron (Chapter 10) found that, in Bolivia, the election of Evo Morales gave some communities a sense of empowerment in their interactions with a mining company, Apex Silver, and that the increased leverage of the community resulting from Bolivia's adoption of ILO Convention 169 regarding the rights of indigenous peoples (even despite the absence of implementation mechanisms) in turn worked to motivate more developed and effective Corporate Social Responsibility (CSR) programs on the part of the company.

A major problem in many host regions, however, results from government dependence on mining revenues. In Chapter 11, Goss points out that in order to bring projects to fruition (and sometimes to secure project financing through international finance institutions), host governments often become partners with private companies in developing a mine, taking majority or minority stakes in a project. However, as Goss points out, 'when local citizens oppose a development, the financial imperatives under which host country governments operate can put them in direct conflict with the citizens they represent' (p. 191).

Focusing on a detailed case study of Trinidad and Tobago (T&T), Shaw (Chapter 15) highlights the complex political economic environment surrounding developing small states, particularly an island state such as T&T. Emphasizing its integration in the global economy through a focus on resource extraction (natural gas destined primarily for the American market), Shaw discusses the various economic and political pressures shaping T&T's development into the twenty-first century, asking how CSR might figure in aligning economic development with social and environmental needs. As is the case with many emerging and small economies competing on a global level, the level of commitment to SEV remains unclear, and the question of whether integration into the global economy enables or disables SEV in these states is an important line of future research.

The organizations, elements, and influences that we have identified for future research considerations include national, regional, and local governments (including self-governing indigenous groups such as First Nation governments in Canada); juridical systems, laws, and regulations pertaining to mining; partnerships with home governments, industry players, IFIs (as creditors), some development NGOs, and affected communities (in principle). We note also that International Financial Institutions (IFIs) and national debt relations are of primary importance for many indebted nations of the global South.

Mechanisms, tools, strategies, and initiatives undertaken by host governments can include regulation of industry impact through legislation and law-making, mandated minimum standards of reporting and verification (including environmental impact assessments), taxation regimes and royalty structures, distribution mechanisms (social investments and/or industry subsidies), bilateral or multilateral (regional) trade agreements, and negotiations in international forums (UN, OECD).

The SEV-enabling dynamics for host governments are varied, but largely dependent on governance capacity. Jurisdiction over business firms operating within national boundaries provides host governments with the ability to enact regulations to hold business firms accountable. Revenues from extraction activities could fund sustainable development/SEV priorities, and stable political environments with strong democratic institutions can empower affected communities and assist in negotiation of economic and SEV priorities. Further, effective distribution policies can empower constituents and build local capacity.

However, the SEV-disabling dynamics experienced by host governments seem by all accounts to be more prevalent. External pressure through relationships with IFIs, industry players, and individual firms to enact permissive regulatory environments to attract investment often excludes SEV priorities. The pervasive threat of capital flight triggers permissive regulation and exacerbates a lack of political will to hold firms accountable. Low regulatory and governance capacity is aggravated by high-corruption environments, and poorly functioning democratic institutions can be easily 'captured' by elite interests. Finally, legacies of colonialism, low education and health, poverty, and violent conflict often compromise local governance capacity and the democratic will-formation necessary to enforce SEV priorities.

Home governments

The role of home governments in mining development overseas has emerged as a key policy issue in recent years, particularly in Canada, which is 'home' to a significant number of large mining companies with global operations. Following a number of high-profile environmental and social disasters connected to Canadian mining companies operating overseas in the 1990s, the Canadian government has come under increased pressure to regulate and hold accountable Canadian mining companies operating in developing nations. Government responses to this public pressure include the creation of a Standing Committee on Foreign Affairs and International Trade focused on Canadian mining companies operating overseas, which in 2005 submitted a parliamentary report to the government that led to a series of roundtable discussions held throughout the country in 2007. The result of the multi-sectoral initiative known as the National Roundtables on CSR (for a more detailed discussion, see Coumans, Chapter 7) was an advisory group report that recommended a stronger role for the Canadian government in monitoring and regulating the overseas activities of Canadian mining companies. Although the Canadian government did not implement all of the recommendations of the advisory report, this issue is still very much alive with the narrow parliamentary defeat in 2010 of a private member's bill focused on regulating the activities on Canadian mining companies abroad, and with the government's release of a 'CSR strategy' document outlining 'softer' regulatory lines (including the creation of an arms-length CSR Counsellor's office for dispute resolution and a CSR Centre of Excellence for development of industry best-practices in CSR) (see Boon, Chapter 4, and Coumans, Chapter 7, for more detailed discussion of these initiatives).

In Chapter 3, Webb provides an overview of Canada's regulatory and legal mechanisms as they relate to Canadian companies operating overseas. He finds that while private litigation is possible, it is extremely rare, and that the most robust legal mechanisms for holding Canadian companies accountable may be shareholder proposals (a mechanism discussed in more detail in the following section on financial actors, and in Chapters 11 and 12, by Goss and Sosa, respectively).

Like host governments, home governments can often find themselves in a conflicted position, caught between the competing demands of economic development on the one hand and social and environmental protection on the other. As Campbell et al. point out in Chapter 5, an initiative of the Canadian government aimed to increase regulatory capacity of the mining industry in Peru (PERCAN, the Peru–Canada Mineral Resources Reform Project) may actually work to undermine Canada's reputation in cases of community–company conflict where PERCAN may be seen as the Canadian government acting in support of Canadian mining companies – a clear conflict of interest that makes government involvement in CSR initiatives fraught with complexities. Further, a tendency to focus primarily on technocratic solutions to complex problems risks alienating important interests that may not easily be defined in technical, scientific, or business terms. Coumans' discussion in Chapter 7 of the involvement of the Canadian International Development Agency (CIDA) in mining CSR projects highlights many similar concerns.

As a category of analysis, home government is most often aligned primarily with a national government, including departments of foreign affairs and international trade, embassies, and consulates acting on behalf of home companies. Further analysis should include a focus on diplomatic relationships between home and host country officials, the role of export credit agencies, and the recourse mechanisms within the juridical system, laws, and regulations of home governments. As elaborated by the authors mentioned above, home governments may partner with host governments to improve regulatory infrastructure, and may also work directly with corporations and industry groups to advocate for trade and/or economic interests. Home governments also directly or indirectly support development and advocacy NGOs and consult with all of the above in policy deliberations.

Mechanisms, tools, strategies, and initiatives available to home governments include bilateral or multilateral trade agreements to protect or encourage national economic interests; policy assistance to host nations to protect image and economic interests with host country; home country laws and regulation to hold companies accountable; consultation and/or diplomatic assistance to improve CSR performance and political relationships between firm and host government; overseas development initiatives (for example, Canadian International Development Agency (CIDA) and International Development Research Centre (IDRC) programs); policy research and/or consultation (examples include the Select Standing Committee on Foreign Affairs report on CSR and extractives, National Roundtables on CSR and Extractives, the CSR Centre of Excellence, and the CSR Counsellor).

Home government SEV-enabling dynamics include political stability and regulatory capacity sometimes lacking in host states; a potential to transcend narrow economic interests of purely economic actors such as banks, industry players, and business firms; economic and human resources to support SEV-focused development; access to technological and intellectual innovation through state-supported university and research and development; a high degree of legitimacy in relations with other system actors; the potential to raise the level of accountability system-wide through binding and enforceable regulations formed in partnership with host governments and industry groups (national policy-making and international agreements); and access to high-level information and the ability to share information with other system actors.

SEV-disabling dynamics could include political priorities and/or economic interests that work against the interests of SEV. Home governments are subject to lobbying influence from powerful industry groups and business firms (dynamics include the threat of capital flight and the need for business tax revenue). Further, horizontal policy incoherence can disable efforts as some departments focus on economic advancement to the exclusion of SEV goals, while other departments work toward SEV enhancement to the exclusion of economic interests. Home countries also face diplomatic and international relations constraints (for example, jurisdictional issues holding firms accountable for abuses in other countries) and short-term political priorities as competing political parties jockey for power.

Communities

Communities affected by mining vary widely in their relationships with other system actors – most notably, companies and host governments. Dashwood (Chapter 2) points out that the concept of stakeholder engagement became important in the mining industry when communities (often aided by NGOs) began to advocate for themselves, sometimes in the process denying companies a 'social license to operate.' Webb (Chapter 3) highlights an increased use of community referenda as a mode of resistance to mining projects in Latin America, with varying levels of success in suspending or terminating the mining project in question.

Communities possess varying degrees of power and influence in negotiations over mining development. Coumans (Chapter 7) documents in detail the efforts of communities in several different development contexts to communicate their self-determined development goals. These goals often do not entail large-scale mining projects, and community goals are frequently dismissed or ignored by companies and government decision-makers alike. Cameron (Chapter 10) points to several factors that, in Bolivia, reduced the capacity for communities to negotiate for long-term benefits from mining projects – in his analysis, high levels of poverty and the desire for mine employment restricted the ability for communities to reject (or realign) mining-led development in favor of other, more sustainable alternatives.

Further analysis of communities should include both those directly affected (for example, communities subject to forced relocation or in direct geographical proximity to a mine) and indirectly affected (for example, affected by an influx of community members relocated from a mining-affected locale, and/or downstream environmental and health impacts). Particular attention must be given to unique histories of any given community as well as to internal power dynamics and conflicting interests between elite and disadvantaged groups. Indigenous communities in particular tend to have different challenges (historical power imbalance and histories of conflict with settler communities; lack of contact in more remote regions; histories of genocide and/or forced relocation; threat of cultural assimilation) and opportunities (may enjoy special status/privilege in certain cases; potential for more cohesive community structure and coordinated action) in different locations. In functioning democracies, the host government should represent and protect community interests. Advocacy and development NGOs often directly partner with communities, and communities may enter partnerships with business firms – these partnerships can be mutually beneficial or manipulative/contentious.

Mechanisms, tools, strategies, and initiatives available to communities may include direct negotiations with business firms to influence the development of projects; impact-benefit agreements with firms; local and regional government representation (political campaigns aimed at local, regional, or national governments), plebiscites, or referenda to gauge support for the project; direct action (including blockades, protests, and information campaigns); and legal action (attempts to sue firm for damages in home country or other jurisdiction if host government fails to adequately regulate the negative impacts of mining).

The SEV-enabling dynamics that may be characteristic of communities include their high degree of legitimacy as direct actors in negotiations with firms and in information campaigns. Communities may possess a longer-term vision for development than other system actors – particularly because they are the ones who must live with the lasting impacts of extractive activities. Communities also generally have a strong connection to place (especially in locations with spiritual and/or cultural significance), which leads them to place a high value on SEV. Community efforts can be enabled by effective political representation and governance capacity at local, regional, and national level, and communities with strong representation and negotiation capacity, as well as effective mechanisms for will-formation (community dialogs) may be able to negotiate directly with firms in a genuinely 'win–win' situation. Further, strong relationships with functional and effective NGOs can help build community capacity – access to information and self-advocacy is crucial in these relationships.

Low levels of power and negotiation capacity in relation to other system actors is a persistent disabling dynamic for communities. This can result in community voices and SEV priorities being ignored in negotiations. Some communities may also be affected by low governance capacity, histories of poverty and social inequality, low levels of education and/or health, and violence and conflict. Internal power imbalances between elite and disadvantaged groups can result in internal divisions, conflict, and the uneven distribution of costs (impacts) and benefits of extractive development. Lack of access to effective remedy in the case of abuses relating to extractive activities is also a strong disabler – at present, communities generally have recourse to very few effective legal mechanisms at national (home or host) or international levels. Communities also suffer from lack of access to information prior to, during, and after extractive activities. As a result, community decisions may be based on inadequate, misleading, or inaccurate information. In the absence of effective resolution processes, conflict can devolve into violence. This is particularly problematic in locations with a history of violent repression where extractives-related grievances add to pre-existing grievances. In such contexts, state repression of dissent appears to support foreign extractives firms.

Advocacy and development NGOs and CSOs

This category of actors is as diverse as communities – thus, specific attention needs to be given to differences in the goals, desires, and capacities of individual organizations categorized here. Coumans (Chapter 7) raises some important questions about the goals and alignments of some development-oriented NGOs that partner with industry actors to promote CSR projects, but at the same time, much of the research and publicity work carried out on behalf of and in partnership with affected communities is done by advocacy NGOs, often also based in 'home' jurisdictions.

NGOs and civil society organizations (CSOs) can be important sources of information about mining-related development issues, and they also serve a crucial function in mobilizing public opinion on critical issues. Dashwood (Chapter 2) points out that large environmental NGOs such as Greenpeace and the World Wildlife Fund (WWF) were key actors mobilizing opinion against global mining operations, often using online information campaigns to raise awareness.

Campbell et al. (Chapter 5) highlight the important role NGOs can play in the construction of political identities in mining-affected areas, pointing to the emergence of rights-based advocacy and the construction of alternatives to mining, including the option of refusal. They highlight the importance of institutional support for such rights-based approaches – support that is currently not addressed in the technocratic approaches to social and environmental management taken by most governments and firms.

Further research on NGOs and CSOs should focus on the different types of support and advocacy offered by different organizations. We identify a range of different organizations involved in mining-related issues, including advocacy NGOs such as MiningWatch Canada, Rights Action, the Halifax Initiative, Observatorio de Conflictos Mineros de América Latina (OCMAL), Amnesty International, CorpWatch, EarthWorks, and Oxfam; and policy and research NGOs such as the International Institute for Sustainable Development (IISD) and the North-South Institute. As described in more detail by Coumans in Chapter 7, development NGOs may partner with companies, communities, and/or governments to engage in direct development projects. CSOs include church and religious groups, which are a significant source of advocacy in Latin America.

Mechanisms, tools, strategies, and initiatives available to NGOs and CSOs include direct pressure on business firms to modify behavior and/or to influence decisions; media campaigns to raise awareness of issues (electronic media are important in this regard); educational/observation delegations to witness and report on impacts of mining; consultation/lobbying with government and/or industry associations; legal action; research and reporting; and direct action (protests, letter campaigns).

The SEV-enabling dynamics characteristic of NGO/CSO actors include the fact that social and/or environmental justice is frequently their primary goal, aligning most actions with SEV. NGOs and CSOs may have a high legitimacy/trust relationship with affected communities, particularly if the organization is based in or has strong connections with the affected area. NGOs are willing to research and publicize negative impacts of mining development, and some organizations have positive working relationships with other system actors (communities, government, companies). NGOs and CSOs may have reliable funding to provide resources for advocacy, and they frequently possess on-the-ground research capacity to document mining development issues as they emerge.

NGO- and CSO-disabling dynamics may include the risk of adversarial/polarized relationships with other system actors (particularly business firms). Some NGOs and CSOs may be unwilling to recognize benefits of mining equally to the harms, and there is a risk of aggravating local conflicts related to mining, especially in split communities. Some NGOs and CSOs may have low legitimacy and trust with other system actors, and they may lack capacity in terms of stable funding and human resources.

Industry associations and other players

Industry associations are particularly active in CSR and standards promotion in the mining industry. Dashwood (Chapter 2) traces the evolution of the International Council on Mining and Metals (ICMM) as the outcome of a series of industry initiatives aimed to address CSR and sustainability in mining. Similarly, Smillie shows in Chapter 6 how the World Diamond Council (WDC) was involved in the creation of the KP, and now that the KP is faltering, how other industry initiatives such as the Diamond Development Initiative International (DDII) and the Responsible Jewellery Council have emerged.

In Chapter 8, Lindsay analyzes three mining industry association codes, finding that despite their depth and detail, all three lack key components of effective regulation – namely, external verification of compliance, full public disclosure, and credible sanctioning mechanisms for non-compliance. Further, in her comparative analysis of industry codes as 'discourse communities,' she argues that these codes may fail to address some of the crucial problems associated with large-scale mining projects as defined by affected communities.

Organizations, elements, and influences characteristic of this actor category include supply chain actors (upstream and downstream), competitors, industry associations, and industry ownership structures. In the mining industry, supply chain actors appear to be less influential than in other industries such as manufacturing and consumer goods. Industry associations are perhaps the most significant actors in this category. Canadian examples include the Mining Association of Canada (MAC), the Prospectors & Developers Association of Canada (PDAC), and the Canadian Institute of Mining, Metallurgy & Petroleum (CIM). International industry associations include the Institute of Materials, Minerals & Mining (IOM3) (UK-based) and the ICMM. Industry associations often partner with host governments, development agencies, and NGOs to develop standards and regulations for the industry.

Mechanisms, tools, strategies, and initiatives available in this category include supply chain agreements to improve efficiency and/or responsibility in suppliers; industry-based self regulation standards and codes such as the MAC's Toward Sustainable Mining (TSM), PDAC's e3 Plus: A Framework for Responsible Exploration, and ICMM's Sustainable Development Framework; industry peer-to-peer learning and information sharing (best practices), research, and development (technology and management level); awards/recognition for high performers; direct lobbying to influence host state regulations; and consultations with states and/or supranational development organizations.

SEV-enabling dynamics in this category include consultation on industry standards with multiple stakeholders (industry, NGO, state, community). Some standards/guidelines are moving toward external verification, and best practice learning and information sharing based on industry experience is standard. Industry associations and other industry players may undertake current research relevant to industry challenges. Industry actors also have the potential to enforce CSR standards through the supply chain, and some like MAC's TSM provide continuous improvement methods and practical advice for all levels of the mining cycle, from exploration to closure.

SEV-disabling dynamics in this category include the following: lobbying emphasis on self-regulation ('soft law') diminishes industry accountability and reduces governance capacity of home and host governments; industry goals (expansion of mining industry, extraction vs recycling, and/or reclamation and reduction of minerals use) are at odds with SEV; industry standards are not enforceable and lack credible retaliation mechanisms for underperformers; non-industry players consider industry-driven standards dubious; firm participation in standards and guidelines initiatives is still relatively low; and skewed power relations affecting consultation and partnership with NGOs, communities, and the state can mean that industry goals and objectives take precedence over SEV goals. As argued by Coumans in Chapter 7, industry initiatives can also work to shift responsibility for SEV away from corporations, obscuring the ways in which the industry is skewed to benefit economic actors while shifting the responsibility for mining's negative impacts onto host governments and communities.

Financial institutions and investors

At a high level, international financial institutions (IFIs) are crucial actors in setting the economic context under which mining development takes place. Campbell et al. discuss in Chapter 5 how the economic reforms undertaken in Peru with the support of the World Bank (WB), the Inter American Development Bank (IADB), and the International Monetary Fund (IMF) opened the country to a sharp influx of foreign investment in the mining sector, but at the same time undermined the capacity of the Peruvian government to effectively regulate the industry.

In Chapters 11 and 12, Goss and Sosa, respectively, assess the potential for various financial mechanisms to influence firm behavior. Goss (Chapter 11) finds that there currently exist few credible financial mechanisms for applying pressure on companies to improve their SEV focus. Large equity holders could in theory influence firm behavior through soft advocacy, but investor boycotts are rarely effective unless implausibly large numbers of shareholders act in concert. Debt holders (primarily banks) are primarily concerned with the ability of the firm to repay its debt, and although the emergence and adoption of the Equator

Principles shows that major banks are responsive to social and environmental concerns, further research needs to be conducted on the efficacy of the Equator Principles in influencing firm behavior.

In her analysis of two cases of shareholder activism, Sosa shows the variable results of this approach to influencing corporate decision making. Drawing from research that points to various factors influencing the saliency of shareholder activism, Sosa's discussion highlights the importance of corporate culture and shareholder legitimacy to explain the different results of similar shareholder advocacy efforts.

Organizations, elements, and influences in the category of financial institutions and investors include private international and national banks, IFIs (primarily the World Bank & IFC), risk insurance agencies, private capital, and major investment funds (some overlap with 'firm' category in terms of certain types of shareholders). IFIs such as WB and IFC partner with host governments to reregulate the mining industry in order to attract investment and indeed, this may be required as condition of a loan to the government. The goals of private banks and investors are closely aligned with industry and business firms – to increase capital. IFIs are in principle development institutions, but their logic is based on the assumption that economic expansion is the primary route to poverty alleviation, development, and national self-sufficiency.

Mechanisms, tools, strategies, and initiatives in this category include standards such as the IFC Performance Standards and the World Bank's Safeguard Policies and Guidelines for Extractives Projects (guidelines on tailings management, security, and human rights, labor rights, indigenous issues); mandated CSR for client firms as a way to protect investment (by diffusing local resistance and/or avoiding regulation that hurts profitability); reporting requirements from some (but not all) finance providers; linking financial backing to CSR/SEV performance (relatively uncommon); and the Equator Principles (major international banks) as a financial industry benchmark for determining, assessing, and managing social and environmental risk in project financing.

SEV-enabling dynamics include: mandated social/environmental guidelines and audits that could hold firms accountable; increased oversight of firm CSR/SEV performance connected to socially responsible investment (SRI); potentially credible retaliation mechanisms if financial backing is linked to CSR/SEV performance; activist and responsible shareholders; and good corporate governance.

SEV-disabling dynamics include the fact that the goals of financial institutions are more closely aligned with that industry than with SEV, and thus they may lack motivation for increased accountability if it risks return on investment; reporting requirements may lack legitimate oversight and/or credible consequences for under-performance; IFI emphasis on Chicago school economics (less government regulation, more free trade) reduces host government capacity to regulate firms and convert economic gain into SEV; power imbalances between IFIs and client (host) governments (whose interests are being served?); private capital and some sovereign funds are less susceptible to international CSR pressure; and wild swings in financial markets and in credit availability.

Reporting and transparency agencies

More research is needed on the efficacy of reporting and transparency organizations in contributing to the protection and maintenance of SEV, particularly given the amount of resources invested (and faith professed) in their development. This category also overlaps considerably with other actor categories, especially given the multilateral participation of the various system actors in the creation of many reporting and transparency agencies.

Smillie's (Chapter 6) discussion of the KP and Hamman et al.'s (Chapter 16) overview of the EITI present some useful insights about reporting and transparency agencies. Smillie notes that the organizational structure of the KP, while necessary and useful in its creation and early application, proved a significant limitation as it was unable to adapt to changing needs in the diamond industry and unanticipated challenges arose. The three fundamental flaws Smillie identifies in the KP are a weak monitoring system, an unworkable consensus decision-making structure, and few penalties for noncompliance aside from suspension.

Hamann et al. argue in Chapter 16 that the Extractive Industries Transparency Initiative (EITI) offers a simple yet effective example of mining company engagement with national-level development. They point out that the implementation of the EITI:

> often results in more far-reaching and complex discussions about the value chain associated with mining. In other words, while the EITI only requires disclosure of payments from companies to the government, it often has the effect (at least in the countries surveyed) of increasing interest in where and how the revenues were generated by the companies, and how the revenues are actually used by the government. (Hamann et al., Chapter 16, p. 274)

More research should be done on the sometimes subtle and unexpected spinoff successes related to reporting and transparency initiatives, as well as their weaknesses and shortcomings. Another useful research initiative would be in compiling an inventory of such organizations and the focus of their activities.

We identify in this category a significant amount of overlap with other actor categories – industry associations produce reporting and transparency guidelines, as do supranational organizations. Mechanisms, tools, strategies, and initiatives in this category include, for example, the GRI's (Global Reporting Initiative) Sustainability Reporting Framework, ISO 14001 and 26000, Transparency International, EITI, Publish What You Pay, and AA 100 Series.

SEV-enabling dynamics may include the availability of accurate and complete information crucial for all system actors, the relatively high legitimacy of initiatives such as the GRI, and 'harder' mechanisms such as the auditing of ISO standards.

SEV-disabling dynamics include the fact that reporting and transparency initiatives are at present based on voluntary participation depending on self-selection (although as Smillie points out in Chapter 6, voluntary can quickly become compulsory in contexts of high participation rates). Further, credible enforcement mechanisms for underperformance have not yet been built into the reporting/transparency infrastructure, third-party verification is still rare, and in many companies and sectors adherence to these standards is considered expensive and time consuming.

Customers

The chapters in this volume, typical of research in the mining industry, did not focus on customers (which could be categorized with industry associations and industry players). However, for the purposes of introducing potentially fruitful lines of analysis, we consider customers in the mining industry to be an important and underdeveloped research area in its own right.

Organizations, elements, and influences belonging to this category could include industrial buyers (overlap with supply chain/industry players) and final consumers, although consumer advocacy and activism is much less common in commodities than in consumer goods. Mechanisms, tools, strategies, and initiatives in this category might include industry-specific oversight and monitoring, information-sharing (for example, the Responsible Gold website), certification and standards such as the KP certification and Council for Responsible Jewellery standards, ethical consumption practices, and supply chain pressure to improve CSR at all stages of extraction/ production. Analysis here should focus on the efficacy of certification efforts and/or information sharing (KP, Responsible Gold, Council for Responsible Jewellery Practices), noting some overlap with industry players and NGO categories.

SEV-enabling dynamics include the potential for well-organized and informed consumer advocacy to have an impact on the reputation of firms, as has been the case in other industries (notably consumer goods manufacturing). This category might also represent also high potential legitimacy with industry players and individual firms.

SEV-disabling dynamics include the following: standards and codes may be co-opted or used strategically by invested actors (for example, states providing oversight may not adhere to standards); access to accurate supply chain information is difficult to obtain in context of mergers and acquisitions, and complex production and distribution of raw materials; consumer apathy, particularly in the case of industrial consumers; and the fact that few consumer initiatives exist in the mining industry.

The firm

A significant amount of research exists on the firm, but most management-oriented approaches (such as those discussed in Chapter 1) use analytic models that de-emphasize the impact of non-firm interactions and actors by centering analysis on the firm itself. Abdala (Chapter 14) provides an interesting case study on one firm initiative promoted by Alcoa in the Juruti municipality of Brazil's Pará State that takes a more inclusive approach to governing sustainable development and CSR initiatives. The Sustainable Juruti initiative is based on a 'pluralistic' decision-making process made up of a diverse set of actors from different groups (both local and non-local) affected by Alcoa's mine. Early indicators show that this initiative is meeting its collectively defined goals, with the added benefit to the firm of reduced conflict surrounding the mine and improved relationships with key stakeholders.

In Chapter 16, Hamann et al. draw from their earlier work on resource governance in Africa, arguing that firms are not only influenced by the institutional and political contexts in which they operate, but that they also crucially influence that context in their own strategies and actions. Firms can interact with host governments in ways that support transparency and effectiveness or in ways that detract from it. The authors draw lessons from three case studies showing how firm interactions with communities, governments, and reporting and transparency organizations (EITI) served to shape the outcomes – both successes and failures – experienced on the ground as they negotiated mining development in complex and conflict-ridden political environments in the African context.

Future research on firms from a SEVGE perspective could include analysis of the interactions between management, employees, unions, shareholders, suppliers, and customers, focusing on the differences in firms ranging in size from small exploration companies with little economic/political influence to large multinationals with significant economic and political influence on other system actors. Research should also take into account the ways in which firms are influenced by economic factors (such as commodity prices and boom/bust cycles), financial factors (access to credit and financial health), ownership factors (M&A, shareholder preferences), corporate governance, size and focus of firm (major firms vs juniors, exploration vs production), and firm partnerships with states, international finance, communities, and NGOs.

Mechanisms, tools, strategies, and initiatives available to firms include CSR programs and policies help obtain/maintain SLO though 'soft law'; impact-benefit agreements with communities (more common in Canada); economies of scale, new technology, and low-cost production to ensure profitability; marketing/PR to improve image (for investors, customers, and host communities and/or states); research and development; direct lobbying of governments (either individually or through industry associations) to influence policy decisions; and partnerships and participation in international organizations (UN) and civil society initiatives.

SEV-enabling dynamics include governance, leadership, and management commitment to SEV (integrated with corporate structure and decision-making); CSR-dedicated capital, human resources, and technological expertise to contribute to SEV; willingness to partner with states, NGOs, and communities and share information and/or expertise; balancing power asymmetries, acknowledging critical rights, negotiating both converging and conflicting interests, and managing relations with stakeholder constituencies (see Covey and Brown, 2001).

SEV-disabling dynamics include over-focus on the financial bottom line that detracts from SEV investment (profit before people and planet); direct lobbying in home and host government to reduce regulatory interference/oversight; strategic and/or short-term CSR programs used for political and financial gain (greenwash, divide and conquer strategies, 'buying' community support); lack of vision and/or horizontal policy incoherence (one department contributing to SEV while other parts of the organization work against SEV); lack of expertise in development and community relations; partnerships skewed by power imbalance and strategic interest manipulation; lack of transparency and poor external communication (especially regarding impacts of operations); lack of accountability for abuses (permissive culture of complicity and/or irresponsibility); intense external competition exerting pressures to underperform in CSR/SEV; and corruption – particularly problematic in relationships with high-corruption environment in host governments.

Media

More research needs to be conducted on the role of media and public opinion in negotiations and debate over mining development. Although not explicitly addressed by the authors in this volume, we present it as a separate category to highlight the need for more attention. The organizations, elements, and influences characterizing this category may include ownership structures of local, national, and international media organizations. Attention should be given to the differences in reporting between advertising-driven media and grassroots, reader-supported media. We note that alternative/grassroots media (especially online) are a key source of information about extractives that may not be reported in mainstream media. Other issues include journalistic ethics, the setting of news priorities by management, and differences in home country vs host country vs international reporting.

Mechanisms, tools, strategies, and initiatives in this actors category include standard news reports and breaking news, in-depth reports (for example, the W5 episode 'Paradise Lost' regarding Goldcorp's operations in Guatemala), and journalistic codes of ethics (reporting accurate information). SEV-enabling dynamics include the fact that transparency and global awareness has made malfeasance difficult to hide. Media may act as impartial observers in cases where information is difficult to obtain and accurate, and fact-checking and balanced information contributes clarity and assists other system actors to understand issues. An investigative approach balanced with willingness to expose both positive and negative outcomes of extractive activities can contribute to informed debate.

SEV-disabling dynamics include the following: newsroom pressures often lead to surface-level reporting on complex issues – only the 'sensational' may be reported; the paucity of mainstream media coverage of important extractives-related issues; media coverage tends toward either local issues or 'big story' international issues; objective and accurate information may be hard to access in some cases; and misinformation can be passed on as information.

Min(d)ing the gaps? Future directions in mining and CSR

Whether it be in the area of research, operations, or policy-making and governance, the complex environment surrounding large-scale resource extraction in developing parts of the world is characterized by gaps – in knowledge and information, in governance capacity and coherence, and often in common understanding and goals. When confronted with such gaps, the various actors involved in resource governance have the option of either seeking to address and redress them in the interests of the common good (defined here as SEV), or alternatively, ignoring and/or seeking to exploit them for private benefit.

A central assumption of this book is that, in order to address and redress the various gaps that disable efforts to contribute meaningfully to SEV in the mining industry, actors with very different goals and worldviews will need to learn to build understanding and to make compromises. The preceding chapters offer some excellent insights about the difficult issues faced by various actors in their efforts to negotiate the complex environment of resource governance, either by seeking to understand, define, and meet their responsibilities, or in choosing to avoid them.

Further research in this area is crucial, but it must be sensitive to the reality that for many of the actors involved in resource governance, lives and livelihoods are at stake. Further research must also take seriously the disabling effects of historical power dynamics between key actors, both at the level of global political economy and in local contexts where advantage and disadvantage can (and often do) collide violently. Thus, it may well be crucial to consider the meaning not only of corporate social responsibility, but also the responsibilities of governments of all levels, development agencies and NGOs, finance institutions, communities, researchers, and policy-makers alike as they interact to form the broad context in which SEV-enhancing development is defined, understood, and used as a basis for action or inaction in the resource extraction industries.

Further research on CSR and SEV in the extractive industries might address the issues explored by previous chapters, including the role of norm-formation and guiding standards; the efficacy of these 'soft' CSR standards and their relationship to 'hard' legal frameworks (national and international); the modes of governance and regulatory interventions of invested actors; the contours of effective oversight and compliance agreements and mechanisms; the enabling and disabling environments that condition choices made and strategies devised by key actors on the ground (including communities, companies, and governments); as well as the roles played by a wide variety of secondary actors in shaping the broader context of resource extraction.

We hope the present work contributes to an expansion in the scope of research on the extractives to take into account these and other emerging dynamics. We believe the complexity and conflict (both actual and potential) characterizing resource extraction and governance will only be exacerbated by the increase in resource scarcity in many sectors (including mining), making finer-grained, more complex, and comprehensive research vital to protecting SEV and the common interests of all actors involved.

Note

1. Sharon Livesey (2002) notes research documenting 40 working definitions of sustainable development in the five years following publication of the Brundtland report.

Appendix: The Latin America–Canada Mining Connection*

According to Lemieux (2005), in 2005 Latin America as a region had a 48 per cent share of global copper production, as well as a 37 per cent share of iron ore, a 26 per cent share of bauxite, a 20 per cent share of zinc, and a 16 per cent share of nickel. In 2010, Latin America's metal and mineral reserves are far from exhausted and mining activity continues to be a critical element of many Latin American economies. In Peru, for example, minerals comprise 75 per cent of export revenue and 19 per cent of Gross Domestic Product (GDP), while in Chile mineral extraction represents 44 per cent of export revenue and 9 per cent of GDP (ECLAC, 2010). Export and tax revenue trends are likely to continue and even accelerate. For example, ECLAC (2010) notes that for countries specializing in the production of natural resources, particularly mining, the recovery in the prices of commodities had a positive impact on fiscal revenues from these sources.

According to ECLAC (2010), exports from Latin America as a whole were expected to increase by about 25.5 per cent in current terms in 2010, equivalent to 12.2 per cent in constant terms. Mining and oil products saw the fastest growth (41.3 per cent in the first three-quarters of 2010), with agricultural and manufacturing products expanding a little more slowly (ECLAC, 2010, p. 52). For countries specializing in the production of natural resources, the 2010 recovery in the prices of commodities had a positive impact on fiscal revenues from these sources (ECLAC, 2010, p. 29). For example, ECLAC reports that 'in Ecuador and Peru, the largest revenue increases [in 2010] reflected the resources yielded by commodities (oil in the former and income taxes paid by mining companies in the latter) and a higher tax take' (ECLAC, 2010, p. 32).

Arellano's (2010) report also makes the observation that despite the global financial and economic crisis of 2008, FDI flows to Latin America and the Caribbean reached a record high in 2008, with services and natural resources sectors being the greatest beneficiaries. The US continued to be Latin America's largest source of FDI (Foreign Direct Investment), followed by Spain and Canada.

Canada is a leader in mining activity, not only in Latin America but globally. According to a Mining Association of Canada (MAC) report (MAC, 2009), as of 2008 Canadian-listed firms operated an estimated 4600 mineral projects in varying states of development worldwide. This investment represents a total direct investment of approximately CAN$67 billion. According to an earlier MAC report (MAC, 2007 quoted in Arellano, 2010), Canada received the biggest slice (19 per cent) of global exploration expenditure in 2006 and exported more than CAN$1 billion, Toronto has become the mining capital of the world and is the global leader for mining finance with 38 per cent of the world's equity raised in 2006, and Vancouver is home to the world's leading cluster of exploration companies and expertise.

Arellano (2010) notes that Canadian FDI in Latin America was driven by two main corporate strategies: natural resource-seeking by mining companies and market-seeking by banks or financial service providers. In 2008, of Canadian FDI to the developing world, 42 per cent went to Latin America; excluding offshore financial centres in the Caribbean, the main recipients of Canadian FDI were Brazil (7.5 per cent), Chile (6.0 per cent), Mexico (4.2 per cent), Argentina (2.1 per cent), Peru (1.5 per cent), and the Dominican Republic (1.0 per cent). Further, the value of investments by Canadian companies in Latin America and the Caribbean more than doubled between 2003 and 2007, from US$18.3 billion to US$37.4 billion.

Latin America's shares of global resources, relatively attractive investment climate, and geographic proximity have made it a natural choice for Canadian mining companies to invest heavily in mining and exploration. According to DFAIT (2009), Canadian mining companies have invested over CAN$60 billion in developing countries, including about CAN$41 billion in Latin America (including Mexico) and almost CAN$15 billion in Africa. Arellano (2010) notes that mining is the only sector in which Latin America is the most prominent region for Canadian FDI. The investments of Canadian mining companies are focused on Mexico, Chile, Brazil, Argentina, and Peru.

The dynamism of current mergers and acquisitions (M&A) activity in the industry complicates the analysis of the 'Canadian' mining sector, since it is very difficult to provide answers to questions such as: When does a Canadian firm cease to be – or start to be – Canadian? Does nationality of the parent company, headquarters, or major shareholders matter in determining whether the firm is or isn't Canadian? These questions become urgent when considering home country regulation: whether a firm is considered Canadian or not would affect the legislation that would apply to it.

This trend toward M&A has escalated for a number of reasons, including ongoing awareness of the finite nature of metal resources, soaring demand, high commodity prices, the appeal of acquisition over new exploration (saves time and has less risk), and the presence of new players in the form of companies from China and India. Given the above conditions, Latin American mines that were previously less attractive are now in high demand, partly because they are perceived as less risky than those in other parts of the world. Canadian companies already operating in the region are also highly sought after for M&A deals due to their traditionally higher growth potential in comparison with their Latin American-owned counterparts. China's Aluminum Corp US$792 million buyout of the Vancouver firm Peru Copper in 2007 (Dolan, 2007), Swiss-based Xstrata PLC's US$18 billion acquisition of Toronto-based Falconbridge Ltd. in 2006, and Anglo-Australian miner Rio Tinto's 2007 US$38.1 billion takeover of Canadian aluminum giant, Alcan Inc. are just some of a number of recent acquisitions of Canadian firms in the region. Larger Canadian companies are also busy buying up smaller Canadian firms. Two such recent instances are the Vancouver-based Goldcorp's purchase of Glamis Gold's operations in Guatemala and Honduras and the US$10.4 billion 2006 take-over of Vancouver's Placer-Dome by Toronto-based Barrick Gold (Rights Action, 2006).

Canadian exploration companies and juniors are also quite active in Latin America. Their level of activity in the region shows an increasing trend (CEPAL, 2010). The business strategy of Canadian junior explorations companies appears to be to obtain capital in the financial markets to use in exploration, and if a financially feasible deposit is found, either sell or be contracted by a major firm, often Canadian, to develop and operate the site.

Given the high level of Canada–Latin America mining interactions, it is not surprising to see a high level of CSR-related activity directed toward operations in the region by not only companies, but also by NGOs (Non-governmental Organizations), and the Government of Canada (See Sagebien et al., 2008; and Boon in this volume).

Note

*This section draws heavily on J. M. Arellano's (2010) briefing paper on Canadian FDI in Latin America as well as on the ECLAC (2010) briefing paper.

Bibliography

Abdala, F. (2010) 'Sustainable Juruti Model: a proposed model for mining and local development', Brazilian Mining Congress, Belo Horizonte, 5 August 2010.
—— (2007) 'Governança Global sobre Florestas', doctoral thesis, University of Brasilia, Brasilia.
ActionAid (2008) *Precious Metal: The Impact of Anglo Platinum on Poor Communities in Limpopo, South Africa* (London: ActionAid).
Ades, A. and R. Di Tella (1999) 'Rents, Competition, and Corruption', *The American Economic Review*, 89, 982–94.
Aiken, M. (2010) 'First Nation Denies Exploration Bid by De Beers', *Kenora Daily Miner and News*, 17 December, www.kenoradailyminerandnews.com/ArticleDisplay.aspx?e=2895331, date accessed 19 February 2011.
Akabzaa, T. (2009) 'Mining in Ghana: Implications for National Economic Development and Poverty Reduction', in Bonnie Campbell (ed.) *Mining in Africa: Regulation and Development* (London, New York: Pluto Press).
Alatas, S. H. (ed.) (1986) *The Sociology of Corruption* (Singapore: Times Books).
Alcazar, W. (2001) *A Public Expenditure Tracking Survey of Peru's Vaso de Leche Program* (Washington, DC: World Bank).
Alcoa (2009) *Mina de Bauxita de Juruti: Balanço de Ações e Resultados Socioambientais* (Juruti: Alcoa).
Alford, R. & R. Friedland (1992) *Powers of Theory: Capitalism, the State, and Democracy* (Cambridge, UK: Cambridge University Press).
Almeida, J. M. G. Jr (1986) *Carajás – Desafio Político, Ecologia e Desenvolvimento* (São Paulo: Editora Brasiliense).
Aliaga D. C. and J. De Echave (1994) *Minería y Reinserción Internacional en el Perú* (Lima: Instituto para el Desarrollo de la Pesca y la Minería).
Anderson, S., R. Dreyfus and M. Perez-Rocha (2010) 'Mining for Profits in International Tribunals', Institute for Policy Studies website, 29 April, www.ips-dc.org/reports/mining_for_profits_in_international_tribunals, date accessed 14 September 2010.
Angel, J. and P. Rivoli (1997) 'Does Ethical Investing Impose a Cost upon the Firm? A Theoretical Examination', *Journal of Investing*, 6 (4), 57–61.
AngloGold Ashanti (AGA) (2005) 'Human Rights Watch Report on AngloGold Ashanti's Activities in the DRC', News release, 1 June 2005, http://www.anglogold.com/NR/rdonlyres/1BC8B7B3-6363-48B3-AE85-B315606CF248/0/2005Jun01_SAreleaseDRC.pdf, date accessed 15 February 2011.
Arellano, J.M. (2010) Canadian foreign direct investment in Latin America. Background paper. Prepared for the Dialogue on Canada-Latin American Economic Relations, 27–28/5/10. The North-South Institute (May), http://www.nsi-ins.ca/english/pdf/FDI%20backgrounder.pdf, date accessed 4 March 2011.
Arellano-Yanguas, J. (2011) 'Aggravating the Resource Curse: Decentralisation, Mining and Conflict in Peru', *Journal of Development Studies*, 47 (4), 617–638.

——— (2008a) 'A Thoroughly Modern Resource Curse? The New Natural Resource Policy Agenda and the Mining Revival in Peru', IDS Working Paper 300, Institute of Development Studies, University of Sussex, Brighton, UK.

——— (2008b) '*Canon minero* and conflicts in Peru', www.sed.manchester.ac.uk/research/andes/seminars/JArellano_PresentationSummary_Seminar3.pdf, date accessed 25 February 2011.

Asociación de Desarrollo Integral San Miguel Ixtahuacan (ADISMI) (2008) 'Repuesta Comunitaria Con Relación al Documento Llamado "Evaluación del Impacto Sobre Los Derechos" por la Transnacional Goldcorp Por Medio De Su Proyecto Marlin', Personal communication between Javier de Leon, the shareholders involved in the HRIA, and Jantzi Sustainalytics, 5 September 2008.

Aste Daffós, J. (1997) *Transnacionalización de la minería peruana. Problemas y posibilidades hacia el siglo XXI* (Lima: Friedrich Ebert Stiftung).

Aste, J., J. De Echave and M. Glave (2003) *Procesos Multi-Actores para la Cogestión de Impactos Mineros en Perú*, Informe Final [final report], Mining Policy Research Initiative, International Development Research Centre, Ottawa, www.idrc.ca/uploads/user-S/11166180311informe_final.pdf, date accessed 15 October 2007.

Auty, R. M. (1993) *Sustaining Development in Mineral Economies: The Resource Curse Thesis* (London: Routledge).

Baker, R. (2005) *Capitalism's Achilles Heel: Dirty Money and How to Renew the Free-Market's System* (Hoboken, NJ: John Wiley & Sons).

Baldacchino, G. and D. Milne (eds) (2006) 'Exploring Sub-National Island Jurisdictions', *Round Table*, 95 (386) (September), 487–627.

Balkau, F. and A. Parsons (1999) 'Emerging Environmental Issues for Mining the PECC Region', *Pacific Economic Co-operation Committee Minerals Forum in Lima, Peru*, 22 April, https://www.u-cursos.cl/ingenieria/2007/1/MI55D/1/material_docente/objeto/123434, date accessed 17 June 2011.

Banerjee, S. B. (2008) 'Corporate Social Responsibility: The Good, the Bad and the Ugly', *Critical Sociology*, 34 (1), 51.

——— (2003) 'Who Sustains Whose Development? Sustainable Development and the Reinvention of Nature', *Organization Studies*, 24 (1), 143.

——— (2001) 'Corporate Citizenship and Indigenous Stakeholders: Exploring a New Dynamic of Organisational-Stakeholder Relationships', *The Journal of Corporate Citizenship*, 1, 39–55.

BankTrack (2010) 'Bold Steps Forward: Towards Equator Principles that Deliver to People and the Planet,' Banktrack website, 14 January, http://www.banktrack.org/download/bold_steps_forward_towards_equator_principles_that_deliver_to_people_and_the_planet/100114_civil_society_call_equator_principles.pdf, date accessed 28 February 2011.

Baptiste, M. (2010a) Press release, 2 November www.nationtalk.ca/modules/news/article.php?storyid=37757, date accessed 19 February 2011.

——— (2010b) 'Will Fish Lake Be Sacrificed for Gold Fever? The Tsilhqot'in Nation's Fight for Its Sacred Lake Is Everyone's Fight', Common Ground web page, 1 November, www.commonground.ca/iss/232/cg232_fishlake.shtml, date accessed 16 February 2011.

Barber, B. (2006) 'Monitoring the Monitor: Evaluating CalPERS' Shareholder Activism', unpublished working paper, Graduate School of Management, University of California, Davis.

Barnea, A. and A. Rubin (2005) 'Corporate Social Responsibility as a Conflict Between Owners', unpublished working paper (Vancouver: Simon Fraser University).
Barreto, M. L. (2001) *Ensaios sobre a Sustentabilidade da Mineração no Brazil* (CETEM.Rio de Janeiro: Centro de Tecnologia Mineral).
Barrick Gold Corporation (2010) 'Reforestation with CIDA in Peru', *Beyond Borders, Responsible Mining at Barrick Gold Corporation*, 3 September, http://barrickbeyondborders.com/2010/09/reforestation-with-cida-in-peru/, date accessed 3 March 2011.
Bastida, E., R. Irarrázabal, and R. Labó (2005) 'Mining Investment and Policy Developments: Argentina, Chile and Peru,' *Centre for Energy, Petroleum and Mineral Law & Policy Internet Journal*, 16, www.dundee.ac.uk/cepmlp/journal/html/Vol16/article16_10.php, date accessed 30 May 2010.
Bauer, R. and P. Smeets (2010) 'Social Values and Mutual Fund Clienteles', unpublished working paper (Maastricht: Maastricht University).
BCE Inc. v. 1976 Debentureholders (2008) 3 S.C.R. 560, 2008 SCC 69.
Bebbington, A. (2007) *Minería y Desarrollo en el Perú: con Especial Referencia al Proyecto Río Blanco, Piura* (Lima: Instituto de Estudios Peruanos).
Bebbington, A., D. Bebbington and L. Hinojosa (2009) 'Extraction, inequalities and territories in Bolivia' (Prepared for the 2009 Congress of the Latin American Studies Association, Rio de Janeiro, Brazil 11–14 June 2009).
Bebbington, A., L. Hinojosa, D. Humphreys Bebbington, M. Luisa Burneo and X. Warnaars (2008a) *Contention and Ambiguity: Mining and the Possibilities of Development*, Brookes World Poverty Institute Working Paper, 57, October 2008.
Bebbington, A., D. Humphreys Bebbington, J. Bury, J. Lingan, J. Pablo Muñoz and M. Scurrah (2008b) 'Mining and Social Movements: Struggles Over Livelihood and Rural Territorial Development in the Andes', *World Development*, 36 (12), 2888–905.
Bedoya, M. (2001) 'Gold Mining and Indigenous Conflict in Peru: Lessons from the Amarakaeri', in C. K. Roy, V. Tauli-Corpuz and A. Romero-Medina (eds) *Beyond the Silencing of the Guns*, pp. 184–210 (Baguio City, Philippines: Tebtebba Foundation).
Belem, G. (2009) 'Mining, Poverty Reduction, the Protection of the Environment and the Role of the World Bank Group in Mali', in B. Campbell (ed.) *Mining in Africa: Regulation and Development* (London, New York: Pluto Press).
——— (2008) 'Quelle gouvernance pour la mise en œuvre du développement durable? L'expérience de l'industrie minière du Mali', PhD thesis, Institut des sciences de l'environnement, Université du Québec à Montréal, Montreal.
Belem, G., E. Champion and C. Gendron (2008) 'La régulation de l'industrie minière canadienne dans les pays en voie de développement: Quel potentiel pour la responsabilité sociale des entreprises?', *McGill International Journal of Sustainable Development Law and Policy*, 4 (1), 51–76.
Beltrão, J. F. and R. C. Domingues-Lopes (2003) 'Indios versus Madereiros: Conflitos sem Trégua', in E. Fontes (ed.) *Contando a História do Pará: Os Conflitos e os Grandes Projetos na Amazônia Contemporânea (sec.xx)* (Belém: E.Motion).
Bergman, L. (2005) 'Curse of Inca Gold', *Frontline World*, Public Broadcasting Service, broadcast October 2005.
Biggs, S. and S. Smith (2003) 'A Paradox of Learning in Project Cycle Management and the Role of Organizational Culture', *World Development*, 31 (10): 1753.

Björnskov, C. and M. Paldam (2004) 'Corruption Trends', in J. Lambsdorff, M. Taube and M. Schramm (eds) *The New Institutional Economics of Corruption – Norms, Trust, and Reciprocity*, pp. 59–75 (London: Routledge).

Blaser, M., H. Feit and G. McRae (eds) (2004) *In the Way of Development: Indigenous Peoples, Life Projects and Globalization* (London, New York, Ottawa, Cairo, Dakar, Montevideo, Nairobi, New Delhi, Singapore: Zed Books in association with International Development Research Centre).

Blore, S. (2006a) *The Lost World: Diamond Mining and Smuggling in Venezuela* (Ottawa: Partnership Africa Canada).

———(2006b) *Triple Jeopardy: Triplicate Forms and Triple Borders, Controlling Diamond Exports from Guyana* (Ottawa: Partnership Africa Canada).

Blore, S. (2005) *The Failure of Good Intentions: Fraud, Theft and Murder in the Brazilian Diamond Industry* (Ottawa: Partnership Africa Canada).

Blowfield, M. and J. G. Frynas (2005) 'Setting New Agendas: Critical Perspectives on Corporate Social Responsibility in the Developing World', *International Affairs*, 81 (3), 499–513.

Boele, R., H. Fabig and D. Wheeler (2001) 'Shell, Nigeria and the Ogoni. A Study in Unsustainable Development: II. Corporate Social Responsibility and "Stakeholder Management" versus a Rights-Based Approach to Sustainable Development', *Sustainable Development*, 9, 121–35.

Boerzel, T. and T. Risse (2010) 'Governance without a State: Can It Work?', *Regulation & Governance*, 4, 113–34.

Boloña, C. (1996) 'The Viability of Alberto Fujimori's Economic Strategy', in E. Gonzales de Olarte (ed.) *The Peruvian Economy and Structural Adjustment: Past, Present and Future* (Miami: North-South Center Press, University of Miami).

Bomsell, O., I. Marques, D. Ndiaye and P. De Sa (1990) *Mining and Metallurgy Investment in the Third World: The End of Large Projects?* (Paris: OECD).

Boon, J. (2009) 'Corporate Social Responsibility (CSR) in the Mineral Exploration and Mining Industry: Perspectives on the Role of 'Home' and 'Host' Governments', MA thesis, Globalization and International Development, University of Ottawa, Ottawa.

Boutilier, R. and I. Thomson (2009) 'Establishing a Social License to Operate in Mining, *Edumine Courses*, http://premium.infomine.com, date accessed 3 October 2009.

Boxembaum, E. (2006) 'Corporate Social Responsibility as Institutional Hybrids', *Journal of Business Strategies*, 23, 1.

Brav, A., W. Jiang, R. Thomas and F. Partnoy (2008) 'Hedge Fund Activism, Corporate Governance and Firm Performance', *The Journal of Finance*, 63, 1729.

Bray, D. (2007a) *Facing Up to Corruption in Nigeria* (London: Control Risks Group).

———(2007b) *Facing Up to Corruption: A Practical Business Guide* (London: Control Risks Group).

Breaking the Silence (2008) 'Maritime Human Rights Group Seeks Answers from Goldcorp', http://frederictonpeace.org/?p=1218, date accessed 16 May 2008.

Bridge, G. (2004) 'Mapping the Bonanza: Geographies of Mining Investment in an Era of Neoliberal Reform', *The Professional Geographer*, 56 (3), 406–21.

Briguglio, L. and E. Kisanga (2004) *Economic Vulnerability and Resilience of Small States* (Commonwealth Secretariat and the University of Malta, London).

British Broadcasting Corporation (BBC) (2008) 'Oil contract scandal shakes Peru', *One Minute News*, 7 October, http://news.bbc.co.uk, date accessed 3 October 2009.
Breuer T., and C. Farrell (2005) *Collaboration between NGOs and the Mining Industry in the Third World*, CHF-PARTNERS in Rural Development, Ottawa, www.chf-partners.ca/proceedings-and-statements/publications/proceedings-and-statements/collaboration-between-ngos-and-the-mining-industry-in-the-third-world.html, accessed 16 February 2011.
Bronfman, A. (2007) *On the Move: The Caribbean since 1989* (Halifax, NS: Fernwood and London: Zed).
Browder, J. O. and B. J. Godfrey (1997) *Rainforest Cities: Urbanization, Development, and Globalization of the Brazilian Amazon* (New York: Columbia University Press).
Brown, A. (2006) 'What Are We Trying to Measure? Reviewing the Basics of Corruption Definition', in C. Sampford (ed.) *Measuring Corruption* (New York: Ashgate Publishing Ltd).
Bury, J. (2008) 'Transnational Corporations and Livelihood Transformations in the Peruvian Andes: An Actor-oriented Political Ecology', *Human Organization*, 67 (3), 307–21.
——— (2004) 'Livelihoods in Transition: Transnational Gold Mining Operations and Local Change in Cajamarca, Peru', *The Geographical Journal*, 170 (1), 78–91.
Business and Human Rights (2010) 'Response of Blackfire Exploration to Allegations of Human Rights Abuses in Connection with Its Mine in Chiapas', Business & Human Right Resource Centre website, 13 September, www.business-humanrights.org/Links/Repository/1002276, date accessed 14 September 2010.
Business Monitor (2009) Special Report: Caribbean 2009–10: Trouble in Paradise, May, Business Monitor International, London.
Business Trinidad & Tobago 2010–2011 (2011) (Port of Spain: Prestige Business Publications).
Business Trinidad & Tobago 2008–2009 (2009) (Port of Spain: Prestige Business Publications).
Business Wire (1998) 'Cambior Secures Dismissal of Omai-Related Class Action', AllBusiness website, 17 August, www.allbusiness.com/legal/legal-services-litigation/6863188-1.html, date accessed 14 September 2010.
Business Wire (2003) 'Manhattan Minerals Corp.: Centromin Peru Ruling on Tambogrande Option Agreement', AllBusiness website, 10 December, www.allbusiness.com/legal/labor-employment-law-alternative-dispute-resolution/5789354-1.html, date accessed 14 September 2010.
Cabrera, M. (2007) *Guía para el manejo de crisis y la comunicación de crisis: Las industrías extractivas y las crisis sociales*, Proyecto de reforma del sector de recursos minerales del Perú (PERCAN)/Ministerio de Energía y Minas, Lima.
Cailteux, K. L. and J. T. Grabill (2010) 'Foreign Tort Suits Under 221 Year-Old Law Ruled Out against Corporations', *The Legal Pulse*, http://wlflegalpulse.com/2010/09/20/, date accessed 17 October 2010.
Calvano, L. (2008) 'Multinational Corporations and Local Communities: A Critical Analysis of Conflict', *Journal of Business Ethics*, 82, 793–805.
Cameron, J. and K. Gibson (2005) 'Alternative Pathways to Community and Economic Development: The Latrobe Valley Community Partnering Project', *Geographical Research*, 43 (3), 274–85.

Cameron, M. A. (1997) 'Political and Economic Origins of Regime Change in Peru: *The Eighteenth Brumaire* of Alberto Fujimori', in M. A. Cameron and P. Mauceri (eds) *The Peruvian Labyrinth: Polity, Society, Economy*, pp. 37–69 (University Park, PA: The Pennsylvania State University Press).
——— (1994) *Democracy and Authoritarianism in Peru: Political Coalitions and Social Change* (New York: St. Martin's Press).
Cameron, M. and C. Hecht (2008) 'Canada's Engagement with Democracies in the Americas', *Canadian Foreign Policy*, 14 (3), 11–28.
Campbell, B. (2009a) 'Introduction', in B. Campbell (ed.) *Mining in Africa: Regulation and Development* (London, New York: Pluto Press).
——— (2009b) 'Guinea and Bauxite-Aluminum: The Challenge of Development and Poverty Reduction', in B. Campbell (ed.) *Mining in Africa: Regulation and Development* (London, New York: Pluto Press).
——— (2009c) 'Conclusion: What Development Model? What Governance Agenda?', in B. Campbell (ed.) *Mining in Africa: Regulation and Development* (London, New York: Pluto Press).
——— (2003) 'Factoring in Governance Is not Enough: Mining Codes in Africa, Policy Reform and Corporate Responsibility', *Minerals & Energy-Raw Materials Report*, 18 (3), 2–13.
Campbell, B., Belem, G. and V. N. Coulibaly (2007) *Poverty Reduction in Africa: On Whose Development Agenda? Lessons from Cotton and Gold Production in Mali and Burkina Faso*, Les cahiers de la Chaire, C.-A. Poissant – Collection recherche No. 2007-01, Montreal, www.ieim.uqam.ca/IMG/pdf/Cahier_2007-01_-_Poverty_reduction_in_Africa.pdf, date accessed 2 March 2011.
Campbell, J. L. (2007) 'Why Would Corporations Behave in Socially Responsible Ways? An Institutional Theory of Corporate Social Responsibility', *Academy of Management Review*, 32 (3), 946.
——— (2006) 'Institutional Analysis and the Paradox of Corporate Social Responsibility', *American Behavioural Scientist*, 49 (7), 925–38.
Campodónico Sánchez, H. (1999) *Las reformas estructurales en el sector minero peruano y las características de la inversión 1992–2008*, Serie Reformas Económicas No. 24, Comisión Económica para América Latina y el Caribe (CEPAL), United Nations, Santiago de Chile.
Campos, E., D. Lien and S. Pradhan (1997) 'Corruption and Its Implications for Investment: Predictability Matters', *World Development*, 27 (6), 1059–67.
Canada Business Corporations Act (1985) R.S., c. C-44, as amended.
Canadian Boreal Forest Agreement (2010) www.canadianborealforestagreement.com/, date accessed 19 October 2010.
Canadian Business for Social Responsibility (CBSR) (2009) *CSR Frameworks for the Extractive Industry* (Canada: CBSR).
Canadian International Development Agency (CIDA) (2007) *Executive Report on the Evaluation of the CIDA Industrial Cooperation (CIDA-INC) Program*, December, Evaluation Division, Performance and Knowledge Management Branch, CIDA, Gatineau, QC, www.acdi-cida.gc.ca/INET/IMAGES.NSF/vLUImages/Evaluations/$file/INC_ExecutiveReport-E.pdf, date accessed 3 March 2011.
Canel, E., U. Idemudia and L. L. North (eds) (2010) 'Rethinking Extractive Industry: Regulation, Dispossession and Emerging Claims', *Canadian Journal of Development Studies*, 30 (1–2) (Special Issue), 1–339.
Cardoso, F. H. and E. Faletto (1979) *Dependency and Development in Latin America* (University of California Press).

Center for International Environmental Law (CIEL) (2009) 'Guatemalan Community Leaders Ask Canadian Government to Investigate Human Rights Violations Committed by Goldcorp Inc. at Marlin Mine', http://www.ciel.org/Hre/Guatemla_Canada_9Dec09.html, date accessed 9 December 2009.

Central American Free Trade Agreement (2005) SICE Foreign Trade Information System website, www.sice.oas.org/TPD/USA_CAFTA/CAFTAfinal/CAFTAind_e.asp, date accessed 14 September 2010.

CERES (2009) *Murky Waters: Corporate Reporting on Water Risk*, Coalition for Environmentally Responsible Economies, www.ceres.org/waterreport, date accessed 22 May 2010.

Chadwick, J. (2005) 'CVRD's Iron Fist: The World's Biggest Producer has Big Plans', *International Mining*, December, 35–43.

Chang, K.-S., B. Fine and L. Weiss (eds) (forthcoming) *Developmental Politics in Transition: The Neo-Liberal Era and Beyond* (London: Palgrave Macmillan).

Chaparro, A. E. (2002) *Actualización de la compilación de leyes mineras de catorce países de América Latina y el Caribe – Volumen I*. Serie Recursos Naturales e Infraestructura No 43, CEPAL (División de Recursos Naturales e Infraestructura), United Nations, Santiago de Chile.

Chaves, E. M. C. (2004) 'Projeto Grande Carajás. Chapter VIII', *Revista Nova Atenas*, 07 (02), 217–24.

Chiapello, E. and N. L. Fairclough (2002) 'Understanding the New Management Ideology: A Transdisciplinary Contribution from Critical Discourse Analysis and New Sociology of Capitalism', *Discourse & Society*, 13, 185–208.

Christian Aid (2009) *Undermining the Poor: Mineral Taxation Reforms in Latin America*, a Christian Aid report, September, Christian Aid, London, www.christianaid.org.uk/resources/policy/tax.aspx, date accessed 16 February 2011.

CFPOA (Corruption of Foreign Public Officials Act) (1998) S.C. c. 24, as amended.

Cleary, D. (1990) *Anatomy of the Amazon Gold Rush* (Iowa City: University of Iowa Press).

Clegg, P. and E. Pantojas (eds) (2009) *Governance in the Non-Independent Caribbean: Challenges and Opportunities in the 21st Century* (Kingston: Ian Randle).

Clemons, R. S. & M. K. Mcbeth (2001) *Public Policy Praxis: Theory and Pragmatism: A Case Approach* (New Jersey: Prentice Hall).

CMMB (2009) 'The Caribbean in a Global Recession', *Investment Quarterly*, 2 (2) (April), 48–68.

Coelho, M.C.N., A. G. Lopes, A. C. da Silva, F. A. O. da Silva, H. Fonseca, I.S. Matos and M.R. de Souza (2002) 'Territórios, cidades e entornos no espaço da mineração em Carajás/Pará - Amazonia Oriental', in S. C Trindade Jr and G. M. Rocha (eds) *Cidade e empresa na Amazônia: Gestão do território e desenvolvimento local* (Belém: Editora Pakatatu).

Columban Fathers (2007) *Mining in the Philippines: Concerns and Conflicts* (West Midlands, UK: Columban Fathers).

Comaroff, J. L. (1998) 'Reflections on the Colonial State, in South Africa and Elsewhere: Factions, Fragments, Facts and Fictions', *Social Identities*, 4 (3), 321–61.

Commission on Sustainable Development Policies (CPDS) (2002) *Agenda 21 Brasileira* (Brasília: Ministério do Meio Ambiente).

Companhia Vale do Rio Doce (2006) Companhia Vale do Rio Doce, http://www.cvrd.com.br (home page), date accessed 12 June 2006.

Contreras, R. and E. Madrid (2006). 'San Cristóbal: Conflictos y negociación es entre comunidades campesinas y una compañía transnacional minera', in T. Bochard (ed.) *Dialogo sostenible* (Sucre, Bolivia: Latinas Editores).

Cooper, A. F., A. Antkiewicz and T. M. Shaw (2006) 'Economic Size Trumps All Else: Lessons from BRICSAM', Working Paper #12, December, CIGI (www.cigionline.org), Waterloo, ON.

Cooper, A. F. and T. M. Shaw (eds) (2009) *The Diplomacies of Small States: Between Vulnerability and Resilience* (London: Palgrave Macmillan for CIGI).

Cooper, A. F. and P. Subacchi (eds) (2010) 'Global Economic Governance in Transition', *International Affairs*, 86 (3) (May), 607–757.

Costa, S. (2008) *Mineworkers' Quality of Life in Remote Communities: A Multiple Case Study in the Brazilian Amazon*, Doctoral thesis (Vancouver: University of British Columbia).

Costa, S. and M. Scoble (2006) 'Mine Accommodation Strategies – Case Studies in Developing Countries', in *Proceedings of the AusIMM First International Mine Management Conference* (AusIMM).

Corporate Register (2009) www.corporateregister.com, date accessed 21 June 2009.

Coumans, C. (2011) 'Occupying Spaces Created by Conflict: Anthropologists, Development NGOs, Responsible Investment and Mining', *Current Anthropology*, 52 (S3), S29–43.

——(2010a) 'Alternative Accountability Mechanisms and Mining: The Problems of Effective Impunity, Human Rights, and Agency', *Canadian Journal of Development Studies*, 30 (1–2), 27–48.

——(2010b) 'Bill C-300: A High Water Mark for Mining and Government Accountability', *MiningWatch Canada Newsletter*, 29 (Autumn), www.miningwatch.ca/sites/miningwatch.ca/files/MWC_newsletter_29.pdf, date accessed 16 February 2011.

——(2009) 'Ongoing Concerns with the Goldcorp Human Rights Impact Assessment', 16 March 2009, http://www.miningwatch.ca/, date accessed 15 October 2009.

——(2008a) 'Realising solidarity: Indigenous peoples and NGOs in the contested terrains of mining and corporate social accountability', in C. O'Faircheallaigh and S. Ali (eds) *Earth Matters: Indigenous Peoples, the Extractive Industries and Corporate Social Responsibility*, pp. 43–66 (Sheffield: Greenleaf Publishing).

——(2008b) 'Re: Fundamental Concerns with the Goldcorp Human Rights Impact Assessment and Erosion of Trust in Canada's Responsible Investment Community's Shareholder Proposal Process', Mining Watch Canada website, 4 December, www.miningwatch.ca/en/letter-shareholder-group-re-human-rights-impact-assessment-goldcorps-guatemala-mine, date accessed 14 September 2010.

——(2002a) *Placer Dome Case Study: Marcopper Mines*, MiningWatch Canada, Ottawa, www.miningwatch.ca/en/placer-dome-case-study-marcopper-mine-marinduque-philippines, date accessed 16 February 2011.

——(2002b) 'The Successful Struggle against STD in Marinduque', in *STD Toolkit* (Ottawa and Berkeley: MiningWatch Canada and Project Underground), www.miningwatch.ca/en/submarine-tailings-disposal-toolkit, date accessed 16 February 2011.

——— (2000) 'Canadian Companies in the Philippines: Placer Dome', in *Undermining the Forests: The Need to Control Transnational Mining Companies: A Canadian Case Study*, pp. 59–67 (London: Forest People's Programme, Philippine Indigenous Peoples Links, World Rainforest Movement).

——— (1995) 'Ideology, Social Movement Organization, Patronage and Resistance in the Struggle of Marinduquenos against Marcopper', *Pilipinas*, 24 (Spring), 38–74.

——— (1994) *Building Basic Christian Communities: Religion, Symbolism and Ideology in a National Movement to Change Local Level Power Relations in the Philippines* (Ottawa: National Library of Canada).

Covey, J. and L. D. Brown (2001) *Critical Cooperation: An Alternative Form of Civil Society – Business Engagement* (Boston: Institute for Development Research).

Cowman, J. (2010) 'The Alien Tort Statute – Corporate Social Responsibility Takes on a New Meaning', *The CSR Digest*, 6 September, www.csrdigest.com/2009/07/the-alien-tort-statute-corporate-social-responsibility-takes-on-a-new-meaning/, date accessed 6 September 2010.

Crabtree, J. (2000) 'Populisms Old and New: The Peruvian Case', *Bulletin of Latin American Research*, 19, 163–76.

Crabtree, J. and J. Thomas (eds) (1998) *Fujimori's Peru: The Political Economy* (London: Institute of Latin American Studies).

CSRAméricas (2009) CSRAméricas home page, Banco Interamericano de Desarrollo, www.csramericas.org/, date accessed 28 February 2011.

Cutler, A. C. (2008) 'Problematizing Corporate Social Responsibility under Conditions of Late Capitalism and Postmodernity', in V. Rittberger and M. Nettesheim (eds) *Authority in the Global Political Economy* (New York: Palgrave Macmillan).

Cyanide Code (International Cyanide Management Code for the Manufacture, Transport and Use of Cyanide in the Production of Gold) (2005), International Cyanide Management Institute, www.cyanidecode.org/, date accessed 14 September 2010.

da Silva Enriquez, M. A. and J. D. Godfrey (2007) 'Social-Environmental Certification: Sustainable Development and Competitiveness in the Mineral Industry of the Brazilian Amazon', *Natural Resources Forum*, 31, 71–86.

Dahan, N., J. Doh and T. Guay (2006) 'The Role of Multinational Corporations in Transnational Institution Building: A Policy Network Perspective', *Human Relations*, 59, 1571–600.

Damonte Valencia, G. (2007) 'Minería y política: recreación de luchas campesinas en dos comunidades andinas', in A. Bebbington (ed.) *Minería, Movimientos Sociales y Respuestas Campesinas; una ecología política de transformaciones territoriales*, 117–71 (Lima: Instituto de Estudios Peruanos and Centro Peruano de Estudios Sociales).

Dashwood, H. (2007a) 'Canadian Mining Companies and Corporate Social Responsibility: Weighing the Impact of Global Norms', *Canadian Journal of Political Science*, 40 (1), 129–56.

——— (2007b) 'Towards Sustainable Mining: The Corporate Role in the Construction of Global Standards', *The Multinational Business Review*, 15, 1 (Spring), 47–65.

——— (2005) 'Canadian Mining Companies and the Shaping of Global Norms of Corporate Social Responsibility', *International Journal* (Autumn), 977–98.

Davidson, J. (1995) 'Enabling Conditions for the Orderly Development of Artisanal Mining with Special Reference to Experiences in Latin America' in *International Roundtable on Artisanal Mining* (Washington DC: World Bank).

Dean, M. (2010) *Governmentality: Power and Rule in Modern Society*, 2nd edn (London: Sage).

Defensoría del Pueblo (2010) *Reporte de Conflictos Sociales no. 78*, October, Defensoría del Pueblo, Lima, Peru.

——(2007) *Informe Extraordinario. Los Conflictos Socio-Ambientales por Actividades Extractivas en el Perú* (Lima: Defensoría del Pueblo, República del Perú).

De Ferranti, D. (2003) *Inequality in Latin America and the Caribbean: Breaking with Tradition?* (Washington, DC: World Bank).

Department for International Development (2010) *Synthesis of Country Programme Evaluations Conducted in Fragile States*, DFID Evaluation Report EV 709, DFIFD, London.

Department of Foreign Affairs and International Trade (DFAIT) (2011) CSR *E-Bulletin/Bulletin électronique sur la RSE*, CSRRSEBTS@international.gc.ca.

——(2010) 'Canada Promotes Socially Responsible Mining in Peru', Government of Canada web page, www.canadainternational.gc.ca/peru-perou/highlights-faits/PERCAN2009.aspx?lang=eng, date accessed 1 October 2010.

——(2009) *Building the Canadian Advantage: A Corporate Social Responsibility (CSR) Strategy for the Canadian International Extractive Sector*, March, www.international.gc.ca/trade-agreements-accords-commerciaux/assets/pdfs/CSR-March2009.pdf, date accessed 22 May 2010.

——(2008) *CSR Survey of Canadian Mining Companies in Peru*, June 23, Embassy of Canada, Lima.

——(2007) *National Roundtables on Corporate Social Responsibility (CSR) and the Canadian Extractive Industry in Developing Countries: Advisory Group Final Report*, DFAIT, Ottawa, www.mining.ca/www/media_lib/MAC_Documents/Publications/CSRENG.pdf, date accessed 3 March 2011.

——(2005) *Government Response to the Fourteenth Report of the Standing Committee on Foreign Affairs and International Trade: Mining in Developing Countries – Corporate Social Responsibility*, 17 October, Department of Foreign Affairs and International Trade, Government of Canada, Ottawa.

Dhir, A. (2010) 'Shareholder Engagement in the Embedded Business Corporation: Investment Activism, Human Rights and TWAIL Discourse', in P. Zumbansen and C. Williams (eds) *The Embedded Firm: Labour, Corporate Governance and Finance Capitalism* (Cambridge, UK: Cambridge University Press).

Diamond Development Initiative International (DDII) (2010) 'Enabling Positive Development for Artisanal Diamond Mining Communities', http://www.ddiglobal.org/, date accessed 7 July, 2010.

Diamond, D. W. (1984) 'Financial Intermediation and Delegated Monitoring', *Review of Economic Studies*, 51, 393–414.

Dias, A.K. and M. Begg (1994) 'Environmental Policy for Sustainable Development of Natural Resources: Mechanisms for Implementation and Enforcement', *Natural Resources Forum*, Vol. 18 (4), 276–86.

Diamond McCarthy (n.d.) 'Province of Marinduque', Current Events, Diamond McCarthy LLP website, www.diamondmccarthy.com/current-events-pom.html, date accessed 2 March 2011.

Dingwerth, K. (2008) 'Private Transnational Governance and the Developing World: A Comparative Perspective', *International Studies Quarterly*, 52, 607–34.

Dolan, K. A. (2007) 'Latin Mining Via Toronto', 2008 Investment Guide International, Forbes.com, 10 December, www.forbes.com/forbes/2007/1210/144.html, date accessed 11 March 2011.

Donaldson, T. and L. E. Preston (1995) 'The Stakeholder Theory of the Corporation: Concepts, Evidence, and Implications', *Academy of Management Review*, 20 (1), 65–91.

Dow Jones Newswire (2010) 'Court Orders Peru to Consult Indigenous Peoples on Mining, Oil Projects', TradingMarkets.com website, 1 September, www.tradingmarkets.com/news/stock-alert/mioie_dj-court-orders-peru-to-consult-indiginous-peoples-on-mining-oil-projects-1145502.html, date accessed 14 September 2010.

Drohan, M. (2010) 'Regulating Canadian Mining Companies Abroad: The Ten-Year Search for a Solution', Policy Brief No. 7, January, Centre for Policy Studies, University of Ottawa, Ottawa.

ECLAC (Economic Commission for Latin America and the Caribbean) (2010) *Preliminary Overview of the Economies of Latin America and the Caribbean*, LC/G.2480-P, Economic Development Division, Economic Commission for Latin America and the Caribbean, United Nations, Santiago, Chile, www.eclac.org/publicaciones/xml/4/41974/2010-976-BPI-Book_WEB.pdf, date accessed 8 March 2011.

——— (2008) *Statistical Yearbook for Latin America and the Caribbean 2007*, http://www.eclac.org, date accessed 25 August 2009.

EIA (US Energy Information Administration) (2010) *U.S. Natural Gas Imports & Exports*: 2009, 28 September, EIA, Washington, DC, www.eia.doe.gov/pub/oil_gas/natural_gas/feature_articles/2010/ngimpexp2009/ngimpexp2009.htm, date accessed 2 March 2011.

——— (2003) The Global Liquefied Natural Gas Market: Status & Outlook, December, DOE/EIA-0637, Energy Information Administration, US Department of Energy, Washington, DC, www.eia.doe.gov/oiaf/analysispaper/global/pdf/eia_0637.pdf, date accessed 28 February 2011.

EITI (2009) 'Supporting Companies', Extractive Industries Transparency Initiative, http://eiti.org/supporters/companies, date accessed 28 February 2011.

EIU (Economist Intelligence Unit) (2009) *Trinidad and Tobago: Country Report*, March, Economist Intelligence Unit, London.

——— (2007) *Trinidad and Tobago: Country Profile*, Economist Intelligence Unit, London.

EMA (2009) 'EMA Appeals Alutrint Judgment', Environmental Management Authority, Republic of Trinidad and Tobago, 7 July, www.ema.co.tt/cms/index.php?option=com_content&task=view&id=170&Itemid=1, date accessed 28 February 2011.

Emmons, G. (2010) '$how Me the Money', *Harvard Business School Alumni Bulletin*, June, www.alumni.hbs.edu/bulletin/2010/june/money.html, date accessed 16 February 2011.

Engel, E., R. M. Hayes and X. Wang (2007) 'The Sarbanes-Oxley Act and Firms' Going-Private Decisions', *Journal of Accounting and Economics*, 44 (1–2), 116–45.

Enriquez, J. (2002) 'Capítulo 4-minería, minerales y desarrollo sustentable en Bolivia', in MMSD South America (ed.) *Mining and Minerals of South America in*

Transition towards Sustainable Development (Montevideo Uruguay: International Development Research Institute).
Enríquez, M.A.R. (2007) 'Mineração: Maldição ou Dádiva? Os dilemas do desenvolvimento sustentável a partir de uma base mineira', doctoral thesis, University of Brasilia, Brasilia.
Ernst & Young (2008) *Corruption or Compliance: Weighing the Costs: 10th Global Fraud Study* (London: Ernst and Young Group Limited).
Escobar, A. (1995) *Encountering Development: The Making and Unmaking of the Third World* (Princeton, NJ: Princeton University Press).
Estrada, D. (2006) ' "Yes" to Gold Mine, but Don't Touch the Glaciers', Inter Press Service News Agency, 15 February, www.ipsnews.net/news.asp?idnews= 32174), date accessed 16 February 2011.
Ethos Institute (2008) 'Best practices from Brazil – CVRD Case Study. Sustainable local development: the challenge of new undertakings in the mining sector', http://www.ethos.org.br (home page), date accessed 14 September 2008.
European Union (2007) CSR Conference: Rhetoric and Realities – Corporate Social Responsibility in Europe, 27 June, Brussels.
European Social Investment Forum (Eurosif) (2008) *2008 European SRI Study*, http://www.eurosif.org, date accessed 10 October 2009.
Fabozzi, F., K. C. Ma and B. Oliphant (2008) 'Sin Stock Returns', *Journal of Portfolio Management*, Fall, 82–94.
Fairclough, N. (2001) 'The Dialectics of Discourse' *Textus*, 14 (2), 231–42.
Fama, E. (1985) 'What's Different About Banking?', *Journal of Monetary Economics*, 15, 29–39.
Falk, R. (1999) *Predatory Globalization: A Critique* (Cambridge, UK: Polity Press).
Farrell, L., E. Mackres and R. Hamann (2009) *A Clash of Cultures (and Lawyers): A Case Study of Anglo Platinum and its Mogalakwena Mine in Limpopo, South Africa* (Cape Town: University of Stellenbosch Business School).
Farrell, T. M. A. (2005) 'Caribbean Economic Integration: What Is Happening Now; What Needs to Be Done', in K. Hall and D. Benn (eds) *Caribbean Imperatives: Regional Governance and Integrated Development*, pp. 177–205 (Kingston: Ian Randle).
Feinberg, R. E. (2011) 'What Role for the Private Sector?', in G. Mace, A. F. Cooper and T. M. Shaw (eds) *Inter-American Cooperation at a Crossroads* (London: Palgrave Macmillan for CIGI).
Figueroa, A., T. Altamirano and D. Sulmont (1996) *Exclusión social y Desigualdad en el Perú*, Organización Internacional del Trabajo (OIT), Instituto Internacional de Estudios Laborales, Lima.
Finnemore, M. (1996) *National Interests in International Society* (Ithica, NY: Cornell University Press).
Finnemore, M. and K. Sikkink (1998) 'International Norm Dynamics and Political Change', *International Organization*, 52, 4 (Autumn), 887–917.
Firger, D. (2010) 'Transparency and the Natural Resource Curse: Examining the New Extraterritorial Information Forcing Rules in the Dodd-Frank Wall Street Reform Act of 2010', *Georgetown Journal of International Law*, 41, 1043.
Fjelstad, O. and J. Isaksen (2008) *Anti-Corruption Reforms: Challenges, Effects and Limits of World Bank Support Background Paper to Public Sector Reform: What Works and Why? An IEG Evaluation of World Bank Support* (Washington, DC: World Bank Independent Evaluation Group).

Flohr, A., L. Rieth, S. Schwindenhammer, and K. D. Wolf (2010) *The Role of Business in Global Governance: Corporations as Norm-Entrepreneurs* (Hampshire, UK, New York: Palgrave Macmillan).

FOCAL (2005) 'External CSR Practice and Investments by Canadian Corporations in Latin America and the Caribbean', September, FOCAL, Ottawa.

Foreign Affairs and International Trade Canada (2010) 'The Review Process', Office of the Extractive Sector CSR Counsellor, www.international.gc.ca/csr_counsellor-conseiller_rse/index.aspx, date accessed 15 November 2010.

Fox, F., H. Ward and B. Howard (2002) *Public Sector Roles in Strengthening Corporate Social Responsibility: A Baseline Study* (New York: The World Bank).

Freeman, R. E. (1984) *Strategic Management: A Stakeholder Approach* (Boston: Pitman).

Freeman, R. E., J. S. Harrison and A. C. Wicks (2007) *Managing for Stakeholders: Survival, Reputations, and Success* (New Haven, CT: Yale University Press).

Frynas, J. G. (2009) 'Corporate Social Responsibility in the Oil and Gas Sector', *Journal of World Energy Law & Business*, 2 (3), 178–95.

——(2004) 'The Oil Boom in Equatorial Guinea', *African Affairs*, 103 (413), 527.

Frynas, J. G. and M. Paulo (2007) 'A New Scramble for African Oil? Historical, Political and Business Perspectives', *African Affairs*, 106 (423), 229–51.

FUNBIO (Brazilian Fund for Biodiversity) (2010) *Sustainable Juruti Fund* (Rio de Janeiro: Funbio).

FGV (Fundação Getulio Vargas) (2009) *Indicadores de Juruti: por onde caminha o desenvolvimento do município* (São Paulo: CES-FGV).

Galbraith, L., B. Bradshaw and M. B. Rutherford (2007) 'Towards a New Supraregulatory Approach to Environmental Assessment in Northern Canada', *Impact Assessment and Project Appraisal*, 25 (1), 27–41.

Gambetta, D. (2002) 'Corruption: An Analytic Map', in S. Kotkin and A. Sajo (eds) *Corruption in Political Transition: A Sceptic's Handbook* (Budapest: Budapest European University).

Gantz, D. (2004) 'The Evolution of FTA Investment Provisions: From NAFTA to the United States – Chile Free Trade Agreement', *American University International Law Review*, 19, 679.

Garvey, N. and P. Newell (2005) 'Corporate Accountability to the Poor? Assessing the Effectiveness of Community-Based Strategies', *Development in Practice*, 15, 389–404.

Gestion (2009) 'Doe Run Daría Mina y Valores en Garantía por el PAMA', *Gestion*, 20 October 2009.

Gibbon, P., S. Ponte and E. Lazaro (eds) (2010) *Global Agro-food Trade and Standards: Challenges for Africa* (London: Palgrave Macmillan).

Gifford, E. J. M. (2010) 'Effective Shareholder Engagement: The Factors that Contribute to Shareholder Salience', *Journal of Business Ethics*, 92, 79–97.

Global Impact Investment Network (GIIN) (2010) 'What Is Impact Investing?' http://www.thegiin.org/, date accessed 25 May 2010.

Global Mining Initiative (2002) *Breaking New Ground: Mining, Minerals and Sustainable Development*, International Institute for Mining and Development report, 9084IIED, www.iied.org/pubs/display.php?o= 9084IIED, date accessed 16 February 2011.

Global Witness (2006) *Digging in Corruption: Fraud, Abuse and Corruption in Katanga's Copper and Cobalt Mines* (London: Global Witness).

Godfrey, B. J. (1982) 'Xingu Junction: Rural migration and land conflict in the Brazilian Amazon', in *Proceedings on the Pacific Coast Council on Latin American Studies Conference* (Pacific Coast Council on Latin American Studies).

Goldcorp, Inc. (2010a) Home Page, www.goldcorp.com, date accessed 25 May 2010.

——(2010b) 'Notice of Annual Meeting of Shareholders and Management Information Circular', www.goldcorp.com, date accessed 26 March 2010.

——(2010c) 'Goldcorp's First Update to the Marlin Mine Human Rights Assessment Report', www.goldcorp.com, date accessed 18 October 2010.

Goldcorp Steering Committee for the Human Rights Assessment of the Marlin Mine website (2010) HRIA Guatemala.com website, May, www.hria-guatemala.com/en/MarlinHumanRights.htm, date accessed 14 September 2010.

Goldstein, A. (2007) *Multinational Companies from Emerging Economies* (London: Palgrave Macmillan).

González, G. (2006) 'Pascua Lama Gold Mine, a Threat to Sustainability', Inter Press Service News Agency, 5 June, www.ipsnews.net/news.asp?idnews=33501, date accessed 16 February 2011.

Gonzales De Olarte, E. (1997) 'Pérou: le blocage des réformes économiques néolibérales', *Problèmes d'Amérique latine*, 25, April–June, 65–85.

——(1996) *El Ajuste Estructural y los Campesinos*, Instituto de Estudios Peruanos (Lima: IEP).

Goss, A. and G.S. Roberts (2011) 'The Impact of Corporate Social Responsibility on Bank Loans', *Journal of Banking and Finance*, 35 (7), 1794–1810.

Government of Canada (2009) Building the Canadian Advantage: A Corporate Social Responsibility (CSR) Strategy for the Canadian International Extractive Sector, www.international.gc.ca/trade-agreements-accords-commerciaux/ds/csr-strategy-rse-stategie.aspx, date accessed 16 February 2011.

Graham, D. and N. Woods (2006) 'Making Corporate Self-Regulation Effective in Developing Countries', *World Development*, 34 (5), 868–83.

Greenwood, M. (2007) 'Stakeholder Engagement: Beyond the Myth of Corporate Responsibility', *Journal of Business Ethnics*, 74, 315–27.

Guha-Sapir, D. and W.G. van Panhuis (2003) 'The Importance of Conflict-Related Mortality in Civilian Populations', *The Lancet*, 361, 2126–8.

Gulati, R., N. Nohria and A. Zaheer (2000) 'Strategic Networks', *Strategic Management Journal*, 21, 203–15.

Gunningham, N., R. Kagan and D. Thornton (2003) *Shades of Green: Business, Regulation, and Environment* (Stanford, CA: Stanford University Press).

Guthrie, J. and L. D. Parker (1990) 'Corporate Social Disclosure Practice: A Comparative International Analysis', *Advances in Public Interest Accounting*, 3, 159–75.

Guttiérez Spelucín, I. A. (2007) 'Community-Company Socio-Environmental Conflicts: Promoting a Culture of Trust and Conflict Prevention through Environmental Security Mainstreaming into Mining Corporations in Peru, MA thesis, Environmental Security and Peace, United Nations University for Peace, New York.

Guyadeen, V. (ed.) (2010) *Centennial Energy Digest: Data from 1988 to 2008* (Port of Spain: T&T Chamber of Industry and Commerce).

Hamann, R. (2004) 'Corporate Social Responsibility, Partnerships, and Institutional Change: The Case of Mining Companies in South Africa', *Natural Resources Forum*, 28 (4), 278–90.

———(2003) 'Mining Companies' Role in Sustainable Development: The "Why" and "How" of Corporate Social Responsibility from a Business Perspective', *Development Southern Africa*, 20 (2), 237–54.

Hamann, R., D. Sonnenberg, A. Mackenzie, P. Kapelus and P. Hollesen (2005) 'Local Governance as Complex System: Lessons from Mining in South Africa, Mali, and Zambia', *Journal of Corporate Citizenship*, 18, 61–73.

Hamann, R., P. Kapelus, D. Sonnenberg, A. MacKenzie and P. Hollesen (2005) 'Local Governance as a Complex System: Lessons from Mining in South Africa, Mali and Zambia', *Journal of Corporate Citizenship*, Summer, 1–18.

Harris, R. and J. Nef (2008) 'Capital, Power and Inequality in Latin America and the Caribbean', in R. Harris and J. Nef (eds) *Capital, Power and Inequality in Latin America and the Caribbean*, pp. 1–23 (New York: Rowman & Littlefield).

Haslam, P. A. (2004) 'The Corporate Social Responsibility System in Latin America and the Caribbean', March, FOCAL, Ottawa.

Haufler, V. (2001) *A Public Role for the Private Sector: Industry Self-Regulation in a Global Economy* (Washington, DC: Brookings Institution Press).

Hecth, S. and A. Cockburn (1990) *The Fate of the Forest: Developers, Destroyers and Defenders of the Amazon* (Bungay: Penguin Books).

Heinkel, R., A. Kraus and J. Zechner (2001) 'The Effect of Green Investment on Corporate Behavior', *Journal of Financial and Quantitative Analysis*, 36 (4), 431–50.

Hellman, J., G. Jones, and D. Kaufmann (2002) *Far From Home: Do Foreign Investors Import Higher Standards of Governance in Transition Economies?* (Washington, DC: World Bank).

Hennessy, H. (2003) 'Gold Mine Fails to Glitter in Peru', BBC News, 3 December, www.minesandcommunities.org/article.php?a=486, date accessed 16 February 2011.

Himley, M. (2008) 'Geographies of Environmental Governance: The Nexus of Nature and Neoliberalism', *Geography Compass*, 2 (2), 433–51.

Ho, C. G. T. and K. Nurse (eds) (2005) *Globalization, Diaspora and Caribbean Popular Culture* (Kingston: Ian Randle).

Hoffman, A. (2008) 'Goldcorp Bested by Mayan Mother', *Globe and Mail*, 10 July, https://secure.globeadvisor.com/servlet/ArticleNews/story/gam/20080710/RMINING10#, date accessed 23 February 2011.

———(1997) *From Heresy to Dogma: An Institutional History of Corporate Environmentalism* (San Francisco: New Lexington Press, 1997).

Hohnen, P. (2009) 'Sustainability: Extracting the Best from Canadian Miners? Canada Is the Latest Country to Announce a CSR Strategy. Paul Hohnen Assesses Its Strengths and Weaknesses', *The Ethical Corporation*, 20 April.

Hong, H. and M. Kasperczyk (2009) 'The Price of Sin: The Effects of Social Norms on Markets', *Journal of Financial Economics*, 93, 15–36.

House of Commons of Canada (2009) 'Bill C-300 An Act Respecting Corporate Accountability for the Activities of Mining, Oil or Gas in Developing Countries', www2.parl.gc.ca/HousePublications/Publication.aspx?Docid=4330045&file=4, date accessed 14 November 2010.

———(2005) *Fourteenth Report of the Standing Committee on Foreign Affairs and International Trade*, 38th Parliament, 1st Session, Ottawa.

Human Rights Impact Assessment – Guatemala (HRIA) (2009) 'Revision of HRIA Objective as Described in the Request for Proposals', http://www.hria-guatemala.com/en/default.htm, date accessed 5 March 2009.

——(2008) 'Memorandum of Understanding between Goldcorp Inc. and the Shareholder Group', http://www.hria-guatemala.com/en/default.htm, date accessed 19 March 2008.

Human Rights Watch (HRW) (2009) 'Kimberley Process: Human Rights Abuse Mars Credibility', http://www.hrw.org/en/news/2009/11/06/kimberley-process-zimbabwe-action-mars-credibility, date accessed 10 January 2010.

——(2005) *The Curse of Gold* (New York: Human Rights Watch).

Humphreys, D. (2001) 'Sustainable development: can the mining industry afford it?', *Resources Policy*, 27, 1–7.

Hutchkins, M. J., C. Walck, D. Sterk and G. Campbell (2007) 'Corporate Social Responsibility: A Unifying Discourse for the Mining Industry?', *Greener Management International*, 52, 17–30.

IBGE (Brazilian Institute of Geography and Statistics) (2010) 'First results of 2010 Census' (Rio de Janeiro: IBGE).

Iguíñiz, J. (1998) 'The Economic Strategy of the Fujimori Government', in J. Crabtree and J. Thomas (eds) *Fujimori's Peru: The Political Economy* (London: Institute of Latin American Studies).

Imai, S., L. Mehranvar and J. Sander (2007) 'Breaching Indigenous Law: Canadian Mining in Guatemala', *Indigenous Law Journal*, 6 (1), 101–39.

Industry Canada (2006) *Corporate Social Responsibility: An Implementation Guide for Canadian Business* (Ottawa: Multimedia and Editorial Services Section, Communications and Marketing Branch, Industry Canada).

Inglehart, R. and C. Wetzel (2005) *Modernization, Cultural Change and Democracy* (New York: Cambridge University Press).

Inmet Mining Corporation (n.d.) 'Industry Involvement', Inmet company website, www.inmetmining.com/sustainability/industryinvolvement/default.aspx, date accessed 13 July 2008.

Instituto Brasileiro de Análises Sociais e Econômicas (1983) *Carajás: O Brasil hipoteca o seu futuro* (Rio de Janeiro: Achiamé).

Instituto Brasileiro de Geografia e Estatística (2007) Ministério do Planejamento, Orçamento e Gestão, http://www.ibge.gov.br (home page), date accessed 10 November 2007.

Instituto Brasileiro do Meio Ambiente e Recursos Naturais Renováveis (1998) Decreto Flona Carajás, Bill Date: 02/02/98, Bill Number: 2.486.

Instituto de Estudios Económicos Mineros (IDEM) (1991) *La importancia económica de la minería en el Perú* (Lima: IDEM).

Instituto Nacional de Estatistica, and The United Nations Development Program (2005) *Estadistico de Municipios de Bolivia* (La Paz, Bolivia: Plural Editors).

Inter-American Commission on Human Rights (IACHR) (2010) 'Precautionary Measure 260-07', http://www.cidh.org/medidas/2010.eng.htm, date accessed 26 May 2010.

International Alert (2005) *Guide to Conflict Sensitive Business Practice* (London: International Alert).

International Council on Mining and the Environment (ICME) (1998) *Sustainable Development Charter* (Ottawa).

International Council on Mining and Metals (ICMM) (2010a) *Good Practice*, 9 (2).
———(2010b) 'Industry Issues: Mining and Economic Development', ICMM website, www.icmm.com/page/4655/mining-and-economic-development, date accessed 3 March 2011.
———(2010c) 'News: 23.10.06–ICMM Support for EITI', ICMM website, www.icmm.com/page/2039/icmm-support-for-eiti, date accessed 3 March 2011.
———(2010d) Position Statement, January, available from www.icmm.com/page/17427/mining-partnerships-for-development-position-statement, date accessed 18 December 2010.
———(2010e) 'Work Programs: Mining: Partnerships for Development', ICMM website, www.icmm.com/page/1409/resource-endowment-initiative, date accessed 18 December 2010.
———(2010f) 'Work Programs: Resource Endowment Initiative', ICMM website, www.icmm.com/page/1409/resource-endowment-initiative, date accessed 18 December 2010.
———(2010g) Sustainable Development Framework, http://www.icmm.com/our-work/sustainable-development-framework, date accessed 4 April 2011.
———(2006a) *The Analytical Framework. The Challenge of Mineral Wealth: Using Resource Endowments to Foster Sustainable Development*, August, ICMM, London, available from www.icmm.com, date accessed 18 December 2010.
———(2006b) *Chile – The Challenge of Mineral Wealth: Using Resource Endowments to Foster Sustainable Development* (London: ICMM).
International Finance Corporation (IFC) (2007) *Participation of Stakeholders: Manual of Best Practices to do Business in Emerging Markets* (Washington, DC: IFC).
———(2006) *Performance Standards on Social and Environmental Sustainability*, www.ifc.org/ifcext/enviro.nsf/Content/PerformanceStandards, date accessed 17 October 2010.
International Development Research Centre (IDRC) (2003a) *Mining Companies and Local Development – Latin America: Chile, Colombia and Peru. Executive Summary* (Ottawa: International Development Research Centre/Mining Policy Research Initiative).
———(2003b) *Mining Policy Research Initiative External Review, Evaluator: David Szablowski*, report prepared for IDRC Programs and Partnerships Branch Management, November, www.idrc.ca/uploads/user-S/118667011611076949170 MPRI_review_12-11-033.doc, date accessed 17 October 2010.
International Labour Office (2010) 'Report of the Committee of Experts on the Application of Conventions and Recommendations', February 2010, http://www.ilo.org/wcmsp5/groups/public/—ed_norm/—relconf/documents/meetingdocument/wcms_123424.pdf, date accessed 15 April 2010.
International Labour Organization (ILO) (1989) Convention No. C169 www.ilo.org/ilolex/cgi-lex/ratifce.pl?C169, date accessed 14 September 2010.
International Monetary Fund (IMF) (2009) 'Regional Economic Outlook: Western Hemisphere: Stronger Fundamentals Pay Off', May, IMF, Washington, DC.
International Organization for Standardization (n.d.), ISO 14000 Essentials, ISO home page, www.iso.org/iso/iso_14000_essentials, date accessed 14 September 2010.
Inquiry of Ministry. Question no. Q-298. By Mr. John McKay (Scarborough-Guildwood). June 9, 2010. Reply by the Minister of International Cooperation Bev Oda.

Ite, U. (2005) 'Poverty Reduction in Resource-Rich Developing Countries: What Have Multinational Corporations Got to Do with It?', *Journal of International Development*, 17 (7), 913–29.

Jamali, D. (2008) 'A Stakeholder Approach to Corporate Social Responsibility: A Fresh Perspective into Theory and Practice', *Journal of Business Ethics*, 82, 213–31.

Jardim, L. (2008) *MST invade Carajás*. Coturno Noturno Blog. http://coturnonoturno.blogspot.com/2008/04/MST-invade-carajs.html, date accessed 30 August 2008.

Jenkins, H. and N. Yakovleva (2006) 'Corporate Social Responsibility in the Mining Industry: Exploring Trends in Social and Environmental Disclosure', *Journal of Cleaner Production*, 14, 271–84.

Jensen, M.C. (2010) 'Value Maximisation, Stakeholder Theory and the Corporate Objective Function', *Journal of Applied Corporate Finance*, 22 (1), 297–313.

Jessop, B. (2004) 'Critical Semiotic Analysis and Cultural Political Economy', *Critical Discourse Studies*, 1 (2), 159–74.

Jiménez, F. (2002) 'Estado, mercado, crisis y restauración liberal en el Perú', in C. Contreras and M. Glave (eds) *Estado y Mercado en la Historia del Perú* (Lima: Pontificia Universidad Católica del Perú).

John, T. and K. John (1991) 'Optimality of Project Financing: Theory and Empirical Implications in Finance and Accounting', *Review of Quantitative Finance and Accounting*, 1, 51–74.

Joiner, D. and G. Wigaraja (2007) 'Measuring Competitiveness in Small States: Introducing the SSMECI', in E. Kisanga and S. J. Danchie (eds) *Commonwealth Small States: Issues and Prospects*, pp. 127–52 (London: Commonwealth Secretariat and CPA).

Jordan, R. (2008) *Conflicto en Minería: Naturaleza, Alcance e Impacto Sobre la Sociedad, la Economía y la Industria (1980 – 2006)* (Informe Temático Sobre Desarrollo Humano and UNDP).

Kahneman, D. and A. Tversky (1979) 'Prospect Theory: An Analysis of Decision under Risk', *Econometrica*, 47 (2), 263–92.

Kapelus, P. (2002) 'Mining, Corporate Social Responsibility and the 'Community': The Case of Rio Tinto, Richards Bay Minerals and the Mbonambi', *Journal of Business Ethics*, 39 (3), 275–96.

Katzenstein, P. J. (ed.) (1996) *The Culture of National Security: Norms and Identity in World Politics* (New York: Columbia University Press).

Keck, M. and K. Sikkink (1998) *Activists Beyond Borders: Advocacy Networks in International Politics* (Ithaca, NY and London: Cornell University Press).

Khanna, P. (2009) *The Second World: How Emerging Powers Are Redefining Global Competition in the 21st Century* (New York: Random House).

Kimberley Process (2003) 'The Kimberley Process Certification Scheme', http://www.kimberleyprocess.com/, date accessed 19 January 2010.

King, A. A. and M. J. Lenox (2000) 'Industry Self-Regulation Without Sanctions: The Chemical Industry's Responsible Care Program', *Academy of Management Journal*, 43 (4), 698–716.

Kingdon, J. (1995) *Agendas, Alternatives and Public Policies*, 2nd edn (New York: HarperCollins).

Kirk, K.S. (1997) 'Iron Ore' in Mineral Commodity Summaries (U.S. Geological Survey).

Kisanga, E. and S. J. Danchie (eds) (2007) *Commonwealth Small States: Issues and Prospects* (London: Commonwealth Secretariat and CPA).

Klare, M. T. (2004) *Blood and Oil: The Dangers and Consequences of America's Growing Dependency on Imported Petroleum* (New York: Metropolitan).

——(2001) *Resource Wars: The New Landscape of Global Conflict* (New York: Owl).

Klippensteins, Barristers and Solicitors (2010a) Webpage for information concerning Copper Mesa lawsuit, www.ramirezversuscoppermesa.com, date accessed 14 September 2010.

Klippensteins, Barristers and Solicitors (2010b) Webpage for information concerning HudBay Minerals Inc. lawsuit, www.chocversushudbay.com/, date accessed 14 December 2010.

Kollman, K. (2008) 'The Regulatory Power of Business Norms: A Call For a New Research Agenda', *International Studies Review*, 10 (3), 397–419.

Kosich, D. (2006) 'Guyana High Court Dismisses $2B Omai Gold Mines Tailings Accident Suit', Mineweb.com, 1 November, www.mineweb.com/mineweb/view/mineweb/en/page15831?oid=13303&sn=Detail, date accessed 14 September 2010.

KPMG (2006) *Global Mining Reporting Survey 2006*, www.kpmg.ca/en/industries/enr/mining/documents/GMS2006.pdf, date accessed 26 March 2010).

Krahman, E. (2003) 'National, Regional, and Global Governance: One Phenomenon of Many?', *Global Governance*, 9, 323–46.

Kurucz, E. C., B. A. Colbert and D. Wheeler (2008) 'The Business Case for Corporate Social Responsibility', in A. Crane, A. McWilliams, D. Matten, J. Moon and D.S. Seigel (eds) *The Oxford Handbook of Corporate Social Responsibility*, pp. 83–112 (Oxford: Oxford University Press).

Kuyek, J. and C. Coumans (2003) *No Rock Unturned: Revitalizing the Economies of Mining Dependent Communities* (Ottawa: MiningWatch Canada).

Labelle, H. (2008) 'The Costs of Corruption', *Compact Quarterly*, 2, http://www.enewsbuilder.net, date accessed 1 October 2009.

Lambsdorff, J. (ed.) (2007) *The Institutional Economics of Corruption and Reform* (New York: Cambridge University Press).

——(2005) *Consequences and Causes of Corruption: What do We Know from a Cross-Section of Countries?*, Working paper V-35-05, University of Passau, Germany.

——(2002) Making Corrupt Deals: Contracting in the Shadow of the Law, *Journal of Economic Behaviour and Organization*, 48, 221–41.

Law, B. (2009a) 'Canadian Mine Accused of Causing Skin Infections', *BBC News*, 11 March 2009.

——(2009b) 'Canada Goldmine Worries Grow', *BBC News*, 30 March 2009.

La Nacion (2010) *'Indígenas Piden Cierre de Mina Canadiense'*, 22 September 2010, http://www.nacion.com/2010-09-23/Mundo/NotasSecundarias/Mundo2531155.aspx, date accessed 30 September 2010.

Lee, C. and D. Ng (2006) 'Corruption and International Valuation: Does Virtue Pay?', Johnson School Research Paper, http://ssrn.com/abstract=934468, date accessed 20 February 2011.

Lemieux, A. (2005) 'Canada's Global Mining Presence', *Canadian Minerals Yearbook*, Natural Resources Canada (Ottawa: Government of Canada), www.nrcan-rncan.gc.ca/mms-smm/busi-indu/cmy-amc/content/2005/08.pdf, date accessed 8 March 2011.

Lemieux, E. (2010a) 'Peru: Challenges of Local Development in Mining Regions', Canadian Foundation for the Americas (FOCAL), September, www.focal.ca/publications/focalpoint/311-september-2010-emilie-lemieux-en, date accessed 14 September 2010.

——(2010b) *Mining and Local Development: Challenges and Perspectives for a Fair and Socially Inclusive Sustainable Development in Canadian Mining Zones in Peru.* Walter & Duncan Gordon Foundation. Unpublished report, June.

Levy, D. L. and A. Prakash (2003) 'Bargains Old and New: Multinational Corporations in Global Governance', *Business and Politics*, 5 (2), 131–50.

Livesey, S. M. (2002) 'The Discourse of the Middle Ground: Citizen Shell Commits to Sustainable Development', *Management Communication Quarterly*, 15 (3), 313.

Machuca, R. M. (2002) 'Cambios de la pobreza en el Perú: 1991–1998. Un análisis a partir de los componentes del ingreso', Investigaciones Breves no. 19, CIES, Lima.

MacMahon, G. and M. Cervantes (2009) *Fraser Institute Annual Survey of Mining Companies 2008* (Vancouver, Canada: Fraser Institute).

Madrid, L. E., Q. N. Guzman, A. E. Mamani, E. D. Madrano and M.R. Nuñez (2002) *Minería y Comunidades Campesinas ¿Coexistencia o Conflicto?* (La Paz, Bolivia: Fundación PIEB).

Mainhardt-Gibbs, H. (2003) *The World Bank Extractive Industries Review: The Role of Structural Reform Programs towards Sustainable Development Outcomes*, Research report no. 1, Extractive Industries Review, World Bank, Washington, DC.

Mamdani, M. (1996) *Citizen and Subject: Contemporary Africa and the Legacy of Late Colonialism* (Princeton: Princeton University Press).

Marín, R. E. A. (2003) 'Conflitos Agrários no Pará', in E. Fontes (ed.) *Coleção Contando a História do Pará: Os Conflitos e os Grandes Projetos na Amazônia Contemporânea (sec.xx)* (Belém: E.Motion).

Marker, B., M. G. Petterson, F. McEvoy and M. H. Stephenson (2005) 'Sustainable Minerals Operations in the Developing World', *Geological Society Special Edition*, 250.

Marshall, I. (2001) *A Survey of Corruption Issues in the Mining and Mineral Sector* (London: International Institute for Environment and Development).

McCool, D. (1995) *Public Policy Theories, Models, and Concepts: An Anthology* (Englewood Cliffs, NJ: Prentice Hall).

McDonald, M. (2010) 'Costa Rican Court Strikes Down Las Crucitas Gold Mine Project', Tico Times Online, 24 November, www.ticotimes.net/News/News-Briefs/Costa-Rican-Court-Strikes-Down-Las-Crucitas-Gold-Mine-Project_Wednesday-November-24-2010, date accessed 25 November 2010.

McGee, B. (2009) 'The Community Referendum: Participatory Democracy and the Right to Free, Prior and Informed Consent', *Berkeley Journal of International Law*, 27, 570–635.

McKay, J. (2009) *Bill C-300: an Act Respecting Corporate Accountability for the Activities of Mining, Oil or Gas in Developing Countries*, http://openparliament.ca/bills/1987/, date accessed 16 February 2011.

McMahon, F. and M. Cervantes (2010) *Survey of Mining Companies 2009/2010: 2010 Mid-Year Update*, Fraser Institute www.fraserinstitute.org/uploadedFiles/fraser-ca/Content/research-news/research/publications/miningsurvey-2010update.pdf, date accessed 16 February 2011.

McPhail, K. (2008) 'Sustainable Development in the Mining and Minerals Sector: The Case for Partnership at Local, National and Global Levels', Bronze Award Essay, International Finance Corporation, World Bank Group, www.ifc.org/ifcext/essaycompetition.nsf/AttachmentsByTitle/Bronze_Mining/$FILE/Bronze_Mining.pdf, date accessed 16 February 2011.

Menz, K. M. (2010) 'Corporate Social Responsibility: Is It Rewarded by the Corporate Bond Market? A Critical Note', *Journal of Business Ethics*, 96, 117–34.

Merton, R. (1987) 'A Simple Model of Capital Market Equilibrium with Incomplete Information', *Journal of Finance*, 42 (3), 483–510.

Midtun, A. (2008) 'Global Governance and the Interface with Business: New Institutions, Processes and Partnerships', *Corporate Governance*, 8 (4), 406–18.

Miller Chevalier (2008) *Latin American Corrupion Survey* (Washington, DC: Miller Chevalier).

Mines and Communities (2010) Website about Blackfire Chiapas mine allegations, Mines and Communities (MAC), 10 March, www.minesandcommunities.org/article.php?a=9973, date accessed 14 September 2010.

Mining Association of Canada (MAC) (2010) Towards Sustainable Mining, http://www.mining.ca/www/Towards_Sustaining_Mining/index.php, date accessed 4 April 2011.

——(2009) *Facts and Figures 2009*, www.mining.ca/www/media_lib/MAC_Documents/Publications/2009/2009_F_and_F_English.pdf, date accessed 8 March 2011.

Mining, Minerals and Sustainable Development (MMSD) (2002) *Breaking New Ground: The Report of Mining, Minerals and Sustainable Development Project* (London: Earthscan).

MiningWatch Canada (2010) 'The Organization of American States' Human Rights Commission, the Region's Most Respected Human Rights Body, Calls on the Guatemalan Government and Goldcorp to Halt Mining', Mining Watch Canada website, 24 May, www.miningwatch.ca/en/oas-human-rights-commission-urges-suspension-mining-activity-goldcorps-marlin-mine-guatemala, date accessed 14 September 2010.

Ministerio de Energía y Minas (MEM) (2010) Resolución Ministerial N° 009-2010-MEM/DM, Peru, Ministerio de Energía y Minas, Lima, www.minem.gob.pe/minem/archivos/file/DGAAM/legislacion/R_M_009_2010_MEM_DM.pdf, date accessed 1 October 2010.

——(2008) Decreto Supremo N° 028-2008-EM, Peru, Ministerio de Energía y Minas, Lima, http://intranet2.minem.gob.pe/web/archivos/ogp/legislacion/ds028-2008.pdf, date accessed 1 October 2010.

——(2007). 'Plan Estratégico de la Oficina General de Gestión Social', Oficina General de Gestión Social, Minsterio de Energía y Minas, Lima.

Mkandawire, T. (2001) 'Thinking about Developmental States in Africa', *Cambridge Journal of Economics*, 25, 289–313.

Modigliani, F. and M. Miller (1958) 'The Cost of Capital, Corporation Finance and the Theory of Investment', *American Economic Review*, 53 (3) 261–97.

Monteiro, M. D. A. (2003) 'Meio século de mineração industrial na Amazônia oriental brasileira: um balanço necessário', in E. Fontes (ed.) *Coleção Contando a História do Pará: Os* .Belém: E. Motion.

Monzoni, M. (2008) *Juurti Sustentável: Proposta de modelo para o desenvolvimento local* (São Paulo: FGV).

Mordant, N. (2003) 'Cambior in Fighting Mode', Mineweb.com, 27 May, www.mineweb.com/mineweb/view/mineweb/en/page15831?oid=11584&sn=Detail, date accessed 14 September 2010.

Morgera, E. (2007) 'Significant Trends in Corporate Environmental Accountability: The New Performance Standards of the International Finance Corporation', *Colorado Journal of International Environmental Law and Policy*, 18, 151.

Mottley, W. (2008) *Trinidad and Tobago Industrial Policy 1959-2008: A Historical and Contemporary Analysis* (Kingston: Ian Randle).

Muradian, R., J. Martinez-Alier and H. Correa (2003) 'International Capital versus Local Population: The Environmental Conflict of the Tambogrande Mining Project, Peru', *Society & Natural Resources* 16 (9), 775–92.

National Archives (2006) *Companies Act 2006*, www.legislation.gov.uk/ukpga/2006/46/contents, date accessed 6 September 2010.

National Roundtables on Corporate Social Responsibility and the Canadian Extractive Industry Advisory Group (2007) *Corporate Social Responsibility (CSR) and the Canadian Extractive Industry in Developing Countries*, Advisory Group Report, Prospectors and Developers Association of Canada, Toronto, www.pdac.ca/, date accessed 15 November 2008.

Natural Resources Canada (NRCAN) (2009) 'Actifs de compagnies canadienne à l'étranger, par pays, 2002–2008', unpublished document, Department of Natural Resources, Ottawa Government of Canada.

———(2008a) 'Actifs de compagnie minières canadiennes à l'étranger et au Canada, 2002–2007', unpublished document, Department of Natural Resources, Government of Canada, Ottawa.

———(2008b) *Minerals Yearbook* (Ottawa, Canada: Natural Resources Canada).

———(2007) *The Role of Industry Associations in the Promotion of Sustainability and Corporate Social Responsibility: Study Findings* (Ottawa: Five Winds International & Strandberg Consulting).

Network in Solidarity with the People of Guatemala (NISGUA) (2009) 'Urgent Action: Crackdown on Local Citizens Opposing Goldcorp's Marlin Mine Escalates in San Marcos, Guatemala', 2 July 2008, http://www.nisgua.org, date accessed 9 July 2009.

Newmont Mining Corporation (2010a) Home Page, www.newmont.com, date accessed 20 January 2010.

———(2010b) 'Report to the Shareholders Status of Implementation of the Community Relationships Review', 23 April, http://www.beyondthemine.com, date accessed 15 July 2010.

———(2010c) 'Environmental/Social Responsibility Standards', www.newmont.com, date accessed 20 October 2010.

———(2009a) 'Community Relationships Review: Global Summary Report', http://www.beyondthemine.com, date accessed 10 October 2009.

———(2009b) 'Newmont Releases Independent Community Relationships Review', 9 March, www.newmont.com, date accessed 20 July 2009.

Neu, P., H. Warsame and K. Pendwell (1998) 'Managing Public Impressions: Enviornmental Disclosures in Annual Reports', *Accounting, Organizations and Society*, 23 (3), 265–82.

Nuñez-Barriga, A. (1999) 'Environmental Management in a Heterogenous Mining Industry: The Case of Peru', in A. Warhurst (ed.) *Mining and the Environment: Case Studies from the Americas*, pp. 137–79 (Ottawa: IDRC).

Nurse, K. (2009) 'Cultural Industries and Cultural Policy in the Context of Globalization: An Agenda for SIDS', in A. F. Cooper and T. M. Shaw (eds) *The Diplomacies of Small States: Between Vulnerability and Resilience*, pp. 244–63 (London: Palgrave Macmillan for CIGI).

———(2007) 'Culture as the Fourth Pillar of Sustainable Development', in Commonwealth Secretariat (ed.) *Small States: Economic Review and Basic Statistics*, Vol. 11, pp. 28–40 (London: Commonwealth Secretariat).

Observatorio de Conflictos Mineros de América Latina (OCMAL) (2009) Pronunciamiento por la Justica Social y Ambiental y por la Paz en Amerindia, http://www.conflictosmineros.net/biblioteca/cat_view/32-estudios-e-informes/35-derechos-humanos-y-criminalizacion, date accessed 4 April 2011.

Odell, C. and A. C. Silva (2006) 'Mining Exploration, Corporate Social Responsibility and Human Rights: Untangling the Facts, Seeking Solutions', Submission to the National Roundtables on Corporate Social Responsibility.

OECD (1997) *Convention on Combating Bribery of Foreign Public Officials in International Business Transactions*, www.oecd.org/dataoecd/52/53/2406809.pdf, date accessed 14 September 2010.

OECD (n.d.) The Paris Declaration on Aid Effectiveness and the Accra Agenda for Action, www.oecd.org/dataoecd/11/41/34428351.pdf, date accessed 16 February 2011.

On Common Ground (2010) 'Human Rights Assessment of Goldcorp's Marlin Mine', http://www.hria-guatemala.com/en/docs/Human%20Rights/OCG_HRA_exec_summary.pdf, date accessed 15 May 2010.

Ondetti, G. (2008) *Land, Protest and Politics: The Landless Movement and the Struggle for Agrarian Reform in Brazil* (University Park: The Pennsylvania State University Press).

O'Neill, J. and A. Stupnytska (2009) 'The Long-Term Outlook for the BRICs and N-11 Post Crisis', Global Economics Paper No. 192, Goldman Sachs Global Economics, Commodities and Strategy Research, New York, www2.goldmansachs.com/ideas/brics/long-term-outlook-doc.pdf, date accessed 2 March 2011.

Ontario Superior Court of Justice (2009) *Plaintiff's Claim no. SC-09-80779-02 (Steven Michael Schnoor vs. Attorney General of Canada and Kenneth Murray Cook)*, 20 February, ONSC, Toronto.

Orihuela, J. C. (2010) 'An Environmental Resource Curse? Governing Air Pollution from Smelters in Chuquicamata and La Oroya', Watson Institute, Brown University, www.watsoninstitute.org/images_news/Orihuela_EnvironmentalResourceCurse.pdf, date accessed 17 October 2010.

Pacific Rim Mining Corp (2008) 'Pacific Rim Files Notice of Intent to Seek CAFTA Arbitration', press release, December, www.marketwire.com/press-release/Pacific-Rim-Files-Notice-of-Intentto-Seek-CAFTA-Arbitration-TSX-PMU-928202.htm, date accessed 14 September 2010.

Pac Rim Cayman LLC v. Republic of El Salvador (2009) ICSID Case No. ARB/09/12 (CAFTA), University of Victoria, Investment Treaty Arbitration, http://ita.law.uvic.ca/alphabetical_list.htm, date accessed 14 September 2010.

Palheta da Silva, J. M. (2004) Poder, Governo e Território em Carajás, doctoral thesis (Presidente Prudente: Universidade Estadual Paulista).

Panama Legislative Assembly (1997) Ley No. 9 Por la cual se aprueba el contrato celebrado entre el Estado y la Sociedad Minera Petaquilla SA.

Pando, J. (1990) *Indigenous Peoples in Latin America*, International Labour Organization website, www.ilo.org/public/english/region/ampro/mdtsanjose/indigenous/cuadro.htm, date accessed 14 September 2010.

Pantin, D. (2010) 'The Caribbean Rentier Economy, State and Society', in B. Meeks and N. Girvan (eds) *The Thought of New World: The Quest for Decolonization*, pp. 107–21 (Kingston: Ian Randle).

——(ed.) (2005) *The Caribbean Economy: A Reader* (Kingston: Ian Randle).

Partnership Africa Canada (2009) *Diamonds and Human Security Annual Review 2009* (Ottawa: Partnership Africa Canada).

Peck, J. and A. Tickell (2002) 'Neoliberalizing Space', *Antipode*, 34 (3), 380–404.

Pedigo, K. and V. Marshall (2009) 'Bribery: Australian Managers' Experiences and Responses When Operating in International Markets', *Journal of Business Ethics*, 87, 59–74.

Peinado-Vara, E. (2004) 'Corporate Social Responsibility in Latin America and the Caribbean', June, Inter-American Development Bank, Washington, DC.

Pegg, S. (2006) 'Mining and Poverty Reduction: Transforming Rhetoric into Reality', *Journal of Cleaner Production*, 14, 376–87.

——(2003) *Poverty Reduction or Poverty Exacerbation? World Bank Group Support for Extractive Industries in Africa*, a report sponsored by Oxfam America, Friends of the Earth-US, Environmental Defense, Catholic Relief Services, and the Bank Information Center, www.edf.org/documents/2737_2737_PovertyRedux2.pdf, date accessed 2 March 2011.

PERCAN (2010) 'Social Issues', PERCAN, Lima, www.percan.ca/Social, date accessed 12 May 2010.

——(2009) 'Brief Description', PERCAN, Lima, www.percan.ca/Plone/BriefDescription/index_html?set_language=en&cl=en, date accessed 24 September 2010.

PERCAN and MEM (Ministerio de Energía y Minas) (2009) 'Project Implementation Plan for PERCAN Extension and Annual Work Plan 2009–2010', Lima.

Perera, A. (2008) 'More Valuable than Gold', Oxfam America website, www.oxfamamerica.org/articles/more-valuable-than-gold/?searchterm=None, date accessed 3 March 2011.

Piedra v. Copper Mesa Corporation (2010) Ontario Court of Justice ONSC 2421, Klippensteins, Barristers and Solicitors website, www.ramirezversuscoppermesa.com, date accessed 14 September 2010.

Philippine Indigenous Peoples' Links and Christian Aid (2004) *Breaking Promises, Making Profits: Mining in the Philippines*, a Christian Aid and PIPLinks report, www.piplinks.org/system/files//philippines_report.pdf, date accessed 16 February 2011.

Pierson, P. (2004) *Politics in Time: History, Institutions and Social Analysis* (Princeton and Oxford: Princeton University Press).

Polanyi, K. (1944) *The Great Transformation* (Boston, MA: Beacon Press).

Pollard, M. (2007) 'Enhancing Small States' Competitiveness in the Global Economy', in Commonwealth Secretariat (ed.) *Small States: Economic Review and Basic Statistics*, Vol. 11, pp. 41–56 (London: Commonwealth Secretariat).

Popplewell, B. (2010) 'Critics Plan to Ask for RCMP Investigation of Calgary Company's Operations in Mexico', *Toronto Star*, 10 March, date accessed 14 September 2010.

Porter, D. and D. Craig (2004) 'The Third Way and the Third World: Poverty Reduction and Social Inclusion in The Rise of "Inclusive" Liberalism', *Review of International Political Economy*, 11 (2), 387–423.

Porter, M. and M. Kramer (2011) 'Creating Shared Value', *Harvard Business Review*, January/February, 62–77.

——(2006) 'Strategy and Society: The Link Between Competitive Advantage and Corporate Social Responsibility', *Harvard Business Review*, 84 (12), 76–92.

Powell, W. and P. DiMaggio (eds) (1991) *The New Institutionalism in Organizational Analysis* (Chicago: Chicago University Press).

Prakash, A. and M. Potoski (2005) 'Green Clubs and Voluntary Governance: ISO 14001 and Firms' Regulatory Compliance', *American Journal of Political Science*, 49 (2), 235–48.

Prefeitura Municipal de Parauapebas (2005) http://www.parauapebas.pa.gov.br (home page), date accessed 16 November 2007.

Presidente de la República (2003) 'Decreto Supremo No. 042-2003-EM' (Lima: El Peruano, Government of Peru, 13 December) pp. 257055–6 (See also www.snmpe.org.pe/pdfs/DECRETO_SUPREMO_N%BA_042-2003-EM.pdf, date accessed 8 September 2010.)

Preville, P. (1997) 'Coming Home to Roost', *Montreal Mirror*, 27 March, www.montrealmirror.com/ARCHIVES/1997/032797/cover.html, date accessed 14 September 2010.

Price, J. (2009) *Foreign Investment in the Latin American Mining Sector: A Risk Assessment* (New York: Kroll).

Prospectors and Developers Association of Canada (PDAC) (2011) *e3Plus: A Framework for Responsible Exploration*, http://www.pdac.ca/e3plus/index.aspx, date accessed 4 April, 2011.

——(2009a) *e3Plus: A Framework for Responsible Exploration, Community Engagement Guide*, http://www.pdac.ca/e3plus/English/toolkits/sr/pdf/e3plus-sr-03-community-engagement.pdf, date accessed 7 February 2011.

——(2009b) *Bill C-300 Position Statement*, August 2009.

——(2007) 'From Theory to Practice: Corporate Social Responsibility and Sustainable Development in Mineral Exploration', short course, 2 & 3 March, PDAC Annual Convention, PDAC Toronto.

Province of Marindunque, Republic of the Philippines (2010) Resolution Reiterating the Declaration of a Fifty (50) Year Large Scale Mining Moratorium in the Province of Marinduque. July 26, 2010, Resolution No. 35 Series 2010, Office of the Sangguniang Panlalawigan, Province of Marindunque, Republic of the Philippines, www.alyansatigilmina.net/files/2010@%20SPRESOLUTION%20@REITERATE%20MINING%20MORATORIUM.pdf, date accessed 2 March 2011.

Public Citizen/Global Trade Watch (2010) 'CAFTA Investor Rights Undermining Democracy and the Environment: Pacific Rim Mining Case', www.citizen.org/documents/Pacific_Rim_Backgrounder1.pdf, date accessed 14 September 2010.

Publish What You Pay Coalition of the Democratic Republic of Congo (2007) 'Final Statement of the Workshop to Build Awareness on the Extractive Industries Transparency Initiative', http://www.publishwhatyoupay.org/en/category/countriesregions/democratic-republic-congo, date accessed 14 February 2011.

Púlsar (2008) 'La Defensoría del Pueblo peruana advirtió en 2007 el conflicto en Moquegua Nota', 25 June 2008, www.agenciapulsar.org, date accessed 15 July 2008.
Ramirez, J. A. (2008) *Analisis de la Interrelacion de la Empresa Minera Pierina con su Entorno Social y Ambiental en Ancash, Peru*, doctoral thesis, Costa Rica: Centro Agronómico Tropical de Investigación y Enseñanza.
Ramsaran, R. (ed.) (2006) *Size, Power and Development in the Emerging World Order: Caribbean Perspectives* (San Juan, T&T: Lexicon Trinidad).
———(ed.) (2003) *The Caribbean in the International Arena: The Implications of Global Instability and Conflict* (San Juan, T&T: Lexicon Trinidad).
Responsible Jewellery Council (2010) Home Page, http://www.responsiblejewellery.com/, date accessed 18 January 2010.
———(2009) 'Core System for RJC Member Certification', http://www.responsiblejewellery.com/certification.html#docs, date accessed 18 January 2010.
Riano, J. and R. Hodess (2008) *2008 Bribe Payers Index* (Berlin: Transparency International).
Rights Action (2008) 'Open Letter from Rights Action to Goldcorp Inc. and Shareholders', www.rightsaction.org/articles/Goldcorp_Open%20Letter_050108.html, date accessed 30 May 2008.
———(2006) 'Mining Mergers and Acquisitions; Impunity and Immunity from Accountability', Rights Action website, 12 September, www.rightsaction.org/urgent_com/Mining_Mergers_091206.html, date accessed 21 February 2008.
Risse, T., S. C. Ropp and K. Sikkink (eds) (1999) *The Power of Human Rights: International Norms and Domestic Change* (Cambridge, UK: Cambridge University Press).
Rist, G. (1997) *The History of Development: From Western Origins to Global Faith* (London: Zed Books).
Rodrigues, M. G. (2003) 'Privatization and Socioenvironmental Conditions in Brazil's Amazonia: Political Challenges to Neoliberal Principles', *Journal of Environmental and Development*, 12, 205.
Rojas, M., B. M'zali, M. F. Turcotte and P. Merrigan (2009) 'Bringing About Changes to Corporate Social Policy through Shareholder Activism: Filers, Issues, Targets, and Success', *Business and Society Review*, 114 (2), 217–52.
Rose-Ackerman, S. (ed.) (1999) *Corruption and Government: Causes, Consequences, and Reform* (Cambridge, UK: Cambridge University Press).
Ruggie, J. (2008) *Protect, Respect, and Remedy: A Framework for Business and Human Rights*. Report of the Special Representative of the Secretary-General on the issue of human rights and transnational corporations and other business enterprises (No. 3), UN Human Rights Council http://www2.ohchr.org/english/bodies/hrcouncil/8session/reports.htm, date accessed 23 March 2011.
———(2004) 'Reconstituting the Global Public Domain – Issues, Actors, and Practices', *European Journal of International Relations*, 10 (4), 499–531.
Sagebien, J. and N. M. Lindsay (2009) Systemic Causes. Systemic Solutions: Embedding CSR in Local/Global Governance Ecosystems. Unpublished working paper.
Sagebien, J., N. M. Lindsay, P. Campbell, R. Cameron and N. Smith (2008) 'The Corporate Social Responsibility of Canadian Mining Companies in Latin America: A Systems Perspective', *Canadian Foreign Policy*, 8 (3), 103–28.

Sagebien, J. and M. Whellams (2011) 'CSR and Development: Seeing the Forest for the Trees', *Canadian Journal of Development Studies*, 31 (3–4), 349–90.

Salas Carreño, G. (2008) *Dinámica Social y Minería – Familias pastores de puna y la presencia del proyecto Antamina (1997–2002)* (Lima: Instituto de Estudios Peruanos).

Samford, S. (2010) 'Averting "Disruption and Reversal": Reassessing the Logic of Rapid Trade Reform in Latin America', *Politics & Society*, 38 (3), 373–407.

Sánchez Albavera, F., G. Ortiz and N. Moussa (1999) 'Panorama minero de América Latina a fines de los años noventa', Serie Recursos Naturales e Infraestructura No 1, CEPAL, División de Recursos Naturales e Infraestructura, United Nations, Santiago de Chile.

Sarrasin, B. (2009) 'Mining and Protection of the Environment in Madagascar', in B. Campbell (ed.) *Mining in Africa: Regulation and Development* (London, New York: Pluto Press).

Schmink, M. and C. H. Wood (1992) *Contested frontiers in Amazônia* (New York: Columbia University Press).

Scherer, A. G. and G. Palazzo (2007) 'Toward a Political Conception of Corporate Responsibility-Business and Society seen from a Habermasian Perspective', *Academy of Management Review*, 32, 1096–120.

Scherer, A. G., G. Palazzo and D. Matten (2009) 'Introduction to the Special Issue: Globalization as a Challenge for Business Responsibilities', *Business Ethics Quarterly*, 19 (3), 327–47.

Schnoor, S. and M. Klippenstein (2010) *Judge Rules that Canadian Ambassador Slandered Documentary Video Maker*, press release, 17 June.

Schubert, S. and T. C. Miller (2008) 'At Siemens Bribery Was Just a Line Item', *New York Times*, 20 December, World Business Section, p. 2.

Schwab, K. (ed.) (2010) The Global Competitiveness Report, 2010–2011, World Economic Forum, Geneva, www3.weforum.org/docs/WEF_Global CompetitivenessReport_2010-11.pdf, date accessed 28 February 2011.

Scott, C. and R. Wai (2004) 'Transnational Governance of Corporate Conduct through the Migration of Human Rights Norms: The Potential Contribution of Transnational 'Private' Litigation', in C. Joerges, I.-J. Sand and G. Teubner (eds) *Transnational Governance and Constitutionalism* (Portland: Hart Publishing).

Scott, P. (2000) 'Reporting in the Mining Sector', *Mining Environmental Management*, 8 (2), 10–12.

Sethi, S. P. and O. Emelianova (2006) 'A Failed Strategy of Using Voluntary Codes of Conduct by the Global Mining Industry', *Corporate Governance*, 6 (3), 226–38.

Shah, S. (2006) 'UK Company Law: Time to Modernize the Rules', *The Ethical Corporation*, March, 36.

Shareholder Association for Research and Education (SHARE) (2010) Shareholder Resolution Database, http://www.share.ca/en/shareholderdb, date accessed 28 May 2010.

———(2009) 'Proxy Voting by Canadian Mutual Funds, 2006-2008: A Survey of Management and Shareholder Resolutions', http://www.share.ca., date accessed 15 April 2010.

———(2007) Submission Brief to the Ontario Expert Commission on Pensions, http://www.pensionreview.on.ca/english/submissions/SHARE.html, date accessed 30 September 2009.

Shaw, T. M. (2010) 'Energy Governance and Global Development: Inter- and Non-state', September, Political Economy of Energy in Europe and Russia (PEEER), University of Warwick, UK.

Shaxson, N. (2007) *Poisoned Wells: The Dirty Politics of African Oil* (New York: Palgrave Macmillan).

Shugart, M. S. (1999) 'Presidentialism, Parliamentarism, and the Provision of Collective Goods in Less-Developed Countries', *Constitutional Political Economy*, 10, 53–88.

Simmons & Simmons (2006) *International Business Attitudes to Corruption: A Survey 2006* (London: Control Risks Group).

Sirkin, H. L., J. W. Hemerling and A. K. Bhattacharya (2008) *Globality: Competing With Everyone from Everywhere about Everything* (London: Headline for BCG).

Slack, K. (2009) *Mining Conflicts in Peru: Situation Critical* (Washington, DC: Oxfam America).

Smillie, I. (2002) *The Kimberley Process: The Case for Proper Monitoring* (Ottawa: Partnership Africa Canada).

Smith, N. J. H., E.A.S. Serrão and P.T. Alvim and I.C. Falesi (1995) *Amazônia – Resiliency and Dynamism of the Land and Its People* (Tokyo: The United Nations University).

Social Investment Organization (SIO) (2009) *Canadian Socially Responsible Investment Review 2008*, http://www.socialinvestment.ca/documents/caReview2008.pdf, date accessed 10 October 2009.

SOCODEVI (2011) 'Reboisement et agroforesterie dans le département de La Libertad', Nos projets, SOCODEVI website, www.socodevi.org/fr/projets/_detail_projet.php?pk_projet=144&continent=1, date accessed 3 March 2011.

South African Human Rights Commission (SAHRC) (2008) *Mining-related Observations and Recommendations: Anglo Platinum, Affected Communities and Other Stakeholders, In and Around the PPL Mine, Limpopo* (Johannesburg: South African Human Rights Commission).

Standing Committee on Foreign Affairs and International Trade (SCFAIT) (2010) *Minutes and Evidence*, 40th Parliament, 3rd Session, March 3, 2010 to present, www2.parl.gc.ca/CommitteeBusiness/CommitteeMeetings.aspx?Language=E&Mode=2&Parl=40&Ses=3&Cmte=FAAE&Stac=3023047, date accessed 14 September 2010.

——— (2005) *Mining in Developing Countries: Corporate Social Responsibility*, 38th Parliament, 1st Session, Fourteenth Report: June, www2.parl.gc.ca/HousePublications/Publication.aspx?DocId=1961949&Mode=1&Parl=38&Ses=1&Language=E, date accessed 14 September 2010.

STCIC (South Trinidad Chamber of Commerce) and UNDP (United Nations Development Programme) (2007) *Mapping Corporate Social Responsibility in T&T: Private Sector and Sustainable Development*, UNDP, Port of Spain.

Stiglitz, J. E. (2006) *Making Globalisation Work* (New York: W.W. Norton & Company Inc).

Stikeman Elliot (1999) 'Quebec Court Decides to Dismiss Proceedings: Tailings Dam collapse to Be Litigated in Guyana', Global Mining Update, April, www.stikeman.com/newslett/minapr99-2.doc, date accessed 14 September 2010.

Stokes, S. C. (1997) 'Democratic Accountability and Social Change: Economic Policy in Fujimori's Peru', *Comparative Politics*, 29 (2) (January), 209–26.

Strange, S. (1996) *The Retreat of the State: The Diffusion of Power in the World Economy* (Cambridge, UK: Cambridge University Press).
Sustainalytics (2010) Jantzi Social Index April 2010 Returns, http://sustainalytics.com/jantzi-social-index-april-2010-returns, date accessed 15 May 2010.
——(2006) 'Jantzi Research Recommends Goldcorp as Ineligible for SRI Portfolios', Sustainalytics website, www.sustainalytics.com/jantzi-research-recommends-goldcorp-ineligible-sri-portfolios, date accessed 14 September 2010.
Szablowski, D. (2010) 'Operationalizing Free, Prior and Informed Consent in the Extractive Sector? Examining the Challenges of a Negotiated Model of Justice', in M. Beck et al. (eds) *Rethinking Extractive Industry: Regulation, Dispossession and Emerging Claims*, special issue of the *Canadian Journal of Development Studies*, 30 (1–2), 111–30.
——(2007) *Transnational Law and Local Struggles: Mining Communities and the World Bank* (Oxford, UK and Portland, OR: Hart Publishing).
Tallontire, A. (2007) 'CSR and Regulation: Towards a Framework for Understanding Private Standards Initiatives in the Agri-food Chain', *Third World Quarterly*, 28 (4), 775–91.
Taub Isenberg, J. (2009) 'Able But Not Willing: The Failure of Mutual Fund Advisers to Advocate for Shareholders' Rights', *Journal of Corporation Law*, 34 (3), 843–93.
Teichman, J. (2004) 'The World Bank and Policy Reform in Mexico and Argentina', *Latin American Politics and Society*, 46 (1), 39–74.
Teivainen, T. (2002) *Enter Economism Exit Politics: Experts, Economic Policy and the Damage to Democracy* (London and New York: Zed Books).
Tereza, I. (2007) *MST faz nova invasão em ferrovia de Carajás da Vale*. Portal Exame, http://portalexame.abril.com/ae/economia/m0142821.html, date accessed 30 August 2009.
Tewarie, B. and R. Hosein (2006) *Trade, Investment and Development in the Contemporary Caribbean* (Kingston: Ian Randle).
Thomas, E. (2010) Devonshire Initiative, www.devonshireinitiative.org/, date accessed 16 February 2011.
Tingay, A. (2004) *Water Quality in the Mogpog River, Marinduque Island. Republic of the Philippines*, November, A. & S. R. Tinguay Pty. Ltd. Environmental Scientists, www.oxfam.org.au/resources/filestore/originals/OAus-WaterQualityMogpogRiver-0904.pdf, date accessed 16 February 2011.
Tinsley, C. R. (1992) 'Mine Financing', in H. Hartman (ed.) *SME Mining Engineering Handbook*, 2nd edn, Vol. 1 (Littleton: SME).
Transparency International (2010a) Corruption Perceptions Index 2010, ISBN 978-3-935711-60-9, available from www.transparency.org/policy_research/surveys_indices/cpi/2010/in_detail#1, Transparency International, Berlin.
——(2010b) 'Global Corruption Barometer 2010', Transparency International website, www.transparency.org/policy_research/surveys_indices/gcb/2010, date accessed 2 March 2011.
——(2009) *Global Corruption Report 2009: Corruption and the Private Sector* (Berlin: Transparency International).
——(2008) Home Page, http://www.transparency.org, date accessed 20 January 2009.

Todo Sobre Majaz (2008) 'Se formó frente regional en defensa del agua y proyecto Alto Piura', www.todosobremajaz.com, date accessed 20 October 2008.

Tostes, J. A. (2007) *Serra do Navio o mito da cidade no meio da selva* (Macapá: SOS Serra do Navio).

TTCSI (Trinidad and Tobago Coalition of Services Industries) (20079–09) 'TTCSI Position on Green Paper: On the Trinidad & Tobago Investment Policy,' www.ttcsi.org/home/green.php, date accessed 28 February 2011.

TTMA (Trinidad & Tobago Manufacturers' Association) (n.d.a) 'Physical Infrastructure', The Trinidad & Tobago Manufacturers' Association web page, www.ttma.com/pillars/social-infrastructure/, date accessed 2 March 2011.

TTMA (The Trinidad & Tobago Manufacturers' Association) (n.d.b) 'Social Infrastructure: Corporate Social Responsibility', The Trinidad & Tobago Manufacturers' Association web page, www.ttma.com/pillars/social_infrastructure/, date accessed 28 February 2011.

United Nations (2009) 'UN Global Compact Participants', 30 June, www.unglobalcompact.org/participantsandstakeholders/index.html, date accessed 28 February 2011.

United Nations Conference on Trade and Development (UNCTAD) (2000a) *Investment Policy Review – Peru*, United Nations, New York and Geneva.

——— (2000b) *Tax Incentives and Foreign Direct Investment – A Global Survey*, ASIT Advisory Studies, No. 16, United Nations, New York and Geneva.

——— (1993) *The Mineral Sector in Peru*, UNCTAD/COM/28, United Nations, New York and Geneva.

United Nations Development Programme (UNDP) (n.d.) 'UNDP Public Private Partnerships', United Nations Development Programme, Trinidad and Tobago, www.undp.org.tt/CSR/index.html, date accessed 28 February 2011.

——— (2010) *World Development Report 2010* (New York: United Nations).

——— (2009) *Human Development Report 2009: Overcoming Barriers. Human Mobility and Development* (London: Palgrave Macmillan for UNDP).

——— (2007) *Objectivos del Desarrollo Milenio: Potosi*, http://www.pnud.bo, date accessed 20 January 2009.

——— (2000) *Human Development Index – Brazil Report* (Brasilia: UNDP).

United Nations Environmental Programme (UNEP) (2009). *From Conflict to Peacebuilding: The Role of Natural Resources and the Environment*.

United Nations Environmental Programme Finance Initiative (UNEP FI) (2010) 'Principles for Responsible Investing. Report on Progress 2010', http://www.unpri.org, date accessed 20 October 2010.

United Nations Environmental Programme Finance Initiative (UNEP FI) and Asset Management Working Group (AMWG) (2005) 'A Legal Framework for the Integration of Environmental, Social and Governance Issues into Institutional Investment', http://www.unepfi.org/fileadmin/documents/freshfields_legal_resp_20051123.pdf, date accessed 10 October 2009.

——— (2006) 'Show Me The Money: Linking Environmental, Social and Governance Issues to Company Value', http://www.unepfi.org/fileadmin/documents/show_me_the_money.pdf, date accessed 20 November 2009.

United Nations Environmental Programme Finance Initiative (UNEP FI), Asset Management Working Group (AMWG) and Mercer (2007) *Demystifying Responsible Investment Performance: A Review of Key Academic and Broker Research on ESG Factors*, http://www.unepfi.org/fileadmin/documents/Demystifying_

Responsible_Investment_Performance_01.pdf, date accessed 20 November 2009.
United Nations Principles for Responsible Investment (UNPRI) (2010) 'Signatories to the Principles for Responsible Investment', http://www.unpri.org/signatories/, date accessed 20 October 2010.
USGS (2006) '2006 Minerals Yearbook: Latin America and the Caribbean', http://minerals.usgs.gov, date accessed 30 September 2009.
Utting, P. (2005) Rethinking Business Regulation: From Self Regulation to Social Control, working paper, United Nations Research Institute for Social Development Technology, Business and Society Programme Paper Number 15, September 2005.
Utting, P. and J. C. Marques (2010) 'Introduction: The Intellectual Crisis of CSR', in P. Utting and J. C. Marques (eds) *Corporate Social Responsibility and Regulatory Governance: Towards Inclusive Development?* (Hampshire, UK and New York: Palgrave Macmillan).
Valeriote, J. (2009) 'Chile's Supreme Court Upholds Indigenous Water Use Rights', *Santiago Times*, 30 November, www.santiagotimes.cl/index.php?option=com_content&view= article&id= 17739:chiles-supreme-court-upholds-indigenous-water-use-rights&catid= 19:other&Itemid= 142, date accessed 14 September 2010.
van Agtmael, A. (2007) *The Emerging Markets Century: How a New Breed of World-Class Companies Is Overtaking the World* (New York: Free Press).
van Dijk, T. A. (1993) 'Principles of Critical Discourse Analysis', *Discourse and Society*, 4 (2), 249–85.
van Duzer, A. (2003) *The Law of Partnerships and Corporations* (Toronto: Irwin Law).
Vecchio, R. (2006) 'Peruvian Protesters Release Workers from U.S.-owned Gold Mine', *Associated Press Newswires*, 4 August 2006.
Veiga, M. M. (1997) Introducing New Technologies for Abatement of Global Mercury Pollution in Latin America (Rio de Janeiro: UNIDO/UBC/CETEM).
Veiga, M. M. and A. R. B. Silva (2003) 'Produção de ouro e impactos do mercúrio na garimpagem da Amazônia', in R. B. E. Trindade and O. Barbosa Filho (eds) *Extração de Ouro - Princípios, Tecnologia e Meio Ambiente* (Rio de Janeiro: O.CETEM/CNPq).
Velasco, L. Jr. (1999) *Privatization: Myths and False Perceptions*, Banco Nacional de Desenvolvimento (Brasília: Ministério do Desenvolvimento, Indústria e Comercio Exterior), http://www.bndes.gov.br/english/studies/privat2.pdf, date accessed 08 June 2006.
Velho, O. G. (1972) *Frentes de expansão e estrutura agrária: Estudo do processo de penetração numa área da Transamazônica* (Rio de Janeiro: Zahar Editores).
Veltmeyer, H., J. Petras and S. Vieux (1997) *Neoliberalism and Class Conflict in Latin America: A Comparative Perspective on the Political Economy of Structural Adjustment* (Houndmills, UK: Macmillan Press).
Vlcek, W. (2008) *Offshore Finance and Small States: Sovereignty, Size and Money* (London: Palgrave Macmillan).
——— (2007) 'Why Worry? The Impact of the OECD Harmful Tax Competition Initiative on Caribbean Offshore Financial Centres', *Round Table*, 96 (390) (June), 331–46.

Vogel, D. (2010) 'The Private Regulation of Global Corporate Conduct', *Business & Society*, 49 (1), 68–87.

——(2005) *The Market for Virtue: The Potential and Limits of Corporate Social Responsibility* (Washington, DC: Brookings Institution Press).

Waddock, S. (2007) 'Corporate Citizenship: The Dark-Side Paradoxes of Success', in G. Flynn (ed.) *Leadership and Business Ethics*, p. 251 (Springer Science & Business Media, New York).

——(2008) 'Building a New Institutional Infrastructure for Corporate Responsibility', *The Academy of Management Perspectives (Formerly the Academy of Management Executive)(AMP)*, 22 (3), 87–108.

Waldie, P. (2010) 'Ottawa Gives Bill Clinton Foundation Special Designation', *Globe and Mail*, 23 November.

Walley, N. and B. Whitehead (1994) 'It's Not Easy Being Green', *Harvard Business Review*, 72 May/June, 46–52.

Webb, K. (2010) *Corporate Social Responsibility at Kinross Gold's Maricunga Mine in Chile: A Multiperspective Collaborative Case Study*, Ryerson Institute for the Study of Corporate Social Responsibility, www.ryerson.ca/csrinstitute/current_projects/Dfait_ch2_KGR_Maricunga.pdf, date accessed 14 September 2010.

——(2005). 'Sustainable Governance in the Twenty-First Century: Moving Beyond Instrument Choice', in P. Eliadis, M. M. Hill and M. Howlett (eds) *Designing Government: From Instruments to Governance*, pp. 242–63, 268–80, 411–21 (Montreal: McGill-Queen's University Press).

——(2004) 'Understanding the Voluntary Codes Phenomenon', in K. Webb (ed.) *Voluntary Codes: Private Governance, the Public Interest and Innovation*, pp. 3–34 (Ottawa: Carleton Research Unit for Innovation, Science and Environment).

——(1999) 'Voluntary Initiatives and the Law', in R. Gibson (ed.) *Voluntary Initiatives: The New Politics of Corporate Greening*, pp. 32–50 (Peterborough, ON: Broadview Press).

Webber, G. (2008) 'Canada', in *Obstacles to Justice and Redress for Victims of Corporate Human Rights Abuse, a Comparative Submission Prepared for Professor John Ruggie*, pp. 35–46, a report by Oxford Pro Bono Publico, University of Oxford, Oxford.

Wei, S. (2000) 'Natural Openness and Good Government', *World Bank Policy Research Working Paper No. 2411*, World Bank, Washington, DC.

Weiss, D. (2009) 'The Foreign Corrupt Practices Act, SEC Disgorgement of Profits, and the Evolving International Bribery Regime: Weighing Proportionality, Retribution, and Deterrence', *Michigan Journal of International Law*, 30, 471.

Weitzner, V. (2010) 'Indigenous Participation in Multipartite Dialogues on Extractives: What Lessons Can Canada and Others Share', in M. Beck et al. (eds) *Rethinking Extractive Industry: Regulation, Dispossession and Emerging Claims*, special issue of the *Canadian Journal of Development Studies*, 30 (1–2), 87–109.

Williams, J. P. (2005) 'The Latin American Mining Law Model', in E. Bastida, T. Wälde and J. Warden-Fernández (eds) *International and Comparative Mineral Law and Policy – Trends and Prospects* (The Hague: Kluwer Law International).

Wise, C. (1997) 'State Policy and Social Conflict in Peru', in M. Cameron and P. Mauceri (eds) *The Peruvian Labyrinth: Polity, Society, Economy* (Pennsylvania: Pennsylvania State University Press).

Wheeler, D., B. Colbert and R. E. Freeman (2003) 'Focusing on Value: Reconciling Corporate Social Responsibility, Sustainability and a Stakeholder Approach in a Network World', *Journal of General Management*, 28 (3) 1–28.

Wollenberg, E., J. Anderson and C. Lopez (2005) *Though All Things Differ: Pluralism as a Basis for Cooperation in Forests* (Jakarta: CIFOR).

World Bank (2008) Home Page, http://go.worldbank.org/, date accessed 20 January 2009.

——(2006) *Global Economic Prospects: Economic Implications of Remittances and Migration* (Washington, DC: World Bank).

——(2004) 'The Costs of Corruption', http://web.worldbank.org/WBSITE/EXTERNAL/NEWS/0,,contentMDK:20190187~pagePK:64257043~piPK:437376~theSitePK:4607,00.html, date accessed 4 October 2009.

——(2003) *Striking a Better Balance: The World Bank Group and Extractive Industries. Final Report*, Extractive Industries Review, World Bank, Jakarta/Washington, DC.

——(1996) 'A Mining Strategy for Latin America and the Caribbean', World Bank Technical Paper no. 345, World Bank, Industry and Mining Division, Industry and Energy Department, Washington, DC.

——(1995a) *Characteristics of Successful Mining Legal and Investment Regimes in Latin America and the Caribbean Region*, World Bank, Industry and Mining Division, Industry and Energy Department, Washington DC, www.natural-resources.org/minerals/CD/docs/twb/Invest_Regime_LAC.pdf, date accessed 1 October 2010.

——(1995b) *Staff Appraisal Report Brazil – Environmental Conservation and Rehabilitation Project*, Report No. 14585-BR, Washington, DC.

——(1992) 'Strategy for African Mining', World Bank Technical Paper no. 181, World Bank, Africa Technical Department Series, Mining Unit, Industry and Energy Division, Washington, DC.

World Commission on Environment and Development (WCED) (1987) *Our Common Future* (New York: United Nations).

World Summit on Sustainable Development (WSSD) (2002) 'The Political Declaration of the World Summit' and 'Plan of Implementation', World Summit on Sustainable Development website, www.worldsummit2002.org/index.htm, date accessed 16 February 2011.

World Vision Canada (2008a) 'Barrick Gold Corporation', World Vision website, Corporate Partners, www.worldvision.ca/GetInvolved/Corporate-Partners/Pages/barrick-gold-corporation.aspx, date accessed 3 March 2011.

World Vision Canada (2008b) 'Corporate Supporters', World Vision website, Corporate Partners, www.worldvision.ca/GetInvolved/Corporate-Partners/Pages/corporate-supporters.aspx, date accessed 3 March 2011.

World Vision Canada (2008c) 'Why Partner with World Vision?' World Vision website, Corporate Partners, www.worldvision.ca/GetInvolved/Corporate-Partners/Pages/Why-Partner-With-World-Vision.aspx, date accessed 3 March 2011.

Xu, Y.-C. and G. Bahgat (eds) (2010) *The Political Economy of Sovereign Wealth Funds* (London: Palgrave Macmillan).

Yakovleva, N. (2005) *Corporate Social Responsibility in the Mining Industries* (Aldershot, UK and Burlington: Ashgate Publishing).

Zadek, S. (2006) The Logic of Collaborative Governance: Corporate Responsibility, Accountability and the Social Contract, Accountability and the Corporate Social Responsibility Initiative, Working Paper No. 17 (Cambridge, MA: John F. Kennedy School of Government, Harvard University).

Zadek, S. and S. Radovich (2006) Governing Collaborative Governance: Enhancing Development Outcomes by Improving Partnership Governance and Accountability. AccountAbility and the Corporate Social Responsibility Initiative, Working Paper No. 23 (Cambridge, MA: John F. Kennedy School of Government, Harvard University).

Zarnikow, D. (2010) 'UN Agency Rep Sent to Chile's South to Discuss Mapuche Rights', *Santiago Times*, 20 August, www.santiagotimes.cl/news/human-rights/19607-un-agency-rep-sent-to-chiles-south-to-discuss-mapuche-rights, date accessed 14 September 2010.

Zinnbauer, D. and R. Dobson (2008) *Global Corruption Report 2008* (New York: Transparency International).

Index

Aboriginal peoples, *see* Indigenous communities
Africa, 10, 11, 25, 217, 253, 299
  Canadian investment in mining, 304
  and CSR, 98, 260–4
  diamond mining, 6, 102–3, 283
  and EITI, 272–4
  governance capacity of national states, 263, 275
  governance system, 260–2, 298
  mining and conflict, 107, 262
  World Bank and regulatory reform, 90
  see also Cameroon; Democratic Republic of Congo (DRC); Ghana; Liberia; Nigeria; South Africa; Zimbabwe
Alcoa, 10, 42, 233, 238, 239, 240, 255, 298
  Sustainable Juruti project, 9, 233, 235, 236–9, 244, 298
Alien Tort Statute (US), 23, 68
AngloGold Ashanti, 110, 264, 270–2, 275
  conflict and legitimacy problems, 271–2, 275
  Human Rights Watch report, 271
  participation in EITI, 272
Anglo Platinum, 264
  Amandelbult platinum mine community benefit-sharing agreement, 268–9, 275
  reputational damage and loss of legitimacy, 267
  response to South African Human Rights Commission report on Mogalakwena mine, 266–7
Anti-corruption, 9, 214, 215, 224, 225–7, 229, 280
  and EITI, 228
  enablers and disablers, 230–1

Apex Silver, San Cristóbal mine (Bolivia), 8, 170, 173–4, 284
  CSR programs, 177
Artisanal mining, 7, 71, 271, 273
  in Brazil, 154, 157, 158, 160–1, 164, 168
  in the diamond industry, 103, 109, 111
Asymmetrical development, 15

Barrick Gold, 18, 36, 127, 131, 132, 144, 150, 169, 228, 305
  CSR project partnership with CIDA, 125–6
  divestment of by Norwegian government, 194
Bill C-300 (Canada), 80, 113, 122, 128, 131, 193
Biodiversity Convention (1992), 38
Blackfire Exploration, 52–3
Bolivia, 8, 18, 171, 175, 183–5, 219, 220–1, 284, 285, 289
  cooperative miners, 172
  election of Morales, 172–3
  mining history, 172–3
Brazil, 6, 7–8, 10, 11, 45, 98, 102, 103, 218, 246, 253, 254, 298, 304
  artisanal mining in, 154, 157, 158, 160–1, 164, 168
  and Kimberley Process, 106, 107
  land conflicts, 154, 157
  military coup, 156, 284
  mining in Amazon region, 155–6
  and mining CSR, 167
  Movimento Sem Terra (Brazilian Landless Workers' Movement), 159–60
  see also Alcoa, Sustainable Juruti project
BRIC countries, 11, 246, 247, 252, 254, 255, 258

342 *Index*

Brundtland report (*Our Common Future*), 39, 302
Business-society relations, 15, 20, 22–3, 24, 133, 135, 140, 152

Cameroon, 264, 272–4
Canadian Boreal Forest Agreement, 82
Canadian government, 7, 62, 74, 78, 79, 80, 81, 82, 85, 94, 98, 101, 116, 121–3, 125, 126, 128, 131, 132, 192, 286–7, 306
  Building the Canadian Advantage: A Corporate Social Responsibility (CSR) Strategy for the Canadian International Extractive Sector (Government of Canada CSR Strategy), 79–81, 98, 124, 129–30
  CSR Centre of Excellence, 80, 81, 287–8
  CSR Counsellor (Marketa Evans), 80, 82, 98, 132, 287–8
  Standing Committee on Foreign Affairs and International Trade (SCFAIT), 101, 121, 286, 288
  and support of mining industry, 123
  see also National Roundtables on CSR (Canada)
Canadian Institute of Mining, Metallurgy & Petroleum, 81, 293
Canadian International Development Agency (CIDA), 121, 123–5, 127, 131, 132, 287, 288
  CSR partnerships with mining companies, 126, 129–30
  in Peru, 85–6
  and transparency, 125
Capacity building, 65, 66, 69, 71–2, 78, 93, 123, 124, 181, 184, 185, 252, 279, 280, 283, 284
Chile, 15, 58, 59, 88, 118, 139, 200, 217, 218, 219, 220, 225, 303, 304
Communities, 5, 8, 17, 52, 64, 71, 75, 96, 99, 101, 115, 139, 163, 166, 170, 172, 173, 175, 182, 184, 185, 187, 240, 243, 281, 292, 294, 299
  capacity to manage benefits and impacts of mining, 50, 180, 181, 184
  complexity of, 265
  conflict with and relationship to mining companies, 90, 97–8, 114, 119, 128–9, 139, 168, 171, 178, 239, 266, 267
  consultation, engagement and partnership with, 44, 57, 130, 163, 166, 168, 209
  and development initiatives/priorities, 128, 130–1
  dialog with companies/government, 240
  effects of poverty/disadvantage, 180
  empowerment through election of Evo Morales (Bolivia), 177–8
  land distribution and the Brazilian Landless Workers' Movement (MST), 159–60
  legitimate representation, 264, 275
  and mining employment, 182, 184
  political identity and relationship to government, 78
  power of, 178–9
  power asymmetries, 97
  problem of dependency, 181
  resettlement, 154
  resistance to mining, 115–16
  right of refusal, 97
  as SEVGE actors, 289–91
Community referendum, 55, 56, 59–61, 116, 207, 289, 290
Companhia Vale do Rio Doce (CVRD, also Vale), Brazil, 156
  Carajás Iron Ore Mining Complex, 157–8
  CSR practices of, 162–4
  privatization of, 157–8
  Vale Foundation, 168
Conflict, 6, 7, 8, 17, 65, 66, 67, 69, 70, 84, 85, 90, 94, 95, 97, 99, 100, 101, 112, 116, 117, 139, 152, 156, 159, 160, 166, 191, 221, 235, 246, 261, 266, 267, 270, 275, 284, 287, 289, 290, 299, 302
  contextual drivers of, 155, 158

Index

Conflict management, 65, 66, 72, 98, 99, 235
Corporate codes of conduct, 65, 248
Corporate irresponsibility, 22, 271
Corporate social responsibility (CSR), 1–2, 12, 20–5, 31–2, 47–8, 64, 69, 73–4, 75, 91, 108, 117, 126, 143, 170, 171, 183–5, 201, 248, 288
  corporate responsibility infrastructure, 22
  and discourse, 134–7, 141
  early CSR movers, 40–2
  and finance, 193–8, 199
  and law, 5, 47–8
  limitations of, 2–3, 19–20, 22–3, 26, 83, 134, 150–1, 171, 175
  as management strategy, 18–19, 20, 27, 66
  reporting, 19, 22, 31, 36, 40–1, 43–4, 61–3, 73, 169, 207, 273, 296–7
  standards and codes, 7, 22, 29, 35, 36, 42, 57, 60, 65, 73, 79, 134, 140–50, 151, 169, 248, 293–4, 295, 301–2
  systemic approach, 2–3, 4, 12–14, 25–8
  and voluntary/self-regulation, 19, 20, 24, 29, 35, 41, 62, 69, 117, 139, 140, 142, 144, 294
Corruption, 9, 16, 17, 18, 29, 73, 76, 78, 106–7, 114, 120, 131, 159, 191, 198, 214–20, 229, 231, 249, 258, 259, 276, 286, 300
  anti-corruption in the mining industry, 224–5
  business costs of, 222–3
  in Latin America, 217–19
  legal aspects (anti-corruption regulation), 52, 53
  as SEVGE disabler, 27, 279–80
  social consequences of, 216–17
  typology of company approaches to corruption, 225–30
Costa Rica, 57, 132, 217
CSR Centre of Excellence, see Canadian government

CSR Counsellor, see Canadian government
CSR, see Corporate Social Responsibility (CSR)

Democratic Republic of Congo (DRC), 104, 105, 112, 260, 261, 263, 264, 270–2, 275, 284
Department of Foreign Affairs and International Trade (Canada) (DFAIT), 65, 80–1, 123
Development, 7, 10, 16, 17, 90, 92, 114, 116–17, 121, 123, 128, 129–30, 165, 239, 244, 259, 265, 266, 276, 277, 289
  developmental state, 246, 247, 249, 255
  pro-poor development, 219, 229, 231
  relationship to CSR, 19, 73, 75
  underdevelopment, 113, 219
Development NGOs, 291–2
  and relationships with mining companies, 123, 125, 127–8, 130–1, 292
Devonshire Initiative (DI), 122–3
Diamond Development Initiative International (DDII), 109, 293
Diamond mining, 6, 102, 106–7, 111–13, 269, 280, 281
  conflict diamonds, 104, 283
  and CSR, 103, 108–11
  and violence, 107
Duty to protect, see Human rights; UN Special Advisor on Business and Human Rights

Ecuador, 18, 49, 151, 219, 220, 221, 303
  Copper Mesa, 49–50
El Salvador, 54–5, 132
Energy Security Through Transparency Act (ESTTA), 191, 196, 200
Environment, 3, 5, 13, 17, 29, 33, 37, 42, 44, 82, 96, 163, 172, 176, 180, 182, 196, 239, 255, 278–9, 282
  damage associated with mining, 32, 35, 36, 39, 286
  environmental regulation, 38, 56–7

Environment – *continued*
  environmental responsibility, 35, 40, 151–2, 162
Equator Principles, 38, 161, 169, 192, 196, 199, 294–5
Export credit agencies, 68, 287
Export Development Canada (EDC), 70, 80, 123, 192, 193
Extractive Industries Transparency Initiative (EITI), 10, 67, 79, 120, 131, 225, 246, 251, 258–9, 264, 275, 296, 299
  as anti-corruption strategy, 228
  limitations of, 258
  success factors, 272–4

Financing (of mining projects), 8, 38, 72, 73, 186, 187, 188–9, 190, 193, 198–200, 285, 295
  financial institutions and lenders/investors, 45, 87, 90, 116, 120, 123, 186, 192, 231, 282, 294–5
  lending conditionalities, 38
  project financing, 161, 192
Foreign direct investment (FDI), 10, 11, 70, 75, 76, 84, 87, 250, 254, 304
Free, prior and informed consent (FPIC), 52, 58, 207
Free trade agreements, 54–5, 94, 250

Ghana, 118, 126, 206, 264, 272–4
Globalization, 15, 17, 23, 28, 92, 140, 234, 245, 247, 248, 265
Global Mining Initiative (GMI), 42
Global norms, 4, 32, 33, 35, 45, 141, 152
Global Reporting Initiative (GRI), 42, 81, 145, 162, 169, 216, 297
Goldcorp, 8, 9, 51–2, 56, 58, 201, 204, 207–10, 211, 300, 305
Governance, 2, 6, 8, 9, 10, 12, 27, 30, 31, 35, 60, 64, 75, 83, 92–3, 121, 138, 139, 152–3, 170, 195, 234, 263, 265, 267, 272, 274, 279, 282, 286, 290, 301–2
  collaborative governance, 23–4
  and ecosystems view, 3, 13, 25–8, 215, 277
  gaps/weakness, 16, 17, 19, 24, 86, 92–3, 118–19, 122, 140, 166, 216, 266, 270
  governance systems, 11, 260–2, 264, 298
  and limited statehood, 275
  modes of governance, 93, 97, 101, 138, 277, 284, 302
  pluralistic/participatory governance, 233–5, 244
  private actors and cultural change, 266
  sustainable governance, 62
Government of Canada, *see* Canadian Government
Government, *see* Home government; Host government
Government social responsibility, 72, 74
Greenwashing, 15, 101, 266, 299
Guatemala, 47, 50, 51, 52, 58–9, 60, 101, 207–10, 218, 300, 305

Home government, 6, 33, 64, 70, 71, 72–3, 74–5, 76, 82, 83, 116, 120, 129, 170, 185, 198, 282, 283, 284, 286–9
  duty to protect, 64, 68
  regulatory jurisdiction, 68, 286
  reputation linked to CSR performance of extractive companies, 67, 78–9, 287
  roles & relationship to CSR, 6, 66–70, 72, 73, 74, 288
Host government, 6, 18, 64, 66, 74, 77, 83, 101, 115, 118–19, 120, 122, 170–3, 185, 190–1, 192, 272, 283–6, 284, 288, 298
  and allocation of mining revenues, 69–70
  governance capacity, 16, 68, 118, 180, 183–4
  joint ventures with mining companies, 190–1
  local and regional government, 75–6

and neoliberalism/structural
adjustment, 16, 76, 91–2, 284,
294, 296
roles and relationship to CSR, 65,
71, 73–4, 76, 77
Huanuni mine (Bolivia), 174
CSR programs, 177
and environmental damage, 174,
179
as a state mine, 179
Human rights, 17, 18, 32–3, 82, 97,
107, 263, 283
abuse of, 101, 112, 116, 119, 126,
264, 271
and corporate responsibilities, 169,
209–10, 265–6
duty to protect, 64, 68
human rights impact assessments
(HRIA), 52, 207–10
responsibilities of governments, 65,
67, 73, 76, 121
standards and guidelines, 81, 140,
161, 295

ILO Convention 169, 58–9, 61, 96,
169, 283, 285
adoption in Bolivia, 172, 178, 185
Impact-benefit agreements, 59, 62,
290, 299
Indigenous communities, 60, 62, 96,
97–8, 101, 119, 129, 139, 171–3,
175, 176, 183, 207, 289
in Brazilian Amazon region, 154,
156, 159–60, 162, 166
consultation with, 163–4
Indigenous Peoples Partnership
Program, 124
legal protection, 57–9
as stakeholders vs. rightholders, 97,
98
Industry associations, 7, 62, 116, 120,
128–9, 133, 292–4
and CSR standards, 140–52
Institutional learning, 29, 235
Inter-American Development Bank
(IABD), 87, 123, 294
International Council on Mining and
the Environment (ICME), 42

International Council on Mining and
Metals (ICMM), 42, 117–20, 128,
131, 143–4, 151, 292
Resource Endowment Initiative,
117–18, 121, 123
Sustainable Development
Framework, 145–9, 293
International Development Research
Centre (IDRC), 66, 288
International Finance Corporation
(IFC), 16, 118–19, 123, 295
Performance Standards, 38, 70, 81,
119, 161, 169, 192, 196
International Monetary Fund (IMF),
75–6, 87, 294

Kimberley Process Certification
Scheme (KPCS, KP), 29, 102,
104–5, 108–9, 112, 280, 281, 293,
296
flaws in organizational structure,
108–9
weakness of in South America,
105–7

Law, 5, 47, 48–50, 55, 57, 58, 60, 61,
68, 87, 89, 92, 108, 153, 285, 288
Legitimacy, 25, 86, 91–3, 135, 234,
271, 275, 281, 298
Liberia, 103, 112, 272
Lobbying, 67, 69, 80, 192–3, 228, 231,
266, 288, 292–3, 299

Media, 11, 23, 96, 107, 141, 281, 292,
300–1
Mining, and environmental
impact/management, 16–17, 18,
32, 33, 36, 39, 42, 114–15, 119,
130, 131, 151, 172, 182, 185,
286
and risk/uncertainty, 18–19, 41, 96,
220–1, 222–3, 270
and social impacts, 17, 96, 116–17,
119
and sustainability/sustainable
development, 4, 31–2, 34, 39,
40–5, 85, 117, 136, 138, 171,
181, 183–5, 193, 194, 235, 237,
265, 278–9

Mining finance, debt financing, 189, 195–6
  flow-through shares, 190
  potential to influence mining firm behaviour/CSR performance, 198–200
  project financing, 189, 192
  risks faced by lenders, 191–2
  through stages of mining development, 189–93, 198
Mining, Metals and Sustainable Development (MMSD), 42, 279
Multilateral Investment Guarantee Agency (MIGA), 119
Multi-stakeholder engagement, 24, 118, 228, 272–3, 274

National Roundtables on CSR (Canada), 66–7, 79, 121–4, 132, 286, 288
Natural Resources Canada (NRCAN), 123
Neoliberalism, 7, 15, 17, 86–8, 130, 136, 141, 152, 284
  and CSR, 75, 137–41
Newmont Mining Corporation, 204, 206
  shareholder resolution and community relations review, 205
Nigeria, 76, 229, 247, 264, 272–4
Non-governmental Organizations (NGOs), 9, 14, 31–3, 37, 39, 52, 60, 67, 79–80, 97, 107, 108–9, 116, 123, 126, 127–31, 151, 164, 210, 212, 291–2

OECD Guidelines for Multinational Enterprises, 81–2, 283
Organizational culture, 266–7

PERCAN (Peru-Canada Mineral Resources Reform Project), 6, 78, 85–6, 94–100, 287
  initiatives and objectives, 95–6
  limitations of, 100–1

Peru, 6, 18, 58–9, 68, 70, 76–8, 83, 99, 100–1, 126, 224, 228, 287, 294, 303
  Antamina, 70, 77–8, 224–5
  canón minero (voluntary fund), 66, 68, 69–70, 74
  Manhattan Minerals (Tambogrande), 56
  mining regulation, 86–90
  neoliberal reforms, 86–8
  Río Blanco, 70, 77
  and social conflict, 84–5
Placer Dome, 36, 43, 114–15, 131, 164, 305
Polanyi, Karl, 137, 141
Policy, 2, 24, 40, 65, 66, 69, 72, 80, 82, 86, 87, 88, 90, 121, 129, 133, 137–8, 152, 165, 172, 207, 227, 268, 273–4, 283, 286, 288, 300, 301
  anti-corruption policy engagement, 228–9
  policy drivers, 79
Political risk insurance (PRI), 192, 196, 223
Poverty, 2, 8, 15, 17, 54, 76, 88, 112–13, 117, 125, 139, 165, 167, 169, 170, 173, 175, 180, 183, 191, 219, 235, 271, 279, 286, 289, 295
  mining as poverty reduction, 118–19, 125
Power, 5, 15, 22, 28, 60, 70, 75, 76, 78, 80, 82, 87, 96, 97, 133–4, 135, 139, 176, 203, 216, 218, 262, 271–2, 282, 289, 290, 294, 296, 299, 300, 301
  balance of in the extractive industries, 68
  community empowerment, 24, 178, 234, 286
Prospectors and Developers Association of Canada (PDAC), 98, 122, 144, 151, 293
  e3 Plus, 28, 145–9
Public opinion, 37, 60, 67, 79, 109, 228, 284, 291
Publish What You Pay initiative, 191, 272, 297

Redress, 29, 49, 50–1, 68–9, 301
Regulation, 23, 29, 38, 39, 58, 68, 69, 73, 75, 78, 86–90, 92, 93, 96, 99, 108, 117, 137, 138, 141, 143, 153, 171, 180, 197, 198, 199, 283–4, 285, 286, 288, 293
  CSR as self-regulation, 24, 35, 139, 140, 142, 144, 294
  'race to the bottom', 76, 198
Relocation (*also* resettlement), of communities, 58, 114, 154, 166, 221, 231, 263, 265–8, 275, 289
Reporting, CSR, 19, 22, 31, 32, 36, 40, 41, 43, 67, 73, 79, 145, 150, 169, 207, 273, 295, 296
Reporting and transparency agencies, 296–7
  see also Extractive Industries Transparency Initiative (EITI); Global Reporting Initiative (GRI); Publish What You Pay initiative; Transparency International
Reputation, 11, 32, 35, 36, 37, 39, 41, 45, 79, 80, 98, 109, 116, 151, 180, 192, 195, 211, 222, 237, 243, 251, 263, 267, 274, 275, 287, 298
Resource curse, 18, 76, 117, 121, 131, 191, 246, 258
Resource nationalism, 18
Responsible investment (RI) (*also* socially responsible investment, SRI), 8–9, 194, 195, 201, 211–12, 295
  business case and, 211
  corporate culture and, 211
  growth of, 202–3
  and shareholder legitimacy, 210
  types of, 203–4
Responsible Jewellery Council (RJC), 110–11, 293
Rio conference (1992 UN Conference on Environment and Development), 39, 42
Risk, 2, 9, 12, 16, 17–18, 20, 28, 38, 41, 78, 92, 96, 100, 139, 140, 150, 166, 189, 192, 193, 195, 196, 199, 201, 203, 221, 222, 226, 227, 229, 230, 243, 263, 270, 279, 305
  contextual drivers of, 158
  industry associations and, 140
Ruggie, John, *see* United Nations, Special Representative of the Secretary General (SRSG) on business and human rights (John Ruggie)

Sanctions, for CSR underperformance, 17, 29, 73, 76, 89, 96, 142, 144, 153, 223, 293
Sarbanes Oxley act, 23, 196
Shareholder activism, 9, 51–2, 201, 295
  ineffectiveness of shareholder boycotts, 193–4
  see also Responsible investment (RI) (*also* socially responsible investment, SRI)
Small island development states (SIDS), 246–7
Social and Environmental Value Governance Ecosystem (SEVGE) model, 3, 4, 13–14, 25–8, 29, 30, 277–301
Social exclusion, 2, 12, 14, 24
Social justice, 151, 152
Social license to operate (SLO), 2, 9, 12, 15, 17, 24, 39, 41, 44, 92, 96, 151, 223, 235, 237, 289, 299
  and development agendas, 116–17
South Africa, 25, 98, 102, 108, 200, 260–1, 263, 275, 276
  legitimate representation structures, 264
South African Human Rights Commission, 266
Stakeholder(s), stakeholder management, 4, 5, 11, 12–14, 20–2, 27, 43–4, 51, 61, 70–1, 97–9, 117, 134, 166, 191–2, 205, 211–12, 231, 240, 264, 267, 275, 279, 289, 299
  complexity of stakeholder relationships, 261–3, 265, 281
State capture, 217, 222, 224, 226, 228

Structural reform (*also* structural adjustment), 15, 75, 86, 93, 94, 138
  see also International Monetary Fund (IMF); World Bank
Supranational and multilateral institutions, 29, 86–7, 123, 176, 192, 282–3
Sustainable development (*also* sustainability), 2, 3, 4, 8, 9, 18, 22, 31–5, 36, 41, 43–5, 85, 98, 129, 166, 170, 171, 175, 181, 183–5, 193, 237, 238, 239, 261, 265, 266, 278–9, 302

Tax, taxation regimes, 7, 23, 70, 79, 84, 90, 103, 109, 117, 159, 173, 181, 189, 190, 223, 224, 229, 237, 285, 303
  avoidance by mining companies, 114, 120, 228
Toronto Stock Exchange (TSX), 49, 50, 197, 199–200, 202
Transnational advocacy networks, 33
Transparency International, 173, 218, 223, 246, 250, 258, 276, 297
Trinidad & Tobago (T&T), and CSR initiatives, 252, 253–4
  as energy-producer, 245, 250
  involvement in EITI, 246, 248, 258–9
  prospects for diversification, 246, 251–3
  relations with US and EU, 248

United Nations, 18, 37, 39, 104, 201, 202, 227, 282–3
  Environmental Programme (UNEP), 37, 202, 282, 283
  Global Compact, 203, 207, 227, 251, 253, 283
  Principles for Responsible Investment (UNPRI), 202–3
  Special Representative of the Secretary General (SRSG) on business and human rights (John Ruggie), 17, 33, 64, 68, 69, 82, 265
US Agency for International Development (USAID), 123

Voluntary Principles on Security and Human Rights, 81, 169

Whitehorse Mining Initiative, 98
Windfall Oils and Mines scandal (Ontario), 197
World Bank, 16, 38, 39, 45, 75, 87, 88, 89, 90, 117, 119, 120, 123, 139, 162, 173, 192, 207, 216, 249, 294–5
  and economic reform, 86–8, 138–9
  Extractive Industries Review (EIR), 279
  Latin American Mining Law Model, 89
  Strategy for Mining in Africa, 90
World Business Council on Sustainable Development, 42
World Commission on Economic Development (WCED), 39
World Diamond Council (WDC), 108, 293
World Heritage Convention, 37, 282
World Vision, 122, 127

Zimbabwe, 107